EIGHTH EDITION

CRIMINAL LAW FOR POLICE OFFICERS

Neil C. Chamelin
*Assistant State Attorney, Second Judicial
Circuit, Leon County, Florida*

Prentice
Hall

Upper Saddle River, New Jersey 07458

Library of Congress Cataloging-in-Publication Data
Chamelin, Neil C.
    Criminal law for police officers / Neil C. Chamelin.—8th ed.
        p. cm.
    Includes bibliographical references and index.
    ISBN 0-13-094101-8
        1. Criminal law—United States. 2. Police—United States—Handbooks, manuals, etc. I.
    Title.

KF9219.3.C45 2002
345.73—dc21   2002020610

Publisher: Jeff Johnston
Executive Editor: Kim Davies
Production Editor: Emily Bush, Carlisle Publishers Services
Production Liaison: Barbara Marttine Cappuccio
Director of Production & Manufacturing: Bruce Johnson
Managing Editor: Mary Carnis
Manufacturing Buyer: Cathleen Petersen
Design Director: Cheryl Asherman
Cover Design Coordinator: Miguel Ortiz
Cover Designer: Marianne Frasco
Cover Image: Nick Koudis/PhotoDisc
Marketing Manager: Jessica Pfaff
Assistant Editor: Sarah Holle
Editorial Assistant: Korrine Dorsey
Composition: Carlisle Communications, Ltd.
Printing and Binding: R.R. Donnelley & Sons

Pearson Education LTD.
Pearson Education Australia PTY, Limited
Pearson Education Singapore, Pte. Ltd.
Pearson Education North Asia Ltd.
Pearson Education Canada, Ltd.
Pearson Educatíon de Mexico, S.A. de C.V.
Pearson Education-Japan
Pearson Education Malaysia, Pte. Ltd.

10 9 8 7 6 5 4 3 2 1
ISBN 0-13-094101-8

To the safety of my family and friends, wherever in the world
they may be
and
to the memory of those lost on September 11, 2001

Contents

CHAPTER 4

The Criminal Act

CHAPTER 5

The Mental Element

CHAPTER 6

Matters Affecting Criminal Responsibility

CHAPTER 7

Assault and Related Crimes

CHAPTER 8

Homicide

CHAPTER 9

Sex Offenses and Offenses to the Family Relationship

CHAPTER

10

Theft

CHAPTER

Robbery

CHAPTER

Burglary and Related Offenses

CHAPTER

Arson

CHAPTER 14

Forgery and Related Offenses

CHAPTER 15

False Imprisonment, Abduction, and Kidnapping

CHAPTER 16

Involving Narcotic Drugs and Alcoholic Beverages

CHAPTER

Extortion, Blackmail, and Bribery

CHAPTER

Offenses by and Against Juveniles

CHAPTER 19

Traffic Offenses

CHAPTER 20

Crimes Affecting Judicial Process

CHAPTER

Crimes Against Public Order

CHAPTER

Organized, White Collar, and Commercial Crimes

Preface

In this eighth edition, I have taken to heart some recommendations of reviewers, past and present, and have expanded the material on juveniles and placed it in a new chapter entitled "Offenses By and Against Juveniles." In addition, new material has been added concerning carjacking, contraband forfeiture, environmental crimes, computer crimes, and other topics, in hopes of bringing the substantive law into focus with changing and evolving developments.

I have added key words and phrases at the beginnings of each chapter and glossaries at the ends to aid in a clear understanding of the materials. As is emphasized in the first chapter when discussing language and communication, the success of this text as a learning aid depends on my ability as a writer to ensure that you understand the words as I mean them.

In this edition, I have expanded the references to states that assume one position or another on legal issues, used actual quotes from statutes in more jurisdictions, added more thought-provoking questions in the text and at chapter ends, and added some additional anecdotes to liven up the subject.

My thanks to the reviewers of this book and their helpful suggestions. They are: Daniel Ford, Cameron University, Lawton, OK; Bob Walsh, University of Houston/Downtown, Houston, TX; Ellen Cohn, Florida International University, Miami, FL; and Daniel S. Campagna, Aurora University, Aurora, IL. A very special thanks goes to the dedicated law students at Florida State University who spent hours researching and verifying materials to assist me in this effort. They are: Brian Stabley, Diane Whiddon, Rob Rogers, and Tequisha Myles.

Historical Background of Criminal Law

▓ KEY WORDS AND PHRASES ▓

Civil law system
Civil wrongs
Common law
Communication
Constitution
Crime
Language

Law
Responsibility
Retributive justice
Stable yet flexible
Stare decisis
Substantive criminal law

1.0 INTRODUCTION

The need for law lies in the history of the human race. In early times, when the first humans appeared on earth, laws were not needed, for conflicts did not arise. But when people began to live in groups, communities, and societies, laws became necessary. People are individuals, whose desires, needs, and wants differ from those of others. These differences cause conflict. Law became necessary as a means of social control, to either alleviate conflicts or to settle them in a manner most advantageous to the group.

As a means of social control, the enforcement of law in early rural societies was usually handled informally by friends, family, and neighbors who could criticize, correct and ostracize those who violated the folkways and mores. In more urban areas, where interpersonal relationships were not close, and where people living next door to each other were strangers, disputes between people and violations of the rules of society had to be handled by more formalized law, law enforcement, and courts.

1.1 THE NATURE OF LAW

The question "What is law?" invites a multiplicity of answers, for **law** is a broad concept with many definitions. For purposes of this book, law can be defined as a group of rules governing interaction. Law is a set of regulations governing the relationships among people and between people and their government.

Law is nothing more than language. Just as a carpenter uses a hammer and saw, a lawyer's tool is the ability to communicate. **Communication** is defined not merely as the conveyance of words but, more properly, as the conveyance of words with the ability to make oneself understood to the listener or reader. The more one studies the complexities of the law, the more one realizes the truth of the statement that **language** is the essence of law. The importance of being able to communicate in terms that mean the same thing to the parties on both the sending and receiving ends of a communication cannot be overemphasized. What the law strives for is uniformity of interpretation. The question then arises, "Why are not the laws so pointedly written that everyone knows exactly what they mean?" This question is relatively easily answered when the problem is seen in proper perspective. Laws must be **stable yet flexible** enough to be interpreted so that they may be molded to fit the problems of a complex and changing society. If every law were written to cover only one specific situation, two major problems would arise. First, the number of laws would increase at least one hundred times over, each law covering too specific a subject. The civil law systems in France and Germany are subject to this weakness. Second, many of these laws would become obsolete so quickly that legislatures would spend most of their time repealing them. However, when laws can be interpreted flexibly, they can be read to include new and unexpected situations as they arise.

The U.S. Constitution operates on the same principle, and we use that document to illustrate this point. The framers of the Constitution could not foresee the problems of the twentieth century. If the Constitution had been prepared only for the problems of 1789, it would have collapsed many years ago. It could not function today because the problems of society in the twentieth century are not the same as the problems encountered by those who framed the Constitution in the eighteenth century. Thus the Constitution was purposely drafted to be stable and yet flexible in the sense that it might be capable of interpretation in light of contemporary problems.

Any number of contemporary problems could be used to illustrate this point. Let us consider a few. At one time, the federal government's right to regulate commerce between the states was limited to prohibiting the erection of such barriers as tariffs. Georgia could not prevent Florida from shipping chickens to Georgia merely to protect Georgia chicken farmers. But certain related commerce problems arose. States were unwilling to prevent certain abuses to members of the labor force. As a result, our federal labor laws came into being. Congress cited the commerce clause of the Constitution as authority for these acts. The U.S. Supreme Court upheld the labor laws so that today wages, working conditions, and hours are regulated. This problem was not foreseen by the framers of the Constitution.

Years ago, segregation was said to be supported by the Constitution. However, due to the changing times and attitudes of the people, in 1954 segregation was said to be in opposition to the Constitution. Finally, the right to counsel, which is found in the Bill of Rights, was said to apply only to federal courts and federal cases. However, in the now-famous Gideon case, the right to counsel was said to apply to state courts and state criminal cases. The same document was used; times had changed.

1.2   CRIME DEFINED

Criminal law is only a small part of the entire legal field. If a state statute requires two witnesses for a valid will, having only one witness will render the will invalid but will not result in criminal charges. A **crime** may be defined as a public wrong. It is an act or omission forbidden by law for which the state prescribes a punishment in its own name. What does this mean? A crime must be a wrong against the public, not merely a wrong against a particular individual. There are many laws, in many jurisdictions, governing the rights and duties of people in their relationships to others. However, only those violations that wrong the public are considered criminal and make up the body of the **substantive criminal law**.

The determination as to whether a particular act is criminal or merely civil in nature is a function of the lawmaking body of each jurisdiction. In tribal times, this decision was made by the people. They considered "criminal" those acts that they felt injured the welfare of the entire community. Today, this function rests with the legislatures of the states.

Crimes differ from **civil wrongs** in many respects, but the sole reason they differ is that the legislature says they differ. In other words, only a fine line distinguishes crimes from civil wrongs, and that line is drawn by the legislature, where and when that body so desires and within the limits of what the public will tolerate.

Crimes are prosecuted by the state in its own name. In a civil case, the action is instituted by the wronged individual. Persons convicted of crimes are punished by fines, imprisonment, or death, whereas defendants who lose civil cases are usually ordered to pay the injured party. A crime is a public wrong, whereas a civil wrong is private in nature, not involving the state as a

TABLE 1–1 Distinction Between Crimes and Civil Wrongs

Crimes	Civil Wrongs
Public wrong	Private wrong
Prosecuted by the state in its own name	Action instituted by the wronged individual
Punished by fine, imprisonment, or death	Normally ordered to pay damages to the injured party
Punishment is prescribed	No set amount of monetary damages

party. Punishment is prescribed and must be prescribed for convictions of criminal acts, but there is no set amount of damages to which a wronged person is entitled in a civil suit. These are only a few of the major differences between crimes and civil wrongs (see Table 1–1), differences that exist solely as a consequence of the legislature's having attached the label *crime* to one act and not the other. This is not to say that the legislature has an "either/or" choice. The lawgivers may choose to declare a particular act both a crime and a civil wrong, as in the case of assault and battery. An act of this nature may be both criminally and civilly wrong, in which case the victim may proceed civilly and the state may prosecute. Both avenues are open, and the outcome in one does not determine the outcome of the proceedings in the other.

Although this explanation is factually correct, it is somewhat mechanical and simplistic. Many other social and political, as well as legal, considerations affect legislative prescriptions of criminal and noncriminal wrongs. For example, a criminal conviction can be considered by the trier of fact (jury or judge in a nonjury trial) in a civil case against the same defendant. On the other hand, an acquittal in a criminal case may not be used by the defendant for his or her own benefit in a subsequent civil case. If a civil case against the defendant ends before the criminal case begins, the outcome of the civil case may not be considered in the criminal proceeding regardless of which side prevailed.

The purpose of the criminal law is twofold. First, it attempts to control the behavior of human beings. Failing this, the criminal law seeks to sanction uncontrolled behavior by punishing the law violator. Within the framework of criminal law, punishment may take one of three forms: fine, imprisonment, or death. The advantages, disadvantages, and effectiveness of imprisonment and death involve some of the most controversial social problems in society today. However, an in-depth study of these problems is beyond the scope of this book.

1.3   EARLY DEVELOPMENT OF CRIMINAL LAW

Criminal law is an offspring of personal vendetta. At some time in the development of each society, when one person injured another, it became the **responsibility** of the victim or the victim's family to seek redress. The community in no way became involved. This led to the theory of **retributive justice**. The Code of Hammurabi, circa 2100 B.C., codified the rules that called for punishment to fit the crime: "An eye for an eye." Eventually, despite the setback of the Dark Ages, societies began treating certain offenses as crimes against the sovereign, and the government began punishing individuals who committed offenses against the public. This practice became the keystone of modern criminal law.

During this developmental period, the law did not take the responsibility of the accused into consideration. That is, the law did not ask why a person had committed a crime or whether such individual was accountable for his/her actions. Defenses such as insanity, justification, excuse, intoxication, infancy, and the like were not considered. The mere doing of the act was all that was required to show the commission of a crime. Today, of course, this is no longer true in most instances. The fact that a person commits a wrongful act does not make that act criminal until the perpetrator is convicted, because of the defenses that may bar conviction. These defenses are discussed in Chapter 6.

The key to the doctrine of responsibility is the legal approach to human psychology. The law is based on the assumption that people act of their own free will. Their fate is not predetermined or predestined. Therefore, the law may hold people accountable for their actions. If the law accepted the concept of determinism, it would hold that people are not responsible for their conduct; everything that they do would be predestined, determined by their early environment and their genetic history. Individuals should not be on trial for their criminal acts but, rather, fate should be on trial. This theory, for obvious reasons, would be totally unacceptable in any society as a legal theory. Many deterministic theories have been propounded throughout the history of criminal law. Some of them have been either wholly or partially rejected. The great Italian sociologist Lombroso felt that he could predict criminality and guilt by measuring the accused's head, ears, nose, or some other area. Fortunately, this theory was rejected. In 1968 an Australian court acquitted a man charged with murder on the theory that he was born with imbalanced chromosomes, and that as a consequence of this physiological deficiency, he was destined to commit crimes for which he could not be held legally accountable. The defendant in this case had the support of several medical professionals. Shortly after the decision of this court, a commission was appointed in the United States to research the feasibility of applying this theory. Their conclusion was that there was no correlation between the existence of XYY chromosomes and criminality.

1.4 LEGAL SYSTEMS AND THE BEGINNING OF THE COMMON LAW

Two major legal systems prevail throughout nine-tenths of the civilized world: the civil law and common law systems. The **civil law system** is the predominant legal system of the civilized world; the common law system is prevalent in England, its dominions, and North America. These systems had their beginnings in completely different ways.

The **common law** began as a result of the habits of individuals and the customs of groups. These habits and customs were so entrenched in society that they became the acceptable norms of behavior. When courts developed, violations of these customs produced the cases heard. The courts began recording their decisions, and judges looking for assistance started following previous court decisions when confronted with new cases. This procedure became known as **stare decisis**—the following of precedents. Thus the customs of the common people became the source of the common law, the law of the common people.

The remainder of the world grew under a different system of law, the civil law. We can trace this system back at least as far as the Roman Empire, where laws were written and codified by the rules of the "state" and imposed on the people. As will be seen, the law of the United States is a combination of common law and civil law. The two systems of law began at opposite ends of the legal spectrum. The common law was developed by the common people and was imposed on the rulers of the country. The civil law was developed by the rulers and imposed on the people. Of course, this is a highly simplified explanation of the development of the legal systems, but it will serve as a useful frame of reference.

1.5 COMMON LAW IN THE UNITED STATES

The English colonists who settled America brought with them a large part of the body of law to which they were accustomed—the English common law. As a consequence of this, and of their political dominancy, this system predominated in the colonies with certain modifications—modifications caused by the feeling that certain English laws were oppressive and that it was these laws the colonists had come to America to escape.

Under the federal–state relationship established by our **Constitution,** each state in the United States is sovereign under a federal government. Consequently, each state is free to decide whether it will select the common law system or the civil law system as the basis for its criminal law. The basic difference between the ways these systems operate is that under the common law, any act that was criminal under the old common law remains criminal today, even though it is not found in statutory form. Under the civil law, all crimes are statutory. In the absence of a statute, there can be no crime.

Presently, 36 states have adopted substantive criminal law codes and at least impliedly have abolished common law offenses including: Alabama, Alaska, Arizona, Colorado, Connecticut, Delaware, Florida, Georgia, Maine, Minnesota, Missouri, Montana, Nebraska, New Hampshire, New Jersey, New Mexico, New York, North Dakota, Ohio, Oregon, Pennsylvania, South Dakota, Tennessee, Texas, Utah, Virginia, Washington, and Wyoming. In the remainder of the states, the common law survives either expressly or by implication.

Even in jurisdictions that have abolished the common law, reference is still made to the common law for definitional purposes. For example, the state of Georgia has abolished all common law offenses. The Georgia statutes make murder a crime, however, and explain the situations under which that crime may be charged. But nowhere in Georgia statutes is the word *murder* defined. The Georgia statute reads "A person commits the offense of murder when he unlawfully and with malice aforethought, either express or implied, causes the death of another human being." Thus it is necessary to look to the common law for a definition of the term.

Today, when most crimes are statutory, how significant is the distinction between the common law system and the civil or statutory law system? The common law states have a distinct advantage in being able to reach back into the common law to find additional offenses that might not be covered by statute in their jurisdictions. Although this is a rare occurrence, these states have the power to look to the common law if an offensive act occurs that is not covered by statute. If the offense was punishable at common law, then it is punishable in those states today. An example of this reaching back to the common law occurred in Pennsylvania when the defendant made numerous obscene telephone calls to the complainant. Apparently, Pennsylvania had no statute governing this type of behavior, so the trial court looked to the common law and found a misdemeanor that in substance was defined as "contriving and intending to debauch and corrupt the morals of the citizens." The court invoked this offense and convicted the defendant. The conviction was affirmed on appeal. This conviction could not have been obtained in a jurisdiction that had abolished the common law offenses if there were no statute making this type of conduct criminal.

Any number of offenses that are the constant subject matter of police investigation today were unknown to the common law. The inventive genius of the criminal mind, accompanying the various stages of historical, industrial, informational, and sociological development, has created new antisocial conduct against which society needs protection. Legislatures, in response to these new pressures, have established new offenses by statute. The list of legislatively established crimes could go on here for a number of pages. We will examine many of them in this book. For an example of legislative response, note that embezzlement, as discussed in Chapter 10, was created by statute and was not a common law crime.

The federal judiciary has no power to exercise common law jurisdiction. This comes about not by choice but by mandate. The federal government has only certain enumerated powers. This means that it can exercise only those powers that have been granted to it by the people. The people have given the Congress the power to enact laws but not to adopt the common law. Therefore, the federal judiciary can exercise authority only over crimes enacted by Congress. The federal judiciary, however, like the states, must look to the common law for definitions to aid in interpreting federal laws.

To illustrate, one of the elements of the common law crime of larceny (see Chapter 10) requires that the thief intend to deprive the rightful owner/possessor of the property (steal). Congress created a crime of larceny when property was taken from a federal installation. The statute made no mention of requiring an intent to steal. The defendant was hunting on a military reserve and came across large brass shell casings which he thought were abandoned and unwanted by the military. He put them in his truck and was later arrested for larceny. He argued that he had no intent to steal because he mistakenly believed the casings had already been abandoned and that no one else wanted possession. The U.S. Supreme Court said that when the word *larceny* is used in the federal statute and Congress did not specifically and forcefully eliminate a common law element, the crime must be interpreted to include the common law elements, including the intent to steal. The defendant's conviction was reversed.

▨▨▨ DISCUSSION QUESTIONS ▨▨▨

1. John has committed an act that under the common law of England would be criminal. The same act is not made criminal by any statute of any state or by federal statute. Can such an offense be prosecuted successfully in the federal courts or in your state? Why or why not?

2. In the study of criminal law, why is it essential to understand the significance of the common law and its effect on the law of the United States?

3. What is meant by the statement that the Constitution must be stable yet flexible?

4. Of what significance is the concept of *responsibility* with regard to the criminal law today?

▨▨▨ GLOSSARY ▨▨▨

Civil law system – one of two major legal systems in the world. All laws are codified and originate with the government.

Civil wrongs – private suits between individuals, usually for money damages. These cases do not involve the government.

Common law – one of two major legal systems in the world; developed out of the customs and habits of the people; the cornerstone of most of the law in the United States.

Communication – the essence of the language of the law; the ability and the skill to ensure that the sender and receiver understand the same thing.

Constitution – the basic governing document of the United States; the foundation for our federal legal system and the relationship between the federal government and the states.

Crime – a public wrong; an act or omission forbidden by law for which the state prescribes a punishment in its own name.

Language – the basis for the functioning of the law.

Law – a group of rules governing interaction; a set of regulations governing the relationships among people and between people and their government.

Responsibility – a person may be held accountable for his or her acts and the consequences of those acts because of the exercise of free will.

Retributive justice – a philosophy that the punishment ought to fit the crime; an "eye for an eye" philosophy.

Stable yet flexible – describes the United States Constitution as a document stable enough to last more than two hundred years, serving as the foundation of our government, yet be flexible enough to be interpreted in light of contemporary social issues.

Stare decisis – a Latin term meaning "following precedents" of earlier court decisions.

Substantive criminal law – the branch of the criminal law dealing with the definitions and elements of crimes.

■ REFERENCE CASES, STATUTES, AND WEB SITES ■

CASES

State v. Branson, 487 N.W.2d 880 (Minn. 1992).

State v. Palendrano, 293 A.2d 747 (Superior Ct. of N.J., 1992).

State v. Dobbins, 171 Ohio St. 40 (1960).

Gideon v. Wainwright, 372 U.S. 335 (1963).

Commonwealth v. Mochan, 110 A.2d 788 (Pa. 1955).

STATUTES

Georgia: O.C.G.A. §. 16-5-1(a) (2000).

Pennsylvania: 18 Pa. C.S.A. §. 5504 (2000).

WEB SITE

www.findlaw.com

Fundamentals of Criminal Law

■ KEY WORDS AND PHRASES ■

Concurrent jurisdiction
Corpus delicti
Constitutionality
Ex post facto
Felony
Grant of power
Lesser and greater included offenses
Limitation on power

Mala in se
Mala prohibita
Misdemeanor
Model Penal Code
Morality
Moral turpitude
Municipal ordinance
Public wrong
Vague and ambiguous statute

2.0 INTRODUCTION

As is true in any endeavor, it is necessary to learn some basics of a field of study before progressing to more complex concepts and applications. The student should be able to distinguish between moral issues and legal issues. Understanding the differences and the underlying reasons gets the reader/student out of a social mind set and into a legal frame of reference necessary to appreciate the more complex legal issues presented in this and later chapters.

2.1 *MORALITY* AND THE LAW

The student of criminal law must be able to approach the subject with an open mind and as objectively as possible. There is often a difference between what is morally wrong and what is legally prohibited. Legal problems should not be settled by resorting to emotion, because people who fall prey to emotion discover many wrong answers. This is not to say that there is necessarily a right or wrong answer to a legal question. If this were true, our system would not require the services of attorneys and judges. The facts of a particular case could be fed into a computer and the "right answer" retrieved mechanically.

An act may be committed that is obviously morally wrong in the eyes of most people but for which there is no legal penalty. Suppose that a young girl has swum too far from shore and is struggling to keep from drowning. A man on the beach observes this activity but takes no action to save the girl even though a rescue attempt would involve no danger to his own safety. Instead, he remains on shore and takes photographs of the drowning girl with the thought of having them published in a national magazine. The girl drowns. Most people would agree that this man's conduct was morally wrong and that there probably is some clear-cut crime here with which he should be charged. Although morally this conduct cannot be tolerated, legally no offense has occurred.

Of course, criminal law, like other areas of law, is not completely devoid of moral considerations. The next section of this chapter is concerned with the classification of crimes. Historically, offenses have been classified according to their severity and the threat they pose to the public welfare. It must be recognized that this categorization was based on the precepts of society at the time of classification. The seriousness of the offenses is really a moral consideration.

2.2 CLASSIFICATION OF CRIMES

Crimes are classified in many ways. Among them is the distinction made between offenses *mala in se* and *mala prohibita.* Crimes *mala in se* are defined as those bad in themselves, morally as well as legally wrong. Murder and rape would be among the crimes classified this way. Historically, all common law crimes were *mala in se.* Statutory crimes were not classified this way by legal philosophers even at common law. *Mala prohibita* crimes are those that are wrong simply because they are prohibited though they involve no **moral turpitude**. Other significant differences between offenses *mala in se* and *mala prohibita* are presented in Chapter 5.

Basically, moral turpitude is depravity or baseness of conduct. Most authorities, including the courts, agree with this. However, whether a particular prohibited act constitutes depravity or baseness depends on the attitude of the people and is usually determined by the courts, based on the facts and circumstances on a case-by-case basis. A criminal act that may be *mala in se* (involving moral turpitude) in one jurisdiction may not be in another.

The most popular common law classification of crimes—into the three categories of treason, felonies, and misdemeanors—is perhaps the most workable. This basic classification system and its distinctions between types of crimes remains the preferred system in most jurisdictions. In fact, most recent legislative modifications in this area have involved distinguishing various degrees of misdemeanors and felonies, as will be described later. Although treason is considered to be the most serious of all crimes because it threatens the very existence of the nation, its rarity of occurrence precludes further treatment in this book.

At common law a **felony** was defined as any crime for which the perpetrator could be compelled to forfeit his property—both real and personal—in addition to being subject to punishment through the procedures of death, imprisonment, or fine. The common law felonies were murder, manslaughter, rape, sodomy, larceny, robbery, arson, and burglary. Statutes today make other crimes felonies, but these were the only felonies of common law. The key to distinguishing felonies from misdemeanors was not the punishment that could be imposed but whether forfeiture was required.

The word *felony*, like **misdemeanor**, is just a label used to define a class of offenses, and these labels do not apply uniformly throughout all jurisdictions. Each jurisdiction is free to call a criminal violation by any name it chooses and to impose such punishment for violations as it desires, providing it does not violate the Eighth Amendment of the U.S. Constitution protecting individuals from cruel and unusual punishments. As a result, there are some differences. For example, in Maryland:

It is a misdemeanor to use a machine gun to commit or attempt to commit a crime. Injuring a race horse is a felony with a maximum penalty of 3 years imprisonment. Abducting a child under 16 years of age is a felony with a maximum of 30 years in prison, but abducting a child under 16 years of age for prostitution or fornication is a misdemeanor carrying a maximum penalty of 8 years; abducting a child under 12 years of age from his home is a felony carrying a 20-year maximum sentence; injuring a railroad car is a misdemeanor with a 5-year maximum penalty; marrying a close relative carries a $1,500 fine and banishment from the state forever.

Today, the law does not require forfeiture of property for committing a felony, and therefore the common law rule is no longer applicable. However, most jurisdictions still maintain the distinction between felonies and misdemeanors. Some jurisdictions distinguish on the basis of where the imprisonment is to take place. Most states distinguish felonies from misdemeanors on the basis of the place of imprisonment or the length of confinement. The remaining jurisdictions seem to use character of the offense or the place of imprisonment or both to make this distinction. In many of those jurisdictions, the criteria are difficult to comprehend. In some states, the length of imprisonment is the deciding factor, whereas in others it is difficult to determine what criteria are used to differentiate between felonies and misdemeanors.

On the federal level, persons convicted of crimes for which imprisonment is imposed are all sentenced to federal prisons. Thus the distinction between felonies and misdemeanors on the federal level cannot be based on the place of imprisonment but rather on the length of imprisonment. A felony under federal law is any crime for which the penalty is death or imprisonment for a term exceeding one year.

Why is it important to know the distinction between a felony and a misdemeanor? Much of the treatment of the accused hinges upon this distinction. The procedural steps that may be taken by law enforcement officers in the performance of their duties depend on whether they are dealing with a felony or a misdemeanor. As is illustrated in detail in Chapter 6, the amount of force an officer may use to apprehend a person for commission of a crime will be based on the type of crime for which the individual is being arrested. In addition, one convicted of a felony will lose his/her civil rights, whereas a convicted misdemeanant will not. Even more importantly, the arrest powers that an officer or private citizen has are governed by the classification of the crime for which the arrest is being made.

Whether a crime is a felony or a misdemeanor is governed by the maximum punishment that can be imposed by the courts for a conviction of the offense and not by the punishment that is actually imposed. For example, if Andy commits grand larceny in a state where a felony is defined as any crime punishable by imprisonment in the state prison, assuming grand larceny carries a maximum penalty of five years in the state prison in that state, Andy is guilty of a felony regardless of the fact that the court may sentence him to spend only six months in the county jail.

All crimes carrying penalties less than those imposed for the commission of felonies are misdemeanors. Thus, by the process of elimination, it may be determined which crimes are misdemeanors in any particular jurisdiction.

2.3 ENACTING AND INTERPRETING STATUTES

Most people automatically think of recent U.S. Supreme Court decisions the instant the word **constitutionality** is mentioned. Constitutional law is a broad subject in its procedural applications and is beyond the scope of this book. The following is a highly simplified explanation of who enacts laws and on what authority laws are enacted, in addition to a short summary of a few basic rules observed when a statute is interpreted.

Two phrases, **grant of power** and **limitation on power**, are essential to an understanding of this subject. The federal government exists because the people of the various states, in whom rests total sovereignty, created it.

The American Law Institute has proposed the **Model Penal Code**. Several states have adopted this code with some qualifications and changes. The Model Penal Code (MPC) creates different classes of crimes.

Under the MPC a crime is a felony if it is so designated, no matter what the penalty. Any crime that creates a punishment exceeding one year is also a felony under the MPC. Further, the MPC creates degrees of felonies. First- and second-degree felonies are those so labeled in the MPC. A crime that is a felony but for which there is no designated degree is a third-degree felony. This creation of degrees is for the purpose of sentencing. Certain fines attach to the various degrees of felonies as a result. A first- or second-degree felony carries a $10,000 fine, a third-degree felony a $5000 fine, unless higher (or lower) amounts are specifically attached to a specific offense. Punishments for the different degrees of felony have a fixed maximum and minimum length of imprisonment ranging from one year to life, to one to five years.

A misdemeanor is one designated as such by the MPC no matter what the penalty is. The MPC also adds a new term, petty misdemeanor, and says that a petty misdemeanor is a crime so designated or one where the penalty is for less than a year. A misdemeanor carries a $1000 fine, and a petty misdemeanor a $500 fine, unless higher (or lower) amounts are attached to a specific offense. As under previous law, an undesignated offense providing a sentence is a misdemeanor under the MPC.

The MPC also creates a new class of offense called violations. A violation usually is named as such and, most importantly, provides no jail sentence. It involves only a fine or forfeiture of bond. Because a violation is not a crime, there is to be no disability or legal disadvantage for someone convicted of a violation. Many of the *mala prohibita* offenses are violations under the MPC. A number of states, even though they have not adopted the MPC in total have adopted its method of classifying offenses. Only New Jersey and Pennsylvania have adopted major portions of the Code without changes. Thirty-three other states have adopted portions of the MPC.

It has only the authority specifically given it by the people of the states. The instrument by which the people granted this authority is the U.S. Constitution. In essence, the people stripped their state governments of certain powers, such as governing commerce, coining money, and conducting wars.

For this reason we call the U.S. Constitution a grant of power. The federal government exercises this authority through acts passed by Congress. To determine the constitutionality of these acts, the courts look to the Constitution to see if the people gave the federal government the power to pass laws on the particular subject with which a congressional act deals. If the power and authority can be found in the Constitution, and if Congress acted within the limits set by the people in dealing with the particular subject of the act, it is constitutional. If Congress was without the authority or acting outside the scope of that authority, the act is unconstitutional.

Turning to the state constitutions, which are called *limitations on power*, it can be seen that any powers not specifically granted to the federal government through the U.S. Constitution are reserved to the states or its people. Thus the states are free to exercise any legislative power not given to the federal government unless there are certain rights that the people of the states wish to reserve to themselves, not allowing even their own state legislatures to have a say in this regard. Thus the people, through their respective state constitutions, limit the powers of state legislatures. They may prohibit their state from having an income tax or prohibiting gambling that would otherwise be permissible. It is for this reason that state constitutions are referred to as *limitations on power.* State constitutions are therefore usually negative in their application, if not in their language. State legislatures may enact laws on any subject for which authority has neither been granted to the federal government nor prohibited to them by the people of the state.

To test the constitutionality of a state statute, the state courts first determine if the authority to legislate in the particular area has been granted exclusively to the federal government. For example, if a state enacted a criminal statute prohibiting counterfeiting of money, the court would find that this power has been exclusively granted to the federal government. In this case, the state would not have the authority to pass such a law and the statute would be unconstitutional.

If this authority has not been granted to the federal government through the U.S. Constitution, the court must then look to the state constitution to determine if the people forbade the legislature from enacting such a law. Finding no such prohibition, the court can declare the statute constitutional. It is a principle of law that courts will make every effort to hold statutes constitutional. Only when there is no way of so holding will they declare them to be unconstitutional.

In this section we have implied that the authority to enact laws on any particular subject rests with either the federal government or the state legislatures. However, there are times when the two levels of government will have **concurrent jurisdiction**. Bank robbery, for example, is a federal offense if the bank is insured by the Federal Deposit Insurance Corporation. The act would constitute robbery under state criminal laws. As a result, there are differences among the states on the same subject matter. Federal legislation must apply uniformly to all states on any subject, whereas each state legislature can enact laws on the same subject pertaining only to its geographical boundaries.

One further principle governs the determination of the constitutionality of statutes. Because people are entitled to know what conduct is prohibited by law, statutes cannot be written in such broad terms as to make unclear the type of conduct prohibited. Such statutes will be held unconstitutional for vagueness.

An illustration of what constitutes a **vague and ambiguous statute** was reported in a 2000 federal case. An Arizona statute criminalized any medical "experimentation" or " investigation" involving fetal tissue from induced abortions unless necessary to perform a "routine pathological examination" or to diagnose a maternal or fetal condition that prompted abortion. The plaintiffs in a suit challenging the constitutionality of the statute included individuals suffering from Parkinson's disease who, because of the statute, were unable to receive transplants of fetal brain tissue, which many medical experts believe holds out promise for eventual amelioration or treatment of the disease. Also challenging that statute were doctors, who feared possible criminal prosecutions if they provided these potentially beneficial services to their patients. The federal trial court and the Circuit Court of Appeals both found the statute to be unconstitutionally vague.

Criminal laws must be written with sufficient definitions so that ordinary people can understand what conduct is prohibited, and they must be written so as not to encourage arbitrary and discriminating enforcement. In this instance, the Court felt that since there was no standard to be followed, there was too much opportunity for "moment-to-moment" and officer-by-officer judgments. The Court ended by saying that while the Constitution does not require "impossible standards of clarity," it does require some clarity so that personal biases and prejudices do not become the standard.

There are some instances, however, where a statute can be very clear and specific and yet be written so broadly as to make otherwise innocent conduct a violation. When that happens, courts will probably find the law invalid. For example, the Utah Supreme Court ruled on a city ordinance that made it illegal "in public or in a public place to solicit another to engage in sex." The ordinance did not mention anything about money or other consideration of value as part of the solicitation. The court said that the ordinance would also prohibit husbands and wives from suggesting sexual intercourse "as they strolled through the park." In Houston, an ordinance made it crime to "assault, strike, or in any manner oppose, molest, abuse, or interrupt any policeman in the execution of his duty." The defendant was convicted under the ordinance for yelling to the police, "Why don't you pick on someone your own size?" The U. S. Supreme Court found the ordinance to be substantially too broad because the law granted unfettered discretion to police to arrest individuals for words or conduct that annoyed or offended them.

To avoid confusion, a distinction must be made between the principle discussed in Chapter 1 that statutes are broadly and flexibly written so as to be capable of interpretation and the constitutional issue of vagueness presented here. Under the broad yet flexible philosophy, the standards of conduct are clearly defined, while situations may differ. When a statute is found to be too vague, the standards of conduct are either unclear or absent.

Our legal system is based on the theory that every benefit of the doubt will be accorded to the defendant in criminal cases. This is like the rule in baseball games that a tie goes to the runner. Based on this principle, we take the view that criminal statutes are to be strictly construed when they work against the accused in a criminal case and liberally construed when they benefit the accused. This means that if the statute works against the defendant, it would be interpreted in its narrowest sense so that only the specified conduct would be included. If, however, the statute works a benefit to the defendant, it is broadly construed so that his or her conduct would be interpreted in the most favorable way.

2.4 *EX POST FACTO* LAWS

The U.S. Constitution, in Article I, Sections 9 and 10, prohibits the passage of *ex post facto* laws either by the federal or the state governments. In brief, an *ex post facto* law is one that alters the laws regarding a particular act in such a way as to be detrimental to the substantial rights of an accused person. This can occur in any of three ways. First, if at the time a person commits an act, that act is not criminal but is subsequently made a crime by legislative action, the person cannot then be prosecuted for its violation. Any attempt to prosecute in such a case constitutes an *ex post facto* application of the law. A person is entitled to know what, if any, violation occurs at the time when he or she commits an act. Thus if the act is not criminal at the time committed, no punishment can be imposed.

> In an 1866 case, the U.S. Supreme Court held that post-Civil War Missouri laws banning former Confederates from the Ministry and requiring lawyers to swear that they had not supported the Confederacy as a condition of practicing law in federal courts, violated the *ex post facto* clause of the Constitution.

The second situation to which the rule applies involves the increasing of punishment for a specific crime after it has occurred. Suppose, for example, that Fred commits grand larceny for which the maximum penalty is five years' imprisonment. Before his trial, the legislature increases the maximum penalty to ten years' imprisonment. Any application of the more severe penalty to Fred would be *ex post facto* because it does affect his substantial rights. He can be punished only to the extent provided at the time he committed the criminal act. The U.S. Supreme Court reinforced this position in a case where a defendant was given a harsher sentence under Florida's sentencing guidelines than he would have received prior to the guidelines. The Court said that since the act was committed before the sentencing provisions, the imposition of a harsher punishment raised an *ex post facto* defense.

Third, the *ex post facto* rule applies to decreasing the state's burden of proof. It is *ex post facto* if the legislature decreases the amount of proof the state will be required to produce to convict for a crime. The reason for this is that it would be detrimental to the rights of the accused if the state could convict the accused more easily than it could have at the time the act was committed. But other changes in the rules of evidence, those that do not materially affect the defendant's rights, can be made. In a recent case, the U. S. Supreme Court found a Texas law violated the constitutional prohibition against *ex post facto*. The court ruled that the defendant was wrongfully convicted of fifteen counts of committing sexual offenses against his stepdaughter. The alleged conduct had taken place between 1991 and 1995, when the girl was 12 to 16 years of age. In 1993, Texas amended a statute authorizing conviction of certain sexual offenses based on the victim's testimony alone; the previous statute required the victim's testimony along with other corroborating evidence to convict. The defendant was convicted based on the victim's testimony alone for all the offenses back to 1991. The Court held that for the alleged conduct occurring before the effective date of the law change in 1993, there had to be corroborating evidence to support the victim's testimony. The conviction of those offenses before that date on the lower burden of proof violated the *ex post facto* prohibition.

If the reverse of any of these situations exists, for instance, as when the punishment is decreased or the burden of proof on the state is increased between the time of the commission of the crime and trial, these changes will benefit the defendant and will not be considered *ex post facto*.

Similarly, if the statute making a certain act criminal has been repealed after the accused committed the act, proceedings against the person must be dropped. The conduct must be criminal both at the time of the act and during the course of proceedings to punish the accused. This is true even if the crime is repealed any time before final appeals have been exhausted. Generally, changes in the law that relate to procedure and jurisdiction and that are to be applied retroactively do not come within the prohibition of *ex post facto* laws.

2.5 STATUS OF MUNICIPAL ORDINANCES

Municipal police officers spend most of their working time enforcing municipal ordinances. This is not criminal law enforcement because violations of ordinances are not crimes. The distinction between crimes and violations of ordinances lies in the definition of crime and in the nature of municipal corporations.

A crime has been defined as a public wrong created by the state and prosecuted by the state in its name. **Public wrong** is interpreted as a wrong affecting the people of the entire state, not just of a particular portion of the state. Municipal ordinances are not enacted by the state legislature, nor are they punished by that lawmaking body. Ordinances are enacted by city councils or commissions, affecting only the municipality, and are punished by the city in its name when violated.

A municipality is a corporation like any other corporation, except that it is a public corporation. Municipalities are created by the state legislature and exist at the whim of that body. Municipalities can be dissolved when and if the legislature so chooses unless they have been established under constitutional home rule.

Like other corporations, municipal corporations are required to have a charter granted by the state. Private corporations have rules for governing internal operations, called bylaws. **Municipal ordinances** are the bylaws of municipal corporations. Violations of some ordinances do not carry penalties and therefore are of little concern to municipal police officers. Violations of other ordinances do carry penalties, and subject the violator to arrest and trial, but these are not crimes. They are more closely associated with civil wrongs. For this reason, violations of ordinances are often called *quasi criminal* in nature. It is difficult to define this term except to say that it lies somewhere between a criminal wrong and a civil wrong. Because the state enacts criminal offenses and because such offenses must be applicable throughout the state on a uniform basis, the state legislature cannot delegate this power.

It would appear that individuals may be fined or imprisoned for violation of a municipal ordinance in the same manner as they might for the commission of a crime. In theory, a fine is treated as civil damages for wronging the municipality. Failure to pay a fine subjects the violator to imprisonment. This is not imprisonment for nonpayment of a debt but imprisonment for failure to obey a lawful court order. A municipal court is empowered to assess a fine. If the fine is not paid, only then can the defendant be imprisoned under the theory that he or she has failed to comply with an order of the court. This theory is followed in the states of Georgia, Louisiana, Minnesota, Missouri, Montana, New York, and Wisconsin.

The real problems thus presented are procedural. For example, is the violator of an ordinance entitled to a trial by jury or a summary (bench) trial (by a judge)? In some states neither felonies nor high misdemeanors may be the subject of municipal ordinances. Other states allow municipalities to "track" or adopt verbatim a state law without any variation. If a city may and does adopt a state statute as an ordinance and the offense is of a grave character, the right to a jury trial attaches. If the city prosecutes the case in the name of the state, all constitutional rights must be afforded. The higher the misdemeanor at common law, the more substantial is the guarantee of a jury trial. This principle is still true under modern law. The U.S. Supreme Court has held that although many factors have to be considered, a sentence exceeding six months is sufficiently severe by itself as to require a jury trial. On the other hand, just because an offense carries a maximum penalty of six months or less does not automatically mean that no jury trial may be warranted.

Another constitutional issue—right to counsel—is also affected by the status of municipal ordinances. The case of *Gideon v. Wainwright* decided by the U.S. Supreme Court in 1963 made the right to counsel obligatory on the states. Until 1972, the ruling in *Gideon* was not applied to minor misdemeanors and ordinance violations requiring the state to provide counsel to indigent defendants. But in 1972 the Supreme Court held that absent a knowing and intelligent waiver, no person may be imprisoned for any offense, whether classified as a petty misdemeanor or felony, where there is a substantial chance of incarceration, unless that person was represented by counsel at the trial. The Court said these so-called trivial matters can result in serious repercussions affecting the career or reputation of the defendant. The net effect is that unless a city wants to pay for attorneys in all cases involving indigent defendants, it should impose only fines on ordinance violators, because no jail time can be imposed without the defendant's having had the opportunity to have an attorney. In fact, a judge has the discretion to decide, often at the behest of the state, before trial that no jail time will be given and thus avoid the need to appoint a lawyer. Once this happens, of course, the judge cannot change the decision.

2.6 CORPUS DELICTI

To fully understand what is necessary to prove a case in court, a law enforcement officer must be aware of the factors that constitute a particular crime. This involves the concept of *corpus delicti*.

Contrary to some popular belief and most comic-strip detectives, the *corpus delicti* is not the body of a victim of a homicide but rather the body of the crime. Every offense consists of distinctive elements, all of which must exist for a particular crime to be proven. If one or more of these elements is missing, the crime thought to have occurred, in fact, could not have been committed. What might have been committed was another crime requiring proof of those elements that do exist. The combination of these elements in a particular offense is called the *corpus delicti*.

As an example, at common law, the *corpus delicti* of burglary consisted of the following elements: breaking, entering the dwelling house of another, in the nighttime, with intent to commit a felony therein. All these elements had to exist before the crime of burglary could properly be charged at common law. If, instead of a dwelling house, a store is broken into, burglary has not been committed because one element of the *corpus delicti* is not present. Instead, another crime has been committed. If the defendant is charged with burglary, the state would be unable to prove that burglary has been committed and the case would be thrown out of court.

As noted above, the *corpus delicti* issue has gained its primary renown in homicide cases. To prove a felonious homicide (e.g., murder), one must prove that a person existed and that this person is now dead at the hands of a human agency. There is no lack of cases where innovative defendants have done ghoulish deeds to get rid of the body, believing that these acts disposed of the *corpus delicti*. However, proof of the *corpus delicti* can be direct or circumstantial. Even when there has been a proper confession, our judicial system is reluctant to convict anyone of homicide without verifying the felonious nature of the death. In one case, a baby who could not crawl was found dead in a creek. The state was able to prove live birth, and the fact that the baby was in the creek meant someone or something had put the baby there. Since the last person seen with the baby was the mother, who was ashamed of having had an illegitimate child, the court felt that the *corpus delicti* was properly established.

A totally missing body can be most dramatic. In a Kansas case, the defendant was accused of killing his grandmother. He was tried and convicted. In summation, the court said that the facts established at the trial were sufficient to establish *corpus delicti* even though the grandmother's body was never found. The facts revealed: (1) that the grandmother had not been heard from after April 5 by her friends and customary associates; (2) the existence of a strained relationship between grandson and grandmother; (3) the carefully constructed alibi consisting of the grandson's fabricated drunkenness and an all-night trip to Colorado; (4) the false-alarm fire of two days before; (5) the grandson's avowed intention to get the ranch "one way or another;" and (6) the grandson's premature knowledge of the real fire and his grandmother's death.

In summary, the state must establish that (1) a criminal law has been violated; (2) the violation was not the result of an accident or misfortune and was not self-inflicted; (3) a human agency caused the violation; and (4) the defendant was the human cause of the violation.

2.7   LESSER AND GREATER INCLUDED OFFENSES

In substantive as well as procedural criminal law, the doctrine of **lesser and greater included offenses** plays an important role. The doctrine affects decisions as to what crimes to charge, what plea negotiations take place, what instructions will be given a jury, what impact decisions will have on double jeopardy law, and what verdicts may be rendered by juries.

To understand the doctrine as it originated at common law, one must think of crimes as a series of chains. The concept will become clearer as the reader progresses through this book, but let us use murder as an example. Murder is the most serious offense that can be committed against the body of a human being. (This will be verified in Chapter 8.) There are, however, other less severe offenses against the body. These include simple assault (the threat to do bodily harm) (covered in Chapter 7), battery (a completed assault; Chapter 7), and manslaughter (a heat-of-passion killing; Chapter 8). Thus the chain of offenses against the body or involving bodily harm are in a chain from the most severe or greater included—murder, which is the greatest in this example, to the least severe—simple assault. Murder, then, has a number of lesser included (less severe) crimes, such as manslaughter, battery, and simple assault (in descending order of bodily harm). On the other hand, simple assault has a number of greater included crimes, which in ascending order of bodily harm include battery, manslaughter, and murder.

There are separate chains for other types of crimes. For example, as the reader progresses through this book, it can be seen that battery is the base crime in the chain leading up to rape and sexual battery and a petit theft is the root of the robbery chain. For each of the major felonies recognized at common law, a misdemeanor was at the base of the chain.

The doctrine tends to become confusing in modern times when so many additional offenses have been created by legislative acts. Thus, there are now additional forms of aggravated assault, such as assault with a deadly weapon and assault with intent to commit murder. Other forms of killing not known at common law have led to the enactment of criminal offenses. Vehicular homicide is an example. To determine whether these statutorily created offenses are in the chain of lesser and greater included offenses, it is necessary to determine whether they are in the same general sections of the statutes (codes) of the other offenses in the chain. For example, if vehicular homicide is grouped with other homicides in the statutes, it is included in the assault and murder chain. If, however, vehicular homicide is grouped with other motor vehicle-related offenses, it will be included in a different chain.

The significance of these distinctions by chain affects the prosecutor's decision as to what charge or charges can be brought. Depending on the provable facts, a defendant can possibly be charged with multiple offenses if they are in separate chains and either be convicted of multiple offenses or at least provide the jury with more options for conviction. Conversely, a defendant cannot be charged with multiple offenses within the same chain, although if the prosecutor charges the greatest included offense that he or she thinks can be proved, the jury is then permitted to find guilt of any lesser included offense committed if the prosecutor fails to prove the crime charged.

The U.S. Supreme Court has held that the conviction of a lesser included offense is an implied acquittal of the greater offense. Thus if a person is tried for murder but is only convicted of assault as a lesser included offense, appeals the conviction, and wins a new trial, the state cannot recharge the murder or any other offense in the chain greater than assault. To allow any other decision would place the defendant in double jeopardy.

▬▬▬ DISCUSSION QUESTIONS ▬▬▬

1. Why is it important for an officer to know and understand the difference between a felony and a misdemeanor?

2. Distinguish between the violation of a state criminal statute and the violation of a municipal ordinance.

3. By what process is a state statute declared constitutional or unconstitutional?

4. Define and describe *ex post facto* laws.

5. What are the four components for establishing *corpus delicti*?
6. Is auto theft a lesser included offense of carjacking? Why or why not?

GLOSSARY

Concurrent jurisdiction – crimes over which both state and federal courts have jurisdiction.

Corpus delicti – the body of the crime made up of the elements that comprise the act.

Constitutionality – testing whether a statute satisfies the requirements to be valid under constitutional principles.

Ex post facto – a law that alters conduct, penalty, or burden of proof in a manner detrimental to the substantial rights of an accused.

Felony – at common law, a crime for which one would forfeit all property in addition to punishment; today, generally, a crime that the punishment for which is for more than a year or incarceration in state prison.

Grant of power – the power given by the people to the federal government through the U.S. Constitution.

Lesser and greater included offenses – a series of offenses, built on a core of elements, which are greater or lesser depending on the addition or subtraction of one element from another offense nearest in the chain.

Limitation on power – the powers reserved to the people by restricting the authority of governments in state constitutions.

Mala in se – bad in itself.

Mala prohibita – bad only because it is prohibited.

Misdemeanor – generally, crimes of a minor nature, the penalty for which is less than that for commission of a felony.

Model Penal Code – A model of substantive criminal laws developed by the American Law Institute, which is adopted in whole or in part by a majority of states.

Morality – a basic philosophy of right and wrong which governs one's thoughts and behavior.

Moral turpitude – refers to conduct which is base or depraved and calls into question the morals and ethics of the actor.

Municipal ordinance – the rules or "by-laws" by which a public municipality operates.

Public wrong – a wrong affecting the people of the entire state, not just of a particular portion of the state.

Vague and ambiguous statute – grounds on which statutes can be held unconstitutional because people cannot understand how to act to be in compliance with the law.

■ REFERENCE CASES, STATUTES, AND WEB SITES ■
CASES

U.S. v. Schulte, 610 F.2d 698 (10th Cir. 1979).

State v. Horton, 139 N.C. 588, 51 S.E. 945 (1905).

State v. Grissom, 840 P.2d 1142 (Kan.1992).

State v. Ulvinen, 313 N.W.2d425 (Minn.1981).

Hough v. State, 929 S.W.2d 484 (Tex.1996).

Rose v. Locke, 423 U.S. 48,96 S.Ct. 243, 46 L.Ed.2d 185 (1975).

Carmell v. Texas, 529 U.S. 513 (2000).

Provo City v. Willden, 44 Crim. L. Rep. 2391, 768 P.2d 455 (Utah 1989).

Miller v. Florida, 482 U.S. 423, 107 S.Ct. 2446, 96 L.Ed.2d 351 (1987).

Forbes v. Napolitano, 247 F.3d 903 (9th Cir. 2000).

Duncan v. Louisiana, 391 U.S. 145, 88 S.Ct. 1444, 20 L.Ed.2d 799 (1963).

Gideon v. Wainwright, 372 U.S. 335, 83 S.Ct. 792, 9 L.Ed.2d 799 (1963).

Argersinger v. Hamlin, 407 U.S. 25, 92 S.Ct. 2006, 32 L.Ed.2d 530 (1972).

City of Houston, Texas v. Hill, 482 U.S. 451, 107 S.Ct. 2502, 96 L.Ed.2d 398 (1987).

RAV v. City of St. Paul, 505 U.S. 377 (Minn. 1992).

Chicago v. Morales, 527 U.S. 41 (1999).

Weaver v. Graham, 450 U.S. 24 (1981).

Lindsey v. Washington, 301 U.S. 397, 57 S.Ct.797, 81 L.Ed. 1182 (1937).

Allison v. Kyle, 66F.3d 71 (1995).

Isaacs v. U.S., 159 U.S. 487 (1895).

Miles v. U.S., 103 U.S. 304 (1880).

Kansas v. Pyle, 216 Kan. 423, 532 P.2d 1309 (1975).

Ball v. U.S., 470 U.S. 856 (1985).

Blockburger v. U.S., 284 U. S. 299 (1932).

Ex Parte Garland, 71 U.S. 333 (1866).

STATUTES

Maryland: §§. 27-1 27-2, 27-61, 2-202, 27-338, 27-373, 27-458.

New Jersey: NJSA 2C:1-1 to 2C:98-4.

Pennsylvania: 18Pa CSA §.101

WEB SITE

www.findlaw.com

CHAPTER 3

Jurisdiction

■ KEY WORDS AND PHRASES ■

Appeal
Appellate jurisdiction
Certiorari
Concurrent or overlapping
   jurisdiction
Jurisdiction
Laws

Original jurisdiction
Personal jurisdiction
Petite policy
Rules of procedure
Subject matter jurisdiction
Territorial jurisdiction
Venue

3.0   INTRODUCTION AND DEFINITION

In designing a system to efficiently handle disputes that arise in a complex society, two decisions must be made. First, we seem to find it essential to write down guidelines for all members of society. These we call **laws**. But to write the law leaves the process only half complete. A place for interpretation and enforcement of laws must be established. Without such a place, law enforcement would depend on the individual likes and dislikes of each person. Tyranny would result.

   Through a slow process, most civilized countries have developed courts, in one form or another, to interpret and enforce their laws. Even these courts are bound by guidelines for the conduct of matters before them. The guidelines, called **rules of procedure**, govern the conduct of all the parties concerned in a case. Under our system, the rules are carefully thought out to assure procedural fairness for everyone involved. One would think that these rules would be all that is needed to ensure a person's rights. But before these rules can effectively come into play, the right of a particular court to handle the case before it must be determined.

The fact that a court has a judge or judges, a place to meet, and officers to enforce its commands does not automatically give that court the power to try a case. The court must have jurisdiction. Without jurisdiction no court can validly try or sentence a person. **Jurisdiction** is defined by the legislature under common law principles and is subject to state or federal constitutional provisions.

Jurisdiction, then, is the power of a court to handle a case. That some courts have the power to handle some matters and are denied this power in other areas is simply a matter of convenience and order. We demonstrate the basic jurisdictional problems in this chapter. Jurisdiction has three aspects: territorial, personal, and subject matter.

3.1 TERRITORIAL ASPECTS OF JURISDICTION

The first principle of **territorial jurisdiction** is that no state can enforce the criminal laws of another state or sovereign. The second principle is that a state can enforce its laws only when those laws have been broken. Normally, the arm of the law of a state cannot reach outside the state boundaries. With these rules in mind, let us consider this problem. John, a citizen of New York, goes into Delaware and murders Sam. John returns to New York and is immediately arrested by New York authorities under a warrant charging him with murder. Because New York's law against murder has not been broken, John cannot be tried on the New York charge, nor can New York enforce the laws of Delaware. Therefore, New York has no jurisdiction.

Suppose that John poisons Sam in Maryland. Sam appears to be dead, and John intended to commit murder. John puts Sam in the trunk of a car and drives to Virginia. While in Virginia, John lops off Sam's head. John hopes to leave the head in one place and the torso elsewhere so that identification will be difficult. For the first time it is realized that Sam was not dead until he lost his head. Can Maryland try John for the murder, or can it try him only for attempted murder? The common law rule was that the crime was committed where the injury occurred. In a homicide case, the place where death happened was immaterial. However, death had to be the proximate result of the act that inflicted harm before this rule applied. In this example, the poisoning of Sam was not the cause of his death. What actually caused death was John's act of chopping Sam into pieces. Thus the common law view of this situation would hold that the crime was committed in Virginia, so only that state could prosecute John for murder. John did, however, attempt to murder Sam in Maryland, and therefore an attempted murder charge would be proper in that state.

Some states have changed the common law rule by enacting statutes providing, in effect, that any crime begun in that state but completed in another is chargeable in the first state. Many of these same states provide that any crime begun in one state but completed in the second state is a chargeable crime in the completing state. Thus if two adjoining states have identical statutes of this type, it is conceivable that a crime begun in state A and completed in state B would be chargeable in both states without any implications of double jeopardy. (See Box p. 29.)

A somewhat analogous problem exists in the multi-element crime. Suppose that we have a four-element crime recognized in each of four states, A, B, C, and D. In state A, John boards a plane flying to state D. John commits the first element of the crime over state A, the second over state B, the third over state C, and the fourth over state D. Who has jurisdiction to try John? If we assume that each state has an identical statute, each would have a territorial infringement of its laws that would allow each to try John. The same rule applies to any crime committed in transit when it cannot be determined in which jurisdiction the harm was actually inflicted.

Prosecuting out-of-state conduct is no easy matter. The drug trade in the United States has led many prosecutors to seek prosecution of persons who have never set foot in their state. For example, Blume, a Florida resident, was charged with conspiracy (see Section 4.9) with a Michigan resident and as an accessory before the fact (see Section 4.11) to drug sales in Michigan. The entire sale took place in Florida. Later, the Michigan resident was apprehended selling in Michigan and gave the police Blume's name as supplier. Blume was convicted in Michigan even though he sought to have the case dismissed, claiming that Michigan had no jurisdiction over his conduct in Florida. The Michigan appeals court noted that a state may reach out-of-state conduct, but there are limits. Those limits are to acts that are intended to have and actually do have a detrimental effect within the state. The court said that even though there were both conspiracy and aiding and abetting in this case, the state had to establish that Michigan was the target state. There had to be shown a specific intent to affect Michigan detrimentally. In this case, the prosecution failed to show that the Michigan resident had to sell the drugs in Michigan and only Michigan. The prosecution also failed to show that Blume would only be paid, or paid more, after the sales in Michigan. There was no evidence that Blume even knew what state would be affected. Thus the court held that Michigan had no jurisdiction. The dissenting judges said that Blume knew that the buyer was from Michigan, which was enough to establish jurisdiction. Although the case is illustrative of the point, not all states have addressed this issue.

The developing law under the Model Penal Code has taken a position on territorial aspects of jurisdiction.

Section 1.03. Territorial applicability

1. Except as otherwise provided in this Section, a person may be convicted under the law of this State of any offense committed by his own conduct or the conduct of another for which he is legally accountable if:

   a. either the conduct which is an element of the offense or the result which is such an element occurs within this State; or

   b. conduct occurring outside the State is sufficient under the law of this State to constitute an attempt to commit an offense within this State; or

   c. conduct occurring outside the State is sufficient under the law of this State to constitute a conspiracy to commit an offense within the State and an overt act in furtherance of such conspiracy occurs within the State; or

   d. conduct occurring within the State establishes complicity in the commission of, or an attempt, solicitation or conspiracy to commit, an offense in another jurisdiction which also is an offense under the law of this State; or

   e. the offense consists of the omission to perform a legal duty imposed by the law of this State with respect to domicile, residence or a relationship to a person, thing or transaction in the State; or

   f. the offense is based on a statute of this State which expressly prohibits conduct outside the State, when the conduct bears a reasonable relation to a legitimate interest of this State and the actor knows or should know that his conduct is likely to affect that interest.

2. Subsection 1a does not apply when either causing a specified result or a purpose to cause or danger of causing such a result is an element of an offense and the result occurs or is designed or likely to occur only in another jurisdiction where the conduct charged would not constitute an offense, unless a legislative purpose plainly appears to declare the conduct criminal regardless of the place of the result.

3. Subsection 1a does not apply when causing a particular result is an element of an offense and the result is caused by conduct occurring outside the State which would not constitute an offense if the result had occurred there, unless the actor purposely or knowingly caused the result within the State.

4. When the offense is homicide, either the death of the victim or bodily impact causing death constitutes a "result," within the meaning of Subsection 1a and if the body of a homicide victim is found within the State, it is presumed that such result occurred within the State.

5. This State includes the land and water and the air space above such land and water with respect to which the State has legislative jurisdiction.

3.2 JURISDICTION OVER THE PERSON

No court can validly try a person for a crime unless that defendant is in the courtroom and is known to the court **(personal jurisdiction)**. This is so because the Bill of Rights insulates us against the obvious tyranny that could be imposed if this were not so.

How does a court get jurisdiction over a person? First, a person can consent to jurisdiction without an arrest. The person can also consent to jurisdiction by waiving the right to complain of an illegal arrest. This type of consent is no good unless the court also has jurisdiction of the subject matter (see Section 3.3).

Second, jurisdiction over the person attaches when an arrest is made and the arrest is proper. But do not be misled. The court has jurisdiction whether the arrest was illegal or legal. The court does not inquire into the manner in which defendants got before the court as long as they are there. It is up to the defendant to raise the issue of the correctness of the arrest. Practically speaking, even if a defendant wins his or her protest on the validity of the arrest, sound prosecutors will have prepared the proper papers to make a valid arrest before the defendant leaves the courthouse. The main reason for objecting to an improper arrest is to prevent the state from using evidence obtained during the arrest.

One problem in this area involves the extradition of an accused person from one state to another. When it is discovered that an accused person who is sought by one state is residing in another, the state seeking to prosecute the accused person has two choices. It can either kidnap the person, or it can use the channels provided by the U.S. Constitution and statutes for extradition. This procedure is handled at the executive level of government, not in the courts. It can be used only in criminal matters. A person can be extradited on a misdemeanor charge as well as a felony.

References in this section to kidnapping as a means of bringing a person before a court to establish jurisdiction over the person are not facetious. The principle that courts will not question how jurisdiction of a person was obtained is well-founded. This is true even if the means used constitute kidnapping. But in such cases, it is also well-established that those who are responsible may be prosecuted for kidnapping or for any other offense committed. In international cases, relationships between countries may suffer on a political level when forcible seizure of a person occurs in a foreign country for return to the United States. This may be true especially when an extradition treaty exists but is circumvented for speed and convenience. (Kidnapping and related offenses are discussed in Chapter 15.)

3.3 JURISDICTION OVER THE SUBJECT MATTER

As essential as the territorial and personal aspects of jurisdiction is the requirement that any court seeking to try an accused must have jurisdiction over the subject matter. Here, again, the legislature, limited by constitutional principles, determines which courts can hear which matters. For instance, no federal court within the continental United States, except in Washington, D.C., can grant a divorce. Even if the parties to a divorce action were to consent to the federal court's handling of such a matter, the judgment given by the federal court would be void. A judgment rendered by their friendly neighborhood butcher would serve just as well.

How do state legislatures confer subject matter jurisdiction? Some states with a simple court system confer jurisdiction based on the felony-misdemeanor distinction. This system is satisfactory as long as the terms are consistent. (See Chapter 2 for a detailed discussion of these terms and the problems they involve.) Other states, because of definition problems, spell out the specific crimes that each court may properly handle. Further, some states grant power to their courts not on the basis of labels or names but on the penalty that may be imposed. Of course, these factors may be used in combination with each other.

The final matter to be considered under **subject matter jurisdiction** involves distinguishing which courts have original jurisdiction, which have appellate jurisdiction, and which have both. **Original jurisdiction** is power to try a case that has never been tried before. Trial courts are courts of original jurisdiction. One cannot be tried in a court unless that court has original jurisdiction. The power to review a case that has been heard in trial court is called **appellate jurisdiction**. Strictly speaking, a trial court does not have appellate jurisdiction. Circumstances have caused legislatures to set up a hierarchy of courts that allows some to have both types of jurisdiction. When granted to general trial courts, appellate jurisdiction is limited to review of cases heard in courts of a lesser nature. One court cannot review its own decision under this power, nor can it review the decisions of equal or higher courts.

Although the right to have one's case reviewed is not a right guaranteed by the U.S. Constitution, it is a privilege granted by all levels of government in the United States. This review privilege takes one of two forms. The first form is the **appeal**. Appeal is the direct review of a case that the statute says an appellate court must hear if all the procedural steps are followed. The second form of review is *certiorari,* discretionary review of a case by a higher court. One attempts to get *certiorari* by a petition. The court looks at the petition, considers it, and then determines whether or not to review the case. When a court denies *certiorari*, it does not mean that everything was done correctly in the lower courts; it simply means that the court does not want to review the case.

3.4 CONCURRENT OR OVERLAPPING JURISDICTION

It is possible for the legislature to confer subject matter jurisdiction on its courts in such a way as to create **concurrent or overlapping jurisdiction**.

If two courts claim jurisdiction in such a situation, which court has jurisdiction? The court that first assumes jurisdiction of the person and begins prosecution will have the exclusive power over the case until it waives its priority by legally and voluntarily abandoning it.

Suppose that Al robs a bank in state X that is insured by the Federal Deposit Insurance Corporation. As he is leaving the bank, he is arrested by federal authorities who were advised of Al's conduct. Al is taken to the county jail, where he is housed awaiting trial. Federal authorities often use state facilities to house federal prisoners who are awaiting trial. When the federal authorities seek to secure the prisoner from custody, the state officials refuse to surrender him, saying that they are going to prosecute him under the state law. Who has jurisdiction? It has been held in such cases that the sovereign first having jurisdiction over the person has the initial right to try him. Thus, the state would be compelled to turn the prisoner over to federal authorities because they made the initial arrest.

If a person commits an act that happens to violate both a federal and a state law, can he or she properly be convicted in both jurisdictions without being able successfully to argue double jeopardy as a defense? The answer to this logical question is yes. Even though the crimes arose out of a single act, that act violated the laws of two separate sovereigns, and each has the authority to try and convict the offender. In the bank robbery illustration in the previous paragraph, the crime would constitute both a federal and a state offense, and both jurisdictions could prosecute. As a practical matter, this is rarely done, unless the crimes are dissimilar in nature or the sovereignty first assuming jurisdiction prosecutes unsuccessfully. In any case, however, both sovereigns may legally prosecute. These instances do not constitute double jeopardy. Double jeopardy occurs when the *same* sovereign tries to prosecute the *same* individual for the *same* act and the *same* crime a second time.

Several states prohibit a prosecution if the defendant has already been tried by another government for a similar offense. These statutes usually bar the prosecution when the federal and state statutes are very much alike and when there are no great differences as to elements or penalties.

Similarly, it is a policy of the Department of Justice that no federal case be prosecuted without the approval of the office of the attorney general after there has been a state prosecution for substantially the same act or acts. This is known as the **Petite policy** named after a case in which the policy was cited.

A similar relationship existed between violations of state criminal statutes and municipal ordinances. If an act violated both a state statute and a municipal ordinance, both jurisdictions had the power to prosecute, and the outcome in one would not affect the decision to prosecute in the other. Thus a person who violated a state statute and a municipal ordinance by committing a single act was subject to punishment in both jurisdictions. Again, as a practical matter, this was rarely done, but it was legally permissible. The U.S. Supreme Court stopped the practice when it ruled that a municipality is an extension of the same sovereignty as the state. Therefore, a single act that violates both state law and municipal ordinances can be prosecuted in one or the other, but not both. The Court held that to rule differently would constitute double jeopardy. (See Section 6.19 for additional materials on this topic.) Consider the following case.

In late 1997, the United States Supreme Court decided a case in which a federal banking regulatory agency had imposed fines against three banking officials who had violated federal banking statutes. The agency also prohibited the defendants from engaging in banking as part of the administrative resolution. The government later criminally indicted the three men for the same transactions that formed the basis for the administrative actions. After much legal maneuvering, the District Court accepted the defendants' claim of double jeopardy and dismissed the indictments. The Court of Appeals reversed, and the Supreme Court decided to hear the case. The Court ruled that Congress clearly intended the sanctions imposed by the administrative agency to be civil rather than criminal in nature and that fact, coupled with other tests, clearly meant this case did not meet the requirement to constitute double jeopardy. The case did not involve the imposition of multiple criminal punishments for the same offense when such occurs in successive proceedings.

3.5   VENUE AND ITS RELATION TO JURISDICTION

Territorial jurisdiction discussed in the preceding section is determined by showing **venue**, the place at which the crime was committed. Beyond that, venue is a procedural matter that directly affects the place at which the trial may be held and from which the jury is selected. Generally, people have a right to be tried where the offense took place unless they waive that right by failing to object to the change of trial location or by asking to have the trial changed to a place where they feel they can get a more fair trial.

3.6   INTERNET JURISDICTION

As has been described, historically, a court's jurisdiction is governed by the geographic boundaries within which the court operates and it is relatively clear, by the established rules and decisions, whether a crime was committed within the territorial boundaries of the court's jurisdiction. Now, we are confronted by events, including crimes, occurring in cyberspace. Internet services pose special jurisdictional problems because information on the Internet is available to anyone in the world. It has no physical, geographical location. Courts normally look to see where the injury or loss occurred, but this can be especially difficult when dealing with offenses occurring on the Internet.

Let's Talk Modern!

The Internet has affected almost everyone and everything. A very good computer hacker operating in Oregon causes funds from a New York bank to be transferred to an account in Charlotte, North Carolina, and from there to another account at a bank in Santa Fe, New Mexico. None of these accounts belongs to the hacker. The hacker does it only for "fun" and receives none of the money. Assuming a crime was committed, who would have what territorial jurisdiction?

Courts have no difficulty finding jurisdiction if a Web site is being operated unlawfully or for an unlawful purpose within the geographic territory of the court, but when a Web site is operated outside the territorial jurisdiction of the court, the issues become more complicated, particularly if the Web site is passive (i.e., user simply views the Web site's information but cannot do things like enter data). In one of the few criminal cases involving the Internet, the defendants were found guilty of disseminating obscene material in violation of federal law. The defendants operated a subscriber-based bulletin board system called Amateur Action Bulletin Board out of California. A federal agent downloaded allegedly obscene images from it to his location in Memphis, Tennessee. In denying an appeal from their conviction, the court stated that the manner in which the images moved does not affect their ability to be viewed on a computer screen in Tennessee or their ability to be printed out in hard copy in that distant location. Therefore dissemination of pornography on the Internet, or any online service, leaves open the prosecution anywhere in the world based on a local standard of obscenity.

The Computer Hacker

Assume the computer hacker in the previous example is identified. Which jurisdictions get the hacker for trial? How do they get personal jurisdiction? In what order?

This legal field is in its infancy. The law will continue to evolve. The reader is encouraged to keep up with changes, as they occur.

■■ DISCUSSION QUESTIONS ■■

1. Susan Spud, a detective with the Middleburg, New Hampshire, police department, has been sent to New York City to work with the detectives' bureau. Susan is to learn the latest detection techniques. While in

New York, Susan sees a woman who is wanted for murder in New Hampshire. Susan asks her to lunch. Susan mentions that she is from New Hampshire but not that she is a police officer. The suspected murderer says she would like to see her family but has no way to get to New Hampshire. Susan tells the woman she is going there and that she will be glad to take her. They leave New York City, and as they cross into New Hampshire, Susan stops the car, tells the woman that she is under arrest, handcuffs her, and takes her back to the town where she is wanted. The woman's attorney argues that the court has no jurisdiction. Can this validly be raised? Why or why not?

2. Bill wants to kill Charlie. Bill, who is standing in state A, looks and sees Charlie standing a few feet from him, but in state B. Bill shoots and kills Charlie. Who has jurisdiction to try Bill? Why?

3. Jane is charged with petty larceny in violation of a section of the municipal ordinances of Senior City. She took city property. After prosecution in the city court, the state seeks to prosecute Jane under its statutes. Among other contentions, Jane argues that she has twice been put in jeopardy for the same offense. Is her double jeopardy plea valid?

GLOSSARY

Appeal – the direct review of a case that the statute says an appeallate court must hear if all the procedural steps are followed.

Appellate jurisdiction – the authority of a court to review a case that has been heard in trial court.

Certiorari – the discretionary review of a case by a higher court.

Concurrent or overlapping jurisdiction – offenses over which both the federal courts and state courts have jurisdiction or their jurisdiction overlaps on the same offenses.

Jurisdiction – the authority of a court to hear and decide a case.

Laws – guidelines for all members of society.

Original jurisdiction – the authority of a court to try a case that has never been tried before.

Personal jurisdiction – the authority of a court to hear and decide a case by having the person charged present or capable of being present.

Petite policy – regulation of the Department of Justice that no federal case can be prosecuted without the approval of the attorney general after there has been a state prosecution for the same act or acts.

Rules of procedure – regulations governing the manner in which a court operates. The rules affect the conduct of law enforcement, attorneys, and judges.

Subject matter jurisdiction – the authority of a court to hear and decide a case by the type or the seriousness level of the offense.

Territorial jurisdiction – the authority of a court to hear and decide a case within a geographical area.

Venue – the geographical place where the crime was committed and where the trial will normally be held.

■ REFERENCE CASES, STATUTES, AND WEB SITES ■

CASES

Petite v. U.S., 361 U.S. 529, 80 S.Ct. 450, 4 L.Ed.2d 490 (1960).

Bartkus v. Illinois, 359 U.S. 121, 79 S.Ct. 676, 3 L. Ed.2d 684 (1959).

Abbat v. U.S., 359 U.S. 187, 79 S.Ct. 666, 3 L.Ed.2d 729 (1959).

Heath v. Alabama, 474 U.S. 82, 106 S.Ct. 433, 88L.Ed.2d 387 (1985).

Hudson v. U.S., 522 U.S. 93, 118 S.Ct. 488, 139 L.Ed.2d 450 (1997).

Waller v. Florida, 397 U.S. 387, 90 S.Ct. 1184, 25 L.Ed.2d 435 (1970). (prosecution by state and municipality constitutes double jeopardy).

Frisbie v. Collins, 342 U.S. 519, 72 S.Ct. 676, 96 L. Ed. 541 (1952).

Bright v. State, 490 A.2d 564 (Del. 1985).

Livings v. Davis, 465 So.2d 507 (Fla. 1985).

Brown v. Nutsch, 619 F.2d 758 (1980).

U.S. v. Wheeler, 435 U.S. 313 (1978).

Frazier v. State, 609 A.2d 668 (1990).

Blanck v. Waukesha County, 48 F. Supp,2d 859, (1999).

WEB SITE

www.findlaw.com

4

The Criminal Act

▓ KEY WORDS AND PHRASES ▓

Accessory after the fact
Accessory before the fact
Assault
Attempt
Causation
Conspiracy
Contraband
Factual impossibility
Foreseeability
Intent
Legal impossibility
Nonconsummation

Omission
Parties to crime
Physical impossibility
Possession
Principal in the first degree
Principal in the second degree
Procuring
Proximate cause
Solicitation
Status
Vicarious liability

4.0 INTRODUCTION: THE CRIMINAL ACT IN GENERAL

Every crime requires an act. The criminal act may take a variety of forms and consequently may lead to a variety of complex legal problems that form the subject of this chapter. As will be seen at the beginning of Chapter 5, the law will not allow a person to be formally punished by society merely for thoughts. Nor does the law generally punish someone who commits an act without some form of evil state of mind. The problem of concurrence of act and intent will be discussed in detail later. These rules are noted here to help the student understand this chapter so that it may be studied in light of the significance of Chapter 5.

4.1 POSSESSION AS AN ACT

At common law, **possession** of an article was considered a very weak act, and therefore possession was usually not chargeable. Because possession is an act, state legislatures have the authority to make it a crime. To be a crime, however, the possession must be coupled with an evil state of mind as required by the legislature. Possession of contraband is an exception to this rule. **Contraband** is defined in criminal law as the possession or trafficking of articles so detrimental to the welfare of society that public policy demands that it be unlawful to possess such articles regardless of the reasons for possession. This will vary with the changing attitudes of society. Because no legal justification for possession exists, the mere possession of articles such as untaxed whiskey (moonshine), numbers slips (bolita), or unprescribed narcotics is illegal, regardless of reason or intent. Reference to local statutes on possession of various articles such as guns or burglary tools should make it clear that the state of mind or mental element, often called **intent**, is stressed throughout these statutes. The reason is a logical one. Consider the plight of an officer who stops a motorist at three o'clock in the morning and finds him in possession of a set of lock picks. After arresting the motorist for possession of burglary tools, the officer finds that the man is a locksmith called out of bed at that early hour by worried parents whose two-year-old child had locked himself in the bathroom. And what about the person arrested for possession of a concealed weapon who later turns out to be a store owner on the way to make a nightly bank deposit and who has a written permit from the police chief to carry a weapon while performing this task? Thus, with the exception of contraband articles, possession can be innocent. It is for this reason that the state of mind is so important in possession cases. Therefore, a person who is unaware that he or she possesses something that is or could be used for criminal purposes is not criminally liable. Generally, possession must be conscious, knowing, and with an evil state of mind before it can be punishable.

A state's intention as to whether possession of an item is to be punished is often determined by looking at the penalty. Minor penalties will be seen as establishing strict liability. Greater penalties will be upheld only if the knowledge requirement is found. Thus, courts will save such statutes from constitutional attacks by requiring that possession include a showing of dominion or control. For example, a Pennsylvania statute punishes, as a misdemeanor, the bringing of liquor or drugs into a prison or mental hospital. Standing alone, every doctor or other health professional entering into such an institution could be guilty of the crime. To save the statute, the Pennsylvania Superior Court ruled that the statute requires that drugs or liquor be brought in with the intent that it come into the illegal possession of an inmate.

Possession is not always clear-cut. For example, common household items can be used as burglary tools, depending on the intent with which they are used. Florida has taken an entirely different approach by viewing the possession statute more as an attempt to commit the crime (see Section 4.7). Thus if the prosecution can prove any overt act tending toward the commission of burglary even though the household item is not used specifically to attempt to gain entry, the possession can be charged. Further, where a state prohibits the use of a handgun in the commission of a crime, can the state successfully prosecute if a defendant is merely carrying a gun that is not used, brandished, or otherwise alluded to by the defendant? Is the mere possession of the concealed weapon chargeable as use? The Maryland Supreme Court came to the apparent logical conclusion that its legislature was seeking to punish the use or threat to use such a weapon during the commission of a crime and that merely possessing the weapon in a concealed manner without showing, using, or referring to it was not chargeable. It should be noted, however, that there was not total agreement by members of the court.

The U.S. Supreme Court interpreted the word *use* in relation to a federal drug statute. The statute more severely punished criminal liability when a gun was used during a drug transaction. The defendant in this case traded his gun for drugs and was subsequently convicted of "using a gun" during a drug transaction. By majority vote, the Court upheld the conviction because Congress did not limit the word *use* to shooting or threatening to shoot. Contrary to the Maryland case, there was possession in this instance coupled with an act involving the gun during the commission of the crime. In a later case, the U.S. Supreme Court said that a conviction for use of a firearm, under the relevant statute, requires evidence sufficient to show employment of a firearm by the accused, a use that makes the firearm an operative factor in relation to the predicate (underlying) offense.

4.2 PROCURING AS AN ACT

Unlike possession, **procuring**, even at common law, was a sufficient act to give rise to criminal liability if coupled with an evil state of mind. The theory was that one had to do some act to procure, whereas possession was more closely connected to a status or condition, without requiring any physical effort on the part of the possessor. Nevertheless, it must be repeated that the act of procuring can also be innocent and must be coupled with an evil state of mind to be criminal. For example, one may procure narcotics, but, if one has a prescription, this is not criminal procurement.

The criminal law references to procurement fall into three categories. Procurement can involve obtaining articles with the intent of using them for criminal purposes. The case law in this category is sparse due to the difficulty of proving the criminal purpose intended. The few cases that can be found arose a number of years ago in the British empire. The category commonly associated with procurement involves prostitution statutes and the prohibition against those who procure, or "pimp," for the prostitute. The third category involves the act of procuring another to commit a crime. Statutes such as these are usually classified under the topic of accessory before the fact, which is dealt with in Section 4.11, or solicitation, found in Section 4.10.

4.3   STATUS AS AN ACT

Status can be defined as a condition or state of being. Attempts have been made to treat a status as an act and punish one for simply being in a particular condition. The U.S. Supreme Court has ruled that a statute making it a crime to be a narcotics addict is invalid because the state of being addicted is insufficient to constitute the act necessary for a crime. This law, therefore, violates the Eighth Amendment's prohibition against imposing cruel and unusual punishment. Being an alcoholic falls into the same class as addiction, but public drunkenness is recognized as an act rather than a status. Why the distinction? A person can suffer from the physiological or psychological illnesses of alcoholism or addiction, whether voluntary or not, without the illness affecting that person's mobility. On the other hand, public drunkenness is a voluntary observable fact that, through its effect on mobility and mental stability, presents a threat to society or to the intoxicated person. This potential for danger is one from which the society can justifiably protect itself.

The question arises as to whether vagrancy is considered an act or a status. State and federal courts have upheld the validity of vagrancy statutes. One court recommended that a vagrancy arrest should not be made unless it is clear that a person is a vagrant of his or her own volition and choice. This statement indicates that one who becomes a vagrant by voluntary or intentional conduct has performed a necessary act sufficient to support a criminal charge.

Rising numbers of homeless people and increasing concern about them are putting a strain on the resources of many law enforcement agencies, particularly in urban areas, where citizen pressure supports incarceration because of the volume of begging and panhandling occurring. The city of Atlanta, in anticipation of the 1996 Summer Olympics, enacted an ordinance allowing arrests for "aggressive begging."

Many homeless people suffer from mental illness but are not sick enough to require long-term care in state mental institutions. In many cities,

Boston being an example, the police no longer arrest these people. Instead, especially on the coldest nights, they "round up" such persons and take them directly to shelters for night lodging.

The reasoning stated in this section has also brought about the evolution of juvenile status offenses discussed in Chapter 3.

4.4 METHODS USED TO COMMIT THE ACT

Ordinarily, when thinking of the commission of a crime, we think of the offender's acting with his or her own hand, by shooting a gun, breaking open a window, or driving an automobile, for instance. The law does not limit criminal liability to cases in which perpetrators act by their own hand. There are three more ways in which an offense may be committed. An offense may be committed through an inanimate agency. Using the mails to send poisoned candy from California to a potential victim in Florida, knowing the victim will eat the candy and die, is an example. In this case, the mail becomes the tool by which the offender produces the desired effect. Do crimes involving the use of computers also fall into this category?

An innocent human agent may also be used to commit a crime. For instance, Able, intending to burglarize Sam's house, obtains the assistance of Chuck, a passerby, to climb a ladder and break open a window on the second floor. Able pretends that the house is his own and he has locked himself out. Chuck agrees because Able explains that he has a bad leg and cannot climb the ladder. Able is just as guilty as if he had climbed the ladder himself and broken into the house. However, Chuck's act was not coupled with an evil intent and thus would not be criminal.

Finally, a nonhuman agent, such as an animal, may be used to commit a criminal act. The case of the organ grinder who trained his monkey to enter homes through partially opened windows and remove wallets and purses from dressing tables serves as an example. It has been held that a large dog that bit a law enforcement officer on the calf and thigh after the owner gave a "sic" command, could be classified as a deadly weapon, thus making the charge against the owner more severe. In another case, a court held that an automobile in which the defendant was a passenger, bumped the victim's hip as the defendant reached out the window and snatched her purse, was found to have been used as a weapon. This increased the charge from robbery to armed robbery, carrying a much more severe penalty.

4.5 CRIME BY OMISSION OR NEGATIVE ACT

A crime may be committed by doing some affirmative act or by doing nothing. Persons may be guilty of a crime for failing to act when and where the law imposes a legal duty on them to act. This rule points out the fact that legal duties do not always correspond to moral or ethical duties. A situation that appears to involve criminal liability for failure to act in a given case, may, in fact, involve no legal guilt. The classic example used to illustrate this point is the case involving the magazine photographer who, while at the beach, observes a young girl in the water crying for help. He makes no attempt to rescue her but instead stands on the shore and takes photographs for his magazine. The photographer is an expert swimmer. In such a case, if he has no legal relationship to the drowning girl, the photographer is in no way liable for the girl's death.

A California appeals court was faced with the potential criminal liability of a doctor who withdrew life support systems from a comatose but not brain-dead patient. The conduct was not classified as an affirmative act but rather was viewed as an **omission** of further treatment. The court defined the issue as being the extent of duty owed at that point by physician to patient and held that a physician has no duty to continue treatment once it has proven to be ineffective, especially where the treatment path is extraordinary, as in this case. But see the material on assisted suicide in Section 8.4.

The legal duty to act generally falls on one who has some influence over the victim or offender through some legal relationship such as marriage, parenthood, or contract, which could have been exerted to prevent the injury. The crime charged in cases of failure to act is the same crime that would be charged had the accused produced the same result by acting affirmatively. An evil state of mind must be proved in either case. In the previous paragraph, doesn't a physician have a legal duty to preserve life? Can he or she make a unilateral decision to omit further treatment?

Although cases are rare, a legal duty to act can arise when none existed. A person who has no legal, familial, or contractual duty can be found liable when he or she attempts a rescue, thus causing others to abandon their rescue attempts, and then acts in a criminally negligent manner (see Section 5.4).

4.6 CAUSATION

One of the most complicated legal problems regarding the criminal act involves the law of causation. The rules of **causation** establish nothing more than a cause-and-effect relationship between the act of the accused and the resulting harm.

The majority of criminal cases involve a direct cause-and-effect relationship. For example, Sam fires a gun at Fred; the bullet strikes Fred and he is injured. In this case, it is not difficult to convince a jury that Sam's act produced the result so the jury will convict Sam of the crime. The situation becomes more complex when, after Sam shot him, Fred was taken to a hospital where he died, not from the bullet wound, but from an infection caused by a doctor's failure to sterilize instruments before removing the bullet from Fred. Another example involves the case in which Fred is mortally wounded by Sam but, an hour before he would have died, Young stabs Fred in the back, causing instant death. Is Sam solely liable for the homicide, or is Young solely liable, or are both liable, or is neither one liable? In cases such as these, investigating police officers would generally be correct in arresting any or all of the participants who could possibly be liable and letting the attorneys and courts determine which, if any, of these people will be held responsible for the criminal result. Of course, the facts in each case must be such as to give the officer probable cause for making an arrest or obtaining an arrest warrant.

Several legal principles apply in situations such as these. First, people are presumed to intend the natural and probable consequences of their acts. This is often referred to as **foreseeability**. Suppose that Bea threatens to shoot Adam, who has a weak heart. Adam is frightened so badly by this threat that he suffers a heart attack and dies. Bea will be held to have foreseen that this might occur as a consequence of her act and might therefore be held liable for Adam's death.

A second rule that applies to these situations is that the accused takes the victim as he finds him or her. For example, if Joe strikes Art without the intent to kill but, unknown to Joe, Art is a hemophiliac and bleeds to death because of the injury inflicted by Joe, Joe can be liable for the death.

In these instances the prosecutor must establish more than a simple physical cause-and-effect relationship; that is, the prosecutor must prove the defendant's act was the **proximate cause** of the injury or harm resulting. The difference between these two concepts can best be explained by example: If Dick shoots a gun at Pete and the bullet strikes and injures him, any fact or circumstance surrounding or contributing to this incident can be considered a physical cause of Pete's injury. Thus the fact that company X manufactured the weapon used by Dick would be a physical cause. The production of the ammunition by company Y would also be a physical cause. Obviously, it would be ridiculous to attempt to hold company X or Y criminally liable for Pete's injury simply on the theory that Pete would not have been injured if these companies had not produced the gun or the ammunition.

Proximate cause exists when a particular act or omission is the direct cause of harm, and when this cause will be recognized by the law as the act responsible for the injury. There are three rules used for determining the existence of proximate cause with modifications existing from jurisdiction to jurisdiction. First, proximate cause is established by showing a direct cause-and-effect relationship: Art shoots at Bob, the bullet strikes Bob, and Bob is injured. No other act interfered in producing this result.

The second method of establishing proximate cause is by showing that the act of the accused set a chain of events in motion that indirectly caused the harm. This is commonly referred to as the "but for" test and is illustrated by the examples in which Bea threatens to shoot Adam, and Adam dies of a heart attack from fear of Bea's threat, or in which Joe strikes a person he does not know is a hemophiliac and the victim subsequently bleeds to death. In each of these instances, but for the act of the accused, the victim would not have suffered the harm. Thus the accused is criminally liable in each case.

Finally, proximate cause can be shown by establishing that the act of the accused placed the victim in a position that substantially increased the risk to the victim of being harmed by some other cause. Thus, where Al shoots Ben in the leg, knocking Ben into the street, where he is run over and killed by a passing automobile, Al's act is the proximate cause of death so Al is criminally liable.

All cases are not always so straightforward. In the example above where Art shoots at Bob, injuring him when the bullet strikes, is Art liable for Bob's death if Bob received grossly negligent medical treatment? Most courts take the position that an independent, intervening, superseding act, if reasonably unforeseeable, can become the proximate cause. If that was shown to be true in this case, Art would not be liable for the death. If, however, the subsequent act is reasonably foreseeable, then it cannot be an independent, intervening, superseding cause.

These examples illustrate only some of the problems that can occur in any given criminal situation. Court decisions throughout the states have discussed and decided many of the questions raised. More than one person may be liable for a single harm even though they were acting independently of each other. (See the example in the sidebar.)

Family Affair in 1913

Two men were fighting. One had no weapon. The other pulled a pocketknife and stabbed his opponent three times. The most serious wound was near the navel, which penetrated the abdominal cavity, causing the intestines to protrude. At that time, the defendant's son came running up and fired a shotgun into the deceased who died shortly thereafter. Both were convicted of murder. The court said that a defendant may be guilty of homicide if the knife wound which he inflicted, contributed to hasten the death or contributed to the death of decedent, although death would not have inevitably followed from the knife wound alone, although there was no preconcert or community of purpose between the stabber and the shooter.

4.7 ATTEMPTS

When can a law enforcement officer properly make an arrest for an attempt to commit a crime? This is not an easy question to answer. Even lawyers have difficulty understanding and applying the complex rules for determining what constitutes an attempt.

As a basis for much of our discussion on this topic, consider the following example. Sam, a wealthy man, invited Red to be his house guest. Red discovered that Sam had a safe in his bedroom and found the combination on a piece of paper in a desk drawer. Red decided to rifle the safe one night. After waiting in his room until he believed Sam was asleep, Red emerged, taking with him the paper with the combination and a revolver. Red's bedroom was on the first floor, and the safe was in Sam's bedroom in the rear of the second floor. Red had taken but two or three steps from his bedroom when he stopped, thinking he saw the outline of a man's head at the end of the hallway. Believing it was Sam and that he, Red, would be apprehended, Red shot at the outline. Sam was actually in the hall at the time but behind Red when he shot. What Red thought was a man's head was actually a post at the foot of the staircase. In this example, has the crime of attempt been committed?

Statutes categorically spell out at least three significant elements of the crime of **attempt:** (1) the doing of an act, (2) tending toward the commission of an offense, but falling short of completion so that the target crime is, in fact, not consummated. The difficulty arises not in stating these elements but in applying them to a specific situation. For example, what is an act? When does an act tend toward the commission of an offense? When is a crime completed?

An attempt is not a part of the target crime, but is a separate offense punishable by statute. As with most other crimes, it requires the combination of an act and an intent. Care must be taken to distinguish an attempt from the concept of intent.

Subelements of Attempt

It is necessary to subdivide the three elements of attempt into component parts, which may be called subelements, for lack of a better term.

Legal Impossibility

Simply stated, **legal impossibility** is if the result the accused sought to achieve would not have been a crime, even if the act attempted was completed, the accused cannot be charged with an attempt. This is only logical, for, unless the accused was trying to commit a crime, there is no attempt. It is immaterial whether or not the accused thought a crime was being committed. If it would have been a crime had the accused completed it, this element is satisfied. On the other hand, if the accused thought he or she was committing a crime, but was mistaken in that belief and no crime would have resulted from the completed act, the accused cannot properly be charged with attempt.

To illustrate: Before the last legislative session, it was a crime to sell liquor in Big City. However, the legislature repealed that law and made it legal for anyone to sell liquor in Big City. Neither the sheriff nor John, a bootlegger, knows of the repeal. John does all other steps necessary to constitute an attempt, but before he can sell the liquor the sheriff arrests him. John cannot be charged with an attempt. Even though his motive was evil, he could not have committed the crime for there was none to commit.

Let us vary the problem. Suppose that it was legal to sell liquor in Big City. The legislature met and made such sales illegal. John, still thinking he could sell liquor in Big City, takes all steps short of actual sale when he is arrested. His belief that his conduct was legal will not relieve him of criminal liability.

If the target crime is one imposing strict liability, not requiring any intent, a New York court said that there can be no attempt of such a crime because an attempt would be a legal impossibility in this context. The target crime mandates certain results even though wholly unintended.

Intent

The intent required of an accused charged with an attempt is a specific intent as distinguished from a general intent or guilty state of mind. To charge for an attempt properly, the prosecution must be able to prove the specific intent an accused had at the time the accused attempted to commit a crime. That specific intent must be the intent to commit the target crime.

Intent is a complex subject and is treated more exhaustively in Chapter 5. For the present, it is enough to realize that it is impossible to look into a person's mind and tell that person's actual intent at the time a crime was committed. The only practical way to determine someone's intent is to infer it by what is done and said. For this reason, the law may imply intent from the facts and circumstances surrounding the attempted commission of a crime. If the specific intent can be shown, the intent element has been satisfied. Thus, in the illustration in the preceding subsection, John's purposeful, attempted sale of liquor, even though he thought it to be legal, would be sufficient to supply the intent required, because he specifically intended to commit the prohibited act of selling liquor. Even though the completed crime may or may not require an intent, the attempt to commit any crime always requires a specific intent.

Nonconsummation

By definition, **nonconsummation** is if the attempt falls short of successfully completing the target crime. If the crime is completed, there is no attempt. Statutes specify that an attempt is an act tending toward, but failing in, the commission of a crime because the accused is somehow prevented from completing the crime or is interrupted before completion. If the crime is completed, the attempt merges into the completed crime and can no longer be charged as a separate substantive offense, because completion removes a necessary element. Statutes in some jurisdictions may change the logical result regarding merger and allow the prosecution to elect to charge either the completed crime or the attempt even though the crime was completed.

The Act

An overt act tending toward, but failing in, the completion of the crime is necessary to an attempt charge. What constitutes an act and when it is sufficient to warrant the charge are difficult problems. It is a universal rule that the conduct must have gone beyond mere preparation. A close analysis of this statement may leave the reader in doubt. To complicate the situation slightly, it can be said that the exact point at which preparation leaves off and an act commences is a matter of degree. Each case must be decided on its own facts.

The common law and the Model Penal Code are subtly different in one respect. Under the common law, courts differentiated preparation from perpetration. Thus emphasis was more directed to "falling short of completion" types of facts with guilt being established long after the planning stages of the target crime. The Model Penal Code emphasizes the "substantial step" aspect of the attempt, and the Code says that it is proven when evidence exists that is "strongly corroborative of the actor's criminal purpose." Then it lists these acts which probably would have been preparation under the common law but which may be considered substantial steps under the MPC. Some of those are:

1. Lying in wait or following the victim

2. Enticing the victim

3. Reconnoitering the place

4. Soliciting an innocent agent

To illustrate, if a person seriously enters into negotiations with another to purchase drugs for the purpose of distribution but does not consummate the deal, can he be charged with attempted possession with intent to distribute? One U.S. Court of Appeals said yes. This was an attempt. The defendant in this case did not know he was dealing with an undercover agent. When he broke off the negotiations, it was because the price was too high. The court used the Model Penal Code's definition of substantial step toward the commission of the crime rather than the common law "preparation/ perpetration" standard. This case amply demonstrates the difference between the two standards.

The act done must be capable of being directly connected with the crime for which an attempt is being charged. For example, if Dan buys a gun with the intent to kill Al and does nothing more, this is not a sufficient act. The mere buying of a gun cannot be directly linked only to the purpose of killing Al, so this constitutes mere preparation. Similarly, if Dan buys a mask and hires a taxi to drive him to a bank he intends to rob, this is mere preparation in most states. In and of itself, it is not illegal to buy a mask and hire a taxi. Because the law will not punish intent alone without some wrongful act, no attempt charge will be upheld. Some states would disagree with this interpretation and would find an attempt to have occurred somewhere in that taxi. We feel this ignores the "point of no return" concept essential to the logical distinction between preparation and the act necessary for an attempt. On the other hand, if Dan walks into the bank with gun in hand and announces his purpose, but is somehow prevented from taking the money, he is guilty of an attempt.

Consider the problems raised by the illustration at the beginning of this section. Did Red, by stepping from his room on the first floor of Sam's home, armed with the revolver and the combination for the safe, satisfy the requirement of an overt act necessary for an attempt, or was his conduct merely preparatory?

Physical Impossibility

What happens if, for some reason, it is physically impossible for the accused to complete the target crime? In an attempt to commit murder, what is the result if the gun is not loaded? This is the problem of physical impossibility. By the great weight of authority, if the physical impossibility is unknown to the accused when the act occurs and, despite this factor, all surrounding circumstances indicate an apparent possibility of committing the crime, an attempt can properly be charged. On the other hand, if the accused knew at the time that the gun was not loaded, the accused cannot be charged with an attempt, for then there would have been no intent to commit the crime.

Another example of physical impossibility involves the situation in which a pickpocket can be charged with an attempt when that pickpocket puts his hand into the victim's pocket and finds the pocket is empty. Here, the accused performed an act with the requisite intent that would have been a crime had it been completed, but, due to a factor that was not known, it was physically impossible to complete the crime.

Refer again to the problem set forth earlier in this section in which Red fired a gun at a post, thinking it was Sam, who was actually standing behind Red. In this instance, due to a fact unknown to Red, it was physically impossible for Red to complete the target crime of murder. Yet, because Red did not know that fact, the physical impossibility will not prevent him from being successfully prosecuted for an attempt.

Factual Impossibility

Closely allied to physical impossibility is the subelement of **factual impossibility,** which refers specifically to the adequacy of the means used to commit the crime. Physical and factual impossibility are often difficult to distinguish and sometimes become confused with legal impossibility.

Different jurisdictions follow different tests to determine the adequacy of the means used. Some courts hold that the means must have, in fact, been adequate before an attempt can be charged. Others hold that the means must have appeared to the defendant to be adequate. Most courts hold that the means need only be apparently adequate under the circumstances. Thus an attempt-to-murder charge is proper when a gun is used, as the means are apparently adequate notwithstanding the existence of a physical impossibility in that the gun was unloaded. On the other hand, if one person tries to kill another, using a child's popgun, the means are obviously inadequate and an attempt cannot be charged.

Drug cases and antifencing (sting) cases in recent years pose some interesting and confusing questions in differentiating legal, physical, and factual impossibility. In several cases, undercover agents have arranged a narcotics buy. Money was exchanged and the seller was arrested. Chemical analysis revealed that the substance purchased was in reality a legal substance (powdered milk). Some courts have allowed a claim of legal impossibility and have released the defendant. Other courts have ruled this to involve a factual impossibility and the defendant was held liable. In some states in which antifencing operations are undertaken, police (without entrapment) have been selling once-stolen goods (now recovered) to those in the business of receiving and concealing stolen goods. Defendants have argued that such recovered goods are not stolen; therefore there is no crime. The status of recovered goods not yet returned to the owners has not been clearly or uniformly defined.

Abandonment of Attempt

Ordinarily, one can withdraw before the commission of a crime and be relieved of criminal liability either because the crime is never committed or because it is committed by co-conspirators after the accused has manifested an intent to withdraw and has actually withdrawn.

The law of attempts is a little different. If the act has gone far enough toward consummation, it is an attempt even though the accused may abandon the intent to complete the crime. If the accused fails to abandon the intent in time, an attempt can still be charged, even though the decision had been made not to complete the crime.

Al, Ben, and Sid walk into Big City Bank and rob it. Once inside with guns drawn, Al decides he does not want any more to do with this crime. He makes his intentions known to Ben and Sid, turns, and walks from the bank. Al has probably successfully withdrawn from the commission of the robbery, which Ben and Sid complete, but at the time of Al's withdrawal the conduct of the group had gone far enough to constitute an attempt, and Al could not withdraw from that. Consequently, Al would be prosecuted for attempted bank robbery, whereas Ben and Sid would be prosecuted for the completed bank robbery.

4.8   ASSAULTS DISTINGUISHED FROM ATTEMPTS

Any discussion of the law of attempts requires a brief mention of the law of assaults. The substantive aspects of assault are covered more fully in Chapter 7. They are referred to here solely to distinguish assaults from attempts.

Actually, there are few dissimilarities between assaults and attempts. **Assaults** are defined as acts tending toward the commission of a crime against the person of another. An attempt is directed toward the commission of any offense prohibited by law. The act leading toward the commission of the completed crime must go beyond mere preparation in both assault and attempt cases.

Most assaults require the accused to have acted with a specific intent, as is required in attempts. Modern aggravated assault statutes in many states are an exception to this rule. The specific intent in an assault case is intent to cause bodily harm or intent to place the victim in fear of bodily harm. It is here that assaults differ from attempts in one important aspect. Assaults may be committed not only by attempting a battery but also by placing the victim in a position where the victim fears imminent bodily harm even though no battery was actually intended by the accused. This difference leads to situations in which an assault charge may be proper but a charge of attempt would be improper.

For example, Art points an unloaded gun at Bill. Bill does not know it is unloaded, but Art does. Under the better point of view, described in Section 4.7, an attempt could not be charged, because the specific intent to complete the crime of battery is absent. Under assault law, however, it would be immaterial whether the accused knew the gun was unloaded. If the victim did not know the gun was unloaded, but thought it to be loaded and feared imminent bodily harm, the crime of assault would be complete.

Many states have enacted statutes covering assaults with intent to commit a felony. It is often difficult to distinguish among these statutes, as they are applied to specific fact situations from the law of attempts that exists in all states. Confusion in this area would only arise in crimes against persons. In many states, even the courts are at a loss to explain the difference. For example, Florida is compelled to recognize the existence of such a distinction by statute. The Florida statutes create the separate offenses of attempt and assault with intent to commit a felony. The courts faced with this problem have been unable to supply factual distinctions but have relied instead on rules of theory that the legislature would not have enacted separate offenses without some valid reason. It appears that the validity of that reason remains secreted in the minds of the state legislators. A Florida court said:

> There is a distinction between an "attempt" to commit an offense . . . and an "assault with the intent to commit" such offense. While there is considerable similarity between the two offenses, they are not in all respects the same. Our legislature has recognized this by enacting two separate statutes . . . It may be that in some cases the conduct of the accused would constitute a violation of both these statutes, while in other cases, this would not be true . . . to hold that the two deal with the identical offense would be to impute to the Legislature the enactment of a useless and unnecessary statute.

Legal treatises distinguish the two. While acknowledging that in the majority of cases there is no distinction, an assault with intent to commit a felony must come much closer to success to be chargeable than a chargeable attempt.

To illustrate this point, let us consider these examples. First, Art, intending to kill Bill, enters Bill's room, thinking Bill is asleep in his bed. Art is armed with a loaded revolver. Although it appears to Art that Bill is in bed, in fact no one is in the bed or in the room, for that matter. Art takes careful aim and fires six shots into the empty bed. Art cannot be charged with assault to commit murder because of the absence of the victim from the scene, thus making the proximity of success too remote for an assault. However, Art can be charged with an attempt to murder, because all elements have been satisfied. The defense of physical impossibility is not available to Art because Bill's absence was unknown to Art.

Changing the facts somewhat, assume that Bill was, in fact, in the bed when Art fired. Assume also that Art missed Bill with all six shots. On these assumptions of fact, Art can now be charged with assault to commit murder because he came so close to succeeding. In fact, his conduct came so dangerously close to completion that no other reasonable outcome was probable. Of course, we do not want to lose sight of the fact that Art could be charged with attempted murder instead of assault with intent to commit murder. The choice is up to the prosecutor.

4.9   CONSPIRACY

Conspiracy means concert in criminal purpose. The crime of conspiracy is defined as the combining of two or more persons to accomplish either an unlawful purpose or a lawful purpose by unlawful means. This crime was a misdemeanor at common law regardless of whether the purpose was to commit a felony or a misdemeanor. Many modern statutes make conspiracy either a felony or a misdemeanor, depending on the nature of the target crime.

Conspiracy is a separate crime and does not depend on commission of the conspired crime. The gist of conspiracy is an agreement to commit a crime or to commit a lawful act by criminal means. In a few jurisdictions, like Indiana, conspiracy cannot be charged unless some overt act tending toward commission of the target crime has occurred. In these states, however, the act required is not the same as for an attempt. The Indiana Statute reads:

§. 35-41-5-2. Conspiracy

a. A person conspires to commit a felony when, with the intent to commit the felony, he agrees with another person to commit the felony.

b. The state must allege and prove that either the person or the person with whom he agreed performed an overt act in furtherance of the agreement.

Most states do not require any act beyond agreement. The Model Penal Code follows this view and does not require an act in first and second degree felonies. Because of the social threat created by persons agreeing to commit crimes, conspiracy by itself was criminal under common law, whether or not the ultimate objective was ever accomplished or even attempted.

Although an intended participant in the commission of a crime may withdraw before the target crime is committed, this is not true of conspiracy. Once the agreement has been made, parties are co-conspirators, and withdrawal to escape a conspiracy charge is impossible. Pennsylvania and New Jersey, among most other jurisdictions, adhere to this policy, which is also the position taken by the Model Penal Code.

A formal written or verbal agreement is not necessary for a chargeable conspiracy. It is sufficient if the parties act in concert, working together understandingly with a single design for the accomplishment of a common purpose. It is not necessary that each conspirator know or see the others. Each conspirator need not know all the plans. Al masterminds a bank robbery and employs Ben, Sid, Dan, and Ed, all specialists in some facet of bank robbery. Ben, Sid, Dan, and Ed are assigned to perform their particular arts, knowing others are involved but not their identities. Only Al knows all the participants. Ben, Sid, Dan, and Ed never meet together before carrying out their assignments, yet they are all liable for conspiracy.

One person alone cannot commit conspiracy because, by definition, it requires at least two people. If one co-conspirator is acquitted of the conspiracy charge, the other must also be acquitted. In such a case, the law says that, if one did not conspire, the other could not have conspired alone. This is true when only two people are involved in the conspiracy. If there are three or more involved, however, one or more may be acquitted as long as there are at least two left who could be convicted.

This rule does not apply if one co-conspirator dies or pleads guilty before the trial of the second. Neither death nor a guilty plea renders the person innocent. This can be done only by acquittal. Similarly, if one is convicted of conspiracy and the case against the second is *nolle prossed*, the conviction against the first will stand because a *nolle prossed* is not equivalent to an acquittal. It is simply a decision by the prosecutor not to prosecute at the present time. The rules are deceptively simple. Local case law must be consulted because of the complex variations from one state to another and often within the same state. A detailed analysis of these variations would consume too many pages for the intended scope of this book, because these are matters of procedure.

Conspiracy does not merge into the target crime even when the crime is consummated. This general rule has a few exceptions, for instance, when a statute specifies that if the crime is completed, conspiracy cannot be charged or when, by its very nature, the target crime is one that can only be accomplished by two people, such as adultery, in which only the participants are involved. A third person, however, may conspire for the commission of adultery.

The elements that must be shown to support a charge of conspiracy are (1) that the individuals charged knew the unlawful purpose of their agreement, (2) that each person intended to be associated with the promotion of that unlawful purpose, (3) that each made clear his or her intent to promote the unlawful purpose, and (4) that each was accepted by the other co-conspirators as a participant in the conspiracy.

Joe, Fred, and Art reach an agreement. On the way to the intended crime scene, Dick joins in. At that point, can Dick be charged with conspiracy? The answer is yes. By adopting the purpose of the agreement, he becomes a party to the agreement. All participants in the conspiracy are responsible for acts of the others that are natural and probable consequences of the original plan. Thus, when Art, Bill, and Cal conspire to commit robbery and arm themselves for that purpose, all are liable for the homicide committed by Art during the robbery, even though killing was not specifically a part of their original plan. This would be true even though one of the co-conspirators was absent from the scene of the crime at the time of the killing.

However, if one co-conspirator acts in a manner completely foreign to the intended plan and its natural and probable consequences, the other co-conspirators will not be liable. This is always a question for the jury. To illustrate: Suppose that Ed, Bill, and Sid conspire to commit and do commit the crime of auto theft. This is not a crime of violence, but, unknown to Bill and Sid, Ed is armed. As the trio drives the stolen car away from the curb, Ed sees his arch enemy, Walt, walking down the street. Before Bill and Sid can react, Ed pulls the revolver from under his coat and shoots Walt, killing him. Bill and Sid could not be held liable for Walt's death, because Ed's conduct was beyond the scope of the conspiracy to commit auto theft.

If someone knows a crime is going to be committed, is that person liable for conspiracy by assisting the conspirators in what would otherwise be an innocent manner? John, the owner and operator of a retail food store, had for some months been selling large quantities of sugar and empty gallon jugs to George and Red. John learned that George and Red were using these raw materials to make and sell moonshine whiskey, but John did not stop selling those materials to them. John received no profit from the sale of the moonshine, nor did he have any connection with the operation of the still. Some months later, George and Red were arrested for making the illicit liquor. Empty sugar bags were found at the site for the still and were traced to John's store. Having learned the facts outlined here, officers arrested John for conspiracy to violate the federal liquor laws. Was the conspiracy charge proper?

The courts have come to an unusual conclusion in cases such as this by holding that, if the target crime is not a violent one involving the threat of death or serious bodily harm, mere knowledge is insufficient to make someone a co-conspirator. On the other hand, if a violent crime is intended, for instance, as when Al sells a machine gun to Ben, knowing Ben intends to use it to commit murder, the attitude of the courts is likely to be different. It should be noted, however, that there are some decisions that appear to fall between these extremes. A drug manufacturer selling an unusually high quantity of a particular drug to a licensed buyer has been shown to have enough knowledge to allow a finding of participation in a criminal enterprise.

4.10 SOLICITATION

Counseling, procuring, or hiring another person to commit a crime was a common law misdemeanor called **solicitation**. Solicitation is a substantive offense. Its existence does not depend on the commission of the crime solicited. It is immaterial whether the person solicited even accepts the offer. As with conspiracy, solicitation does not proceed far enough to be an attempt; yet the courts have no difficulty in saying that soliciting another to commit a crime is enough of an act to be punishable.

The justification for making solicitation a crime, as for conspiracy, is public policy. A Connecticut court stated the philosophy:

> The solicitation to another to a crime is as a rule far more dangerous to society than the attempt to commit the same crime. For the solicitation has behind it an evil purpose, coupled with the pressure of a stronger intellect upon the weak and criminally inclined....

There is little doubt that solicitation is chargeable in cases in which the crime solicited is a felony. Courts are split, however, as to whether or not solicitation to commit a misdemeanor is a crime. Most authorities agree that the better view is that solicitation to commit a misdemeanor should be a crime.

4.11 PARTIES TO CRIME: PARTICIPATION

Because an act is required for the commission of every crime, a person's mere presence at the scene of a crime during its commission will not support a criminal charge. The fact that a person is present, even though mentally concurring in the commission of the criminal act, does not make that person liable for the crime. To be liable, a person must participate in some manner. In the rest of this chapter, we will strive to make clear just when someone does participate in committing a crime, in what capacity that person will be liable, and the punishment to which such person will be subject.

At common law, parties to crime were grouped into broad categories called principals and accessories. Each of these groups was further divided into **principals in the first degree**, **principals in the second degree**, and **accessories before and after the fact**.

At common law, all parties to the commission of treason or to the commission of misdemeanors were treated as principals no matter how they participated. An exception to this rule was made for anyone who helped a misdemeanant after the crime was committed. Because of the petty nature of misdemeanors, the law did not treat such persons as participants in the offense. Only in felonies were these distinctions important.

Principal in the First Degree

At common law, a principal in the first degree is the actual perpetrator of the crime, that is, the person who, with his or her own hand or through some inanimate agency, some innocent human agent, or some nonhuman agent, committed the crime.

Principal in the Second Degree

A principal in the second degree at common law is one who was either actually or constructively present at the commission of the crime and who aids and abets in its commission even though not the actual perpetrator. Mere presence at the commission of the crime would not justify charging a person with the crime. A person must aid and abet. This means that the person must assist the perpetrator either by doing some affirmative act or by providing advice. Aiding and abetting also involves criminal intent that must be proven before a conviction could be obtained.

Suppose that a citizen helps an undercover police officer commit a crime as an aider and abettor but not as a principal in the first degree. Can the citizen be convicted as an accomplice, even if the reason for the crime was to arrest the very citizen who aided and abetted the officer? Alaska, in an unusual opinion, felt that the conviction should stand. The defendant tried to argue that the officer's defense of "public authority justification" should also protect the defendant. The court said if such a defense exists in a case, the defense applies only to the officer, is personal to the officer, and thus may not be vicariously asserted.

A person might be a principal in the second degree without rendering physical assistance. It is not necessary, therefore, that the principal in the second degree be actually present when the crime is committed. Only constructive presence is required. Constructive presence meant being sufficiently near to render some type of assistance if necessary. For example, Al, Ben, and Frank conspired to break and enter a store owned and operated by George to steal goods. They agreed that, to facilitate the crime and lessen the danger of detection, Al would entice George from the store to a house about two miles away and keep him there while Ben and Frank broke into the store and stole the goods. Under these facts, Al would be liable for the crime as a principal in the second degree under common law because, despite his distance from the scene of the crime, Al was aiding and abetting during its commission and was, therefore, constructively present.

Accessory Before the Fact

An accessory before the fact at common law is someone who, although not actually or constructively present, or aiding or abetting in the commission of a crime, is a participant by performing prior acts of procuring, counseling, or commanding. This is the basis for the distinction between being an accessory before the fact and soliciting. For a person to be an accessory before the fact, the crime must be committed. If no crime is committed, no accessory charge can be made, and the only alternative is to charge solicitation.

Participation Under Modern Law

The foregoing distinctions between principals in the first and second degrees and accessories before the fact, although still essential, have caused considerable procedural difficulties for our courts. At common law, the penalty for persons convicted as principals or accessories before the fact was the same. Statutes have complicated this otherwise simple procedure by providing different penalties for each type of participation. As a result, if an indictment fails specifically to spell out the level of participation, the indictment is invalid. Although at common law, a principal in the second degree could be tried and convicted before a principal in the first degree was convicted, this was not true with respect to someone charged as an accessory before the fact. This was the second major problem created by distinguishing the participants. An accessory before the fact could not be convicted unless a principal was convicted first. The common law based this on the idea that there could be no accessory relationship unless the identity of the principal was known. Illogically, this was true even though it could be shown that a crime was committed and that the accused was directly related to that offense as an accessory before the fact.

Conversely, even though a principal was identifiable and had been convicted before the accessory's trial, the accused accessory before the fact could still attempt to prove that the principal should not have been convicted. The accused accessory was permitted to do this by introducing facts that were not produced at the trial of the principal.

Recognizing the complexity of the problems often caused by their statutes, the legislatures of many states have taken steps to correct them. Some states have abolished all distinctions between principals in both degrees and accessories before the fact, labeling them all "principals" without reference to degree. In these states, it is still important to define a participant by resorting to the common law distinctions to prove an accused's participation. But the penalty and trial problems do not exist in these states. Among the states that have selected this alternative are Alabama, California, and Connecticut.

Other states have retained the distinctions but eliminated the requirement that the principal be convicted before an accessory before the fact can be convicted. In these states, being an accessory before the fact is a separate offense. The prosecution need only prove that a crime has been committed and that the accused participated as accessory before the fact. Colorado, Florida, Massachusetts, Mississippi, North Carolina, South Carolina, and Tennessee follow this course.

A third group of states have returned to the common law approach of standardizing the penalties for all three categories but have eliminated the procedural problems mentioned here. Among the states in this category are Delaware and New Jersey.

Capability to Commit Crime

What is the importance of the distinctions between principals and accessories? Besides differences in penalties and the use of these classifications to prove participation, there is one further loophole. This involves certain persons' capabilities to commit crimes.

A classic illustration of this point raises the question: Can a woman forcibly rape a man? The answer to this is no. However, criminal liability may attach to a woman involved in a rape. Can a woman be convicted of rape? The answer to this question is yes. The opposite answers to these two questions can be reconciled only by understanding the subject of parties to crimes. The fact that a female cannot be the actual perpetrator, that is, a principal in the first degree, does not affect her liability as a principal in the second degree or as an accessory before the fact under common law. If she procures a man to commit rape or if she aids and abets in the commission of a rape, she is just as guilty of the crime as the actual perpetrator.

As you can see from this example, the purpose of classifying parties to a crime is to prevent a person from escaping criminal liability when the person is not the actual perpetrator of a crime but has done more than merely observe its commission.

Accessory After the Fact

The previous discussion made no mention of the fourth common law participant. Accessories after the fact should not be referred to in connection with the problems raised by accessories before the fact and principals. The liability of an accessory after the fact does not arise until after the crime is completed. Because of this, all states retain the classification of accessories after the fact despite what they have done to the other participant classifications. Because an accessory after the fact does not commit the crime, the penalties, which are usually less severe than those involved in the precommission and commission stages of the crime, are easily justified.

An accessory after the fact is defined as one who personally receives, relieves, comforts, or assists another, knowing that the other person has committed a felony. To convict someone as an accessory after the fact, it must be shown that a felony was, in fact, committed and that it was completed at the time the supposed accessory rendered assistance. It must also be shown that the accused knew a felony has been committed by the person he or she assisted. The last, and possibly most significant, factor to be shown is that assistance was rendered to the felon personally.

Although the common law required a felony to have been committed, it is possible today under the statutes of a few jurisdictions to be an accessory after the fact to a misdemeanor. This appears to be the rule in Arkansas, Colorado, Georgia, and New Hampshire. This is contrary to the common law, which distinguished between principals and accessories only in felony cases.

Misprision of Felony

Mere failure to disclose the commission of the felony is not personal assistance making one liable as an accessory after the fact unless there is a statute requiring disclosure. However, one could be convicted at common law for the separate offense of misprision of felony, a misdemeanor. This involved acts such as concealing the commission of a felony or nonpersonally assisting the felon. Concealing stolen property or concealing the commission of a felony without agreeing to do so would make one liable for misprision of felony at common law. However, because modern statutes cover acts such as this that do not constitute p ersonal assistance, the misprision of felony offense has fallen into disuse and, although there are some state cases applying the common law, the offense exists by statute today only on the federal level. It is interesting to note that I have found only one case in which a conviction for violation of this statute has ever been affirmed on final appeal. Bank robbers took some of the bank money to pay a professional bondsman who arranged for their bonds. The defendants discussed in the presence of the bondsman that the money came from the robbery. The bondsman did nothing. He was convicted of misprision of felony. The conviction was affirmed on appeal.

If an accused concealed a felony by agreeing with the felon to do so, without going so far as to give personal assistance and become liable as an accessory after the fact, the accused would be chargeable with a common law crime called compounding a felony. In substance, compounding a felony consisted of agreeing to conceal a felony, or agreeing not to prosecute, or agreeing to withhold evidence. The agreement was an essential element of this offense. Compounding a felony remains a statutory offense in many states.

§. 153 of the California Penal Code provides, Every person who, having knowledge of the actual commission of a crime, takes money or property of another, or any gratuity or reward, or any engagement, or any promise thereof, upon any agreement or understanding to compound or conceal such crime, or to abstain from any prosecution thereof, or to withhold any evidence thereof....is punishable as follows:....

§. 9A.76.100, Revised Code of Washington provides, "A person is guilty of compounding if:

a.   He requests, accepts, or agrees to accept any pecuniary benefit pursuant to an agreement or understanding that he will refrain from initiating a prosecution for a crime: or

b.   He confers, or offers or agrees to confer, any pecuniary benefit upon another pursuant to an agreement or understanding that such other person will refrain from initiating a prosecution for a crime."

At common law, the only person exempt from prosecution as an accessory after the fact was the spouse of an accused. The bases for this exception were maintenance of the confidential relationship between spouses, preservation of the unity between husband and wife recognized by common law, and recognition of the presumed bond of love and affection. Under modern statutes, most states have extended the exempted class to include certain relatives of the accused, recognizing human frailties caused by such relationships. For example, Florida exempts parent, grandparent, child, grandchild, brother, and sister.

4.12 VICARIOUS LIABILITY

When and under what circumstances should one person be held liable for the criminal actions of another? As described so far in this chapter, there is no problem assigning liability when one person procures, commands, or counsels criminal activity. However, should an employer be held liable for criminal acts committed by an employee? Is it fair to punish one for the acts of another person? To punish an otherwise innocent person for another's acts is called **vicarious liability**. Reluctantly, courts now accept legislative recognition and the creation of vicarious liability.

The common law, with possibly only two exceptions, refused to allow an employer to be charged for the unauthorized crimes of an employee. The two exceptions were the common law misdemeanors of libel and nuisances committed by an employee while acting within the scope of the employment. Because of safety interests involved, many modern statutes are reasonably clear that an employer may be held liable for a misdemeanor committed by an employee. Hazardous business activities engaged in by an employer can result in vicarious liability in most states if an employee engages in a misdemeanor that jeopardizes those safety interests. (See also Section 22.12 on environmental crimes.)

Many traffic laws impose liability on an owner for regulatory violations of the traffic code, such as parking violations, regardless of who operated or parked the vehicle, but do not impose vicarious liability for out-of-the-presence crimes, such as vehicular homicide. Generally, those vicarious liability statutes that have been upheld are those that impose no jail or prison sentence but rather, impose only a fine for a conviction. The taking of money is seen as less of a problem than the taking of one's liberty.

▨▨▨▨ DISCUSSION QUESTIONS ▨▨▨▨

1. D struck B with his fist. B was drunk at the time. The blow caused the death of B, who would not have died had he been sober at the time, according to the testimony of the medical examiner. Is D liable for the death of B? Explain.

2. X, a wealthy man, invited Q to be his house guest. Q discovered that X had a safe in his bedroom, and Q located the combination on a piece of paper in a desk drawer. Q decided to rifle the safe one night. Waiting in

his room until he believed X was asleep, Q emerged, taking with him the paper with the combination and a revolver. Q's bedroom was on the first floor, and the safe was in X's bedroom in the rear of the second floor. Q had taken but two or three steps from his bedroom when he stopped, thinking he saw the outline of a man's head at the end of the hallway. Believing it was X and that he, Q, would be apprehended, Q shot at the outline. X was actually in the hall at the time but behind Q when he shot. What Q thought was a man's head was actually a post at the foot of the staircase. Among other things, Q is charged with (1) attempted theft and (2) attempted homicide. Discuss the propriety of these two charges.

3. A, B, and C conspired to break and enter a store owned and operated by D for the purpose of stealing goods. They agreed among them that, to facilitate the crime and lessen the danger of detection, A would entice D from the store to a house about two miles away and detain him there while B and C broke into the store and stole the goods. Under these facts, in what capacity will A be liable for the resulting crimes, if at all? Explain.

4. X, the owner and operator of a retail food store, had for some months been selling large quantities of sugar and empty gallon jugs to B and C. X learned that B and C were using these raw materials to make and sell moonshine whiskey, but did not stop selling those items to them. X received no profit from the sale of the moonshine, nor did he have any connection with the operation of the still. Some months later, B and C were arrested for making the illicit liquor. Empty sugar bags were found at the site of the still and traced to X's store. Upon learning these facts, officers arrested X for conspiracy to violate the federal liquor laws. Was the charge of conspiracy proper? Explain.

5. Wilkinson, while standing in the rear of a pickup truck, fired his rifle across private land at what he believed to be a grazing deer. In fact, it was a decoy deer deployed by law enforcement to curtail poaching that had been going on in the area. When approached by the game warden, Wilkinson stated, "I know I screwed up. It was too good to be true. I should not have shot that deer." Is an attempt charge proper under common law? How about in your jurisdiction?

GLOSSARY

Accessory after the fact – one who personally receives, relieves, comforts or assists another, knowing the other person has committed a felony.

Accessory before the fact – one who commands, counsels, or procures others to commit a felony but who is not present and does not aid or abet in the commission of the crime.

Assault – acts tending toward the commission of a crime against the person of another.
Attempt – an act tending toward but falling short of completion of a crime.
Causation – cause and effect relationship.
Conspiracy – the combining of two or more persons to accomplish an unlawful purpose or a lawful purpose by unlawful means.
Contraband – any item, the possession of which is unlawful.
Factual impossibility – the means used in an attempt are inadequate to complete the crime.
Foreseeability – to be held accountable for the natural and probable consequences of one's conduct.
Intent – an evil state of mind or, desiring a particular result while committing an illegal act.
Legal Impossibility – the result, in an attempt, if the completed act would not be a crime.
Nonconsummation – in an attempt, failure to complete the target crime.
Omission – failure top do something the law requires be done.
Parties to crime – participants; generally, only important in felonies where participants are divided into principals and accessories.
Physical impossibility – a condition, in an attempt charge, unknown to the actor, that prevents the completion of the crime.
Possession – physical control over something or the right to physically control that thing.
Principal in the first degree – the actual perpetrator of a felony.
Principal in the second degree – one who is actually or constructively present and who aids and abets in the commission of a felony but who is not the actual perpetrator.
Procuring – the act of obtaining illegal items or services.
Proximate cause – legally recognized act or omission that produces a resulting harm.
Solicitation – counseling, procuring, or hiring a person to commit a crime.
Status – a condition or state of being.
Vicarious liability – the liability or accountability of one person for the acts of another.

▬ REFERENCE CASES, STATUTES, AND WEB SITES ▬
CASES

Smith v. U.S., 508 U.S. 223, 113 S.Ct. 2050, 124 L.Ed.2d 138 (1993).

Robinson v. California, 370 U.S. 660, 82 S.Ct. 1417, 8 L.Ed.2d 758 (1962).

Connecticut v. Ernest Schleifer, 99 Conn. 432, 121 A. 805 (1923).

State v. Ferguson, 302 S.C. 269 (1999).

People v. Furlong, 79 N.E. 978 (N.Y. 1907).

People v. Grant, 377 N.E.2d 4 (Ill. 1978).

Bailey v. U.S., 516 U.S. 137 (1998).

Muscarello v. U.S., 524 U.S. 125 (1998).

Henderson v. State, 65 So. 721 (Ala.1913).

Commonwealth v. Williams, 557 A.2d 30 (Penn.1989).

Thomas v. State, 531 So.2d 708 (Fla.1988).

Harris v. State, 626 A.2d 946 (MD 1993).

Wynn v. State, 313 Md. 533 (MD 1988).

People v. Bailey, 549 N.W.2d 325 (Mich.1996).

Gargan v. State, 436 P.2d 986 (Kan.1968).

State v. Bloomer, 618 N.W.2d 550 (IA 2000).

State v. Wilkinson, 724 So.2d 614 (Fla. 1998).

Jenkins v. State, 747So.2d 997 (Fla.1999).

Morris v. State, 722 So.2d 849 (Fla. 1998).

STATUTES

Federal: 18 U.S.C.S. §. 924(c)(1).

Indiana: §. 35-41-52

Pennsylvania: 18 Pa.CS §. 903 and §. 5123.

New Jersey: N.J. Stat. §. 2C:5-2

California: Cal. Pen. Code §. 153

Georgia: O.C.G.A. §. 16-10-90

Wyoming: Wyo. Stat. §. 6-5-203

Washington: Re. Code Wash. (ARCW) §. 9A.76.100

WEB SITE

www.findlaw.com

The Mental Element

■ KEY WORDS AND PHRASES ■

Intent

Mens rea

Motive

Negligence

Recklessness

Specific intent

Transferred intent

5.0 INTRODUCTION: THE MENTAL ELEMENT

We saw in Chapter 4 that for conduct to be criminal, some act, either affirmative or negative, must occur. In this chapter we discuss the need for showing that the act was committed with a certain state of mind.

As a general rule, most crimes require a combination of an act and an intent. When both are required, the act and intent must be simultaneous. If the act occurs without the intent required, later formation of the intent will not make the injuring party criminally liable. For example, if John by accident and with no criminal responsibility kills Fred, John cannot be tried for a crime even if John later decides he is glad he killed Fred. The result would be the same if John planned to kill Fred on Tuesday but abandoned that intent and on Wednesday, by accident and without a criminal state of mind, kills Fred. Therefore, the prosecution must not only prove that the criminal state of mind existed but that it existed at the time the injurious act occurred.

There is no specific length of time that the intent must exist before the act as long as both exist and do so concurrently. The intent may be formed at the time of the act, or it may be formed over a longer period of time. John may see his hated enemy Fred and in that instant intend to and, in fact, kill Fred. Or John may plan for weeks the method of killing Fred and merely await the proper moment for execution. The most difficult time problem associated with intent is the continuing intent fiction employed in criminal law. In Chapter 3 we saw an example of this. Suppose John poisons Sam in Maryland. Sam appears to be dead and John intended to commit murder. John puts Sam in the trunk of a car and drives to Virginia. While in Virginia, John lops off Sam's head. John hoped to leave the head in one place and the torso elsewhere so that identification would be difficult. Until John cut his head off, Sam was not dead.

John intended to kill Sam by poisoning in Maryland. Unknown to John, Sam did not die of poisoning but, rather, died of the wounds inflicted by decapitation in Virginia. The act causing death occurred in Virginia. However, did the necessary mental element exist at that time? By using the continuing intent fiction, courts generally hold that John would be liable for Sam's death in Virginia. The reason for this is that John intended Sam's death. The method by which this was to be accomplished has no bearing on the intent. Thus, John's intent continued from Maryland to Virginia where Sam's death occurred. The mental element in criminal law may take any one of several forms—all often mislabeled **intent**. Intent in its legal sense is only one of several forms. In the following sections of this chapter we discuss the various mental elements.

Proof in court is the object of any discussion of the mental element or state of mind. It must be recognized that it is impossible to look into a person's mind in the literal sense. It would be much easier if we could see a picture of the state of mind that existed when the injuring act occurred. Because we cannot, we must resort to objective tests to determine the apparent subjective mental state. This is accomplished by permitting the fact finders, the jury or judge, to presume or infer the accused's state of mind at the time the act was committed. This inference is based on what the accused did and said and on all other circumstances surrounding the act. Admittedly, this "Monday morning quarterbacking" is imperfect, but it is the best system available today. There are constitutional, moral, and philosophical questions as to whether we ever want to be able to read a person's mind perfectly.

Perhaps one of the best-known principles of constitutional law is that a person is presumed innocent until proven guilty. Proof offered by the prosecution must rise to a very high standard. The proof offered must be "beyond and to the exclusion of any reasonable doubt"—a greater standard than required in a civil suit, which requires only a "preponderance of the evidence," or a 50.1 percent to 49.9 percent split or, as it has been said, "half plus a feather." Once the prosecution has presented evidence that the act was committed, that the accused committed it, and all the surrounding circumstances have been shown, the finder of fact, as mentioned in the last paragraph, may presume that the necessary mental element existed because each person is presumed to intend the natural and probable consequences of the acts he or she commits. Of course, this presumption is rebuttable. The accused can offer contradictory evidence that if believed or reasonably thought to be true, can destroy the presumption and require that the accused be acquitted.

Although one can abandon the intent before committing the act in most instances, the accused cannot escape criminal sanction by repenting after completing the act. The crime is a breach of society's standard and any attempt to make the injured party whole or to apologize will be of no avail as far as guilt is concerned. This is not to say that it may not in some way mitigate or lessen the sentence imposed. But this is solely the function of the court in most cases.

5.1 INTENT

Intent is the primary form of the mental element the prosecution seeks to prove through evidence found by the investigator. Two points of confusion often arise. There is a distinction between the word *intent* in its legal sense and intention. When we speak of doing something intentionally, we refer to voluntary conduct with a specific objective in mind. Although in a number of instances criminal intent can be proven by showing such voluntary conduct with a specific objective, criminality may also be proven by showing either voluntary or involuntary harm resulting merely from a determination to act in a certain way. For example, if John shoots at Fred voluntarily, John is said to have acted intentionally. If John commits this act to injure Fred, he is said to have the requisite criminal intent. If John intentionally shoots at Sam but the bullet strikes Fred, a person John had no intention of hitting, John will nonetheless be liable for his act because of his intent to do harm.

The second point of confusion arises over the meaning of **motive** and the part it plays in criminal law. Very often, motive is thought to be the same as intent. It is not. Watching some courtroom television programs would lead one to believe that proof of motive is essential. Motive, good or bad, is never an essential element of a crime. Neither the presence nor the absence of a motive ever has to be proven at a trial regardless of the nature of the case. In no crime does motive constitute a part of the *corpus delicti* (Chapter 2).

Motive can be defined as those desires that compel or drive a person to intend to do something. Motive involves judgment, but we do not judge a person's motive for doing something. An individual may have the best of motives, as when someone kills a known assassin or mass killer to get rid of that person. Society may have been done a service, but the person is nonetheless guilty of a crime unless the act falls within the bounds of the standard defenses to homicide. Neither do we convict a person for doing an innocent act merely because there existed an evil motive.

This is not to say, however, that motive has no use whatsoever in a criminal case. Motive can play an important role in determining the existence or absence of intent. It can be one of those surrounding circumstances that help to prove intent, because it is impossible to determine exactly what thoughts were in the mind of the accused when committing the offense. It is important to remember, however, that motive is only one factor that can be utilized in finding intent. Motive alone is never sufficient to prove intent.

An example of the difference between motive and intent may help to illustrate: Mr. Smith is lying deathly ill in a hospital bed. He is in misery. Death within a short time is inevitable, but the doctors are doing everything possible to keep him alive. Mr. Smith begs his son to do something to put him out of his misery. The son, realizing his father's pain, procures some sleeping tablets and gives them to his father. The father takes the entire bottle and dies. In this example, his son had good motives. He could no longer bear to see his father suffer. Still, his intent was to kill, and he is criminally liable for the death of Mr. Smith despite his good motives.

To illustrate further: In a Texas case, the defendant shot and killed a man to prevent a theft at night. Under a Texas statute, this killing would have been justified. However, there existed a long-standing dispute between the defendant and the decedent. Why did he shoot? Was it to prevent theft or to settle the score? The court was very clear that it was not about to grant a license to kill.

People may be responsible for unintended consequences of their conduct. As long as the accused commits an unlawful act accompanied by a wrongful state of mind, the accused is held to intend the result of that conduct. This is true even when the specific result the accused achieves is not the specific result intended, as long as the actual harm done would normally flow from such an act under rules set out by law. This is the basis of the rule that a person is responsible for the natural and probable consequences of his or her act. If a person intends to commit the act, that person is held to intend all probable consequences.

A corollary rule derived from the concept of responsibility for unintended consequences is the doctrine of **transferred intent**. Suppose John fires a gun at Sam, intending to kill him. The bullet misses Sam and strikes Bill, who is standing a short distance away. Bill dies as a result. Although John did not want to cause Bill's death, he is responsible for that death, because the law will transfer his homicidal intent from Sam to Bill.

California limits transferred intent to the "shoot at one, hit another" type of case. Thus, if a person shoots into a car to kill one person and actually injures two, that court will not allow the transferred intent doctrine to be used to convict the defendant of two attempted murders. Not all courts agree with this position but will follow the more traditional rule that the original intent will transfer to any number of unintended victims because the specific intent was to kill *someone*. Among others, Maryland follows the traditional rule.

5.2 TYPES OF INTENT *(MENS REA)*

Intent (referred to as *mens rea* in law, is divided into two categories: general intent and specific intent. Some crimes require only proof of general intent or general evil state of mind. Other crimes require proof of a specific state of mind or intent to commit a specific crime.

General Intent

General criminal intent, is all that is required in most crimes. All that has to be shown is that the accused acted with a malevolent purpose—that the accused knew the conduct was wrong. Unless more is required, evidence that will convince a jury that the defendant had this evil state of mind at the time of the act will be enough to raise the presumption that must be overcome by the defense. A specific intent, when required, must be proved and cannot be presumed.

Specific Intent

Some crimes require proof of a **specific intent.** It must be shown that the actor desired the prohibited result. A specific intent cannot be presumed but must be proved like any other element of crime.

For example, John opens a door to someone's house and enters it to escape the rain. Has he broken and entered the dwelling house of another with intent to commit a crime? No. The prosecution would be unable to prove the required specific state of mind by presumption. The jury would not be permitted to presume that merely because John was in another's house, he was there to commit a crime. However, they could presume from the evidence of his being in the house that he trespassed, because criminal trespass requires only a showing of general intent. It would then be up to John to prove a proper defense to his action, if he has any, to escape liability for his act. There are some exceptions to the rules discussed under the specific crimes noted in following chapters. We will attempt, in our discussion of crimes, to indicate whether it is a specific or general intent crime. The reader is cautioned to note that certain crimes that required only a general intent at common law require a specific intent under modern statutes. We will attempt to point out these changes, but local statutes should be consulted to determine the type of intent required in the reader's home state.

States are constantly trying to help prosecutors win their cases by creating presumptions to take the place of proof of the intent. A California statute said that embezzlement would be presumed where a rental car is not returned within five days after the expiration of the lease. A jury instruction was given in a case based on this presumption. The U.S. Supreme Court felt that the instruction potentially relieved the state of having to prove each element beyond a reasonable doubt and therefore caused the jury to bypass the intent issue. The case was sent back to California to determine if a rational jury could have found the intent proven despite the instruction; if so, the error was harmless, but if not, the case had to be reversed.

5.3 RECKLESSNESS

A less wrongful state of mind than intent is that called **recklessness**. Committing a prohibited act with a reckless state of mind is nonetheless criminal and will often justify holding an individual criminally liable even though intent cannot be proved.

A reckless state of mind implies that one acts, not intending harm, but with complete disregard for the rights and safety of others, causing harm to result. If John voluntarily commits an act without intending to hurt anyone, but his conduct is so dangerously done and he is not concerned with the consequences, he is acting with a reckless state of mind. The accused must foresee that his voluntary act would possibly or probably cause the result and that he acted unreasonably under circumstances he knew about. The defendant's actual state of mind must be shown. Suppose that John, an adult, decides to do some target practice with his new .22-caliber target pistol in front of his house on Christmas morning in a busy residential neighborhood. There are children playing in the street, but John is determined to test his new weapon. He fires once at the tire of a passing car and misses. The bullet strikes and kills a child playing across the street. John intended to hurt no one. His reckless state of mind would nevertheless make him criminally liable, but probably for a lesser degree of homicide than if he had acted with the intent to kill. The key to recklessness is the actor's indifference to consequences.

5.4 NEGLIGENCE

The final type of mental element we discuss is **negligence**. We will attempt to show what type of negligence will render one criminally liable and what type will not.

Elements

There are four elements of negligence: a standard of care, breach of that standard, proximate cause, and harm or injury produced. In any discussion of negligence, it would appear that only conduct is being examined. Although it is difficult to separate or compartmentalize the state of mind and the conduct involved in negligence, it is enough to say that negligence, which involves acts of omission as well as acts of commission, will, in any number of instances, render one guilty of a crime. Negligent acts are usually voluntary but do not rise to the level of intended conduct.

Standard of Care

The first element, standard of care, concerns everyone's legal responsibility to act or refrain from acting in ways dangerous to others. Every person owes a duty to every other person not to infringe upon someone else's welfare or safety. When one person harms another, it must first be determined whether or not a duty was owed. If a duty was owed, this element is satisfied. The duty may be fixed either by statutory or common law. Take, for example, the duty owed to a child by a parent. The duty may arise when someone who owes no duty to another performs an act directed to that other person. This is best illustrated by an example: A passerby has no duty to save a stranger in distress. Once the passerby begins to save the stranger, he has a duty to act with reasonable care.

Breach of Standard

Once the duty is found, it becomes essential to determine if there was a breach of that duty. Did the person owing the duty meet the standard of care? This factor involves what is known to the law as the "reasonable person test." This test is applied by asking whether the accused exercised the same amount of care that a reasonable, prudent person exercising ordinary caution would have used under similar circumstances. If the accused did not act this way, the duty has not been fulfilled. Unfortunately, there is no model to which we can turn for this standard. The reasonable man is a legal fiction. The standard may vary from jurisdiction to jurisdiction and from case to case because the trier of fact examines the circumstances of each case and determines what the reasonable man would have done under similar circumstances. The reasonable man criterion will vary according to the skill of each person. In a case involving need for medical attention, a doctor would be required to exercise more care than would an ordinary passerby.

Proximate Cause

If a duty has been discovered and if it is thought that the duty has been breached, it is necessary to determine if that breach was the proximate cause of the alleged injury. This determination is based on all the rules of causation discussed in Chapter 4.

Harm or Injury

Of course, there cannot be liability unless an injury or actual harm has occurred. The harm, however, need not be severe. The slightest injury to person or property will support this element of negligence.

Civil vs. Criminal Negligence

All these elements define both civil and criminal negligence. The elements of civil negligence, often called simple or ordinary negligence, are the same as the elements of criminal negligence. Criminal negligence is often called gross or culpable negligence, but all these terms are interchangeable. The difference between civil and criminal negligence is solely a matter of degree and is usually a question to be settled by the fact finder. The distinction will depend on the facts of the particular case and not on any hard and fast legal principles. The guidelines the jury uses to distinguish civil and criminal negligence are based not on the severity of the harm or injury that results but, rather, on the severity of the breach itself.

Negligence differs from recklessness in one important aspect. In recklessness, conduct is governed by the actual state of mind of the accused. In negligence it is possible that the accused is actually not aware of the consequences of that conduct even though a reasonable person would be. The same conduct can render one both civilly and criminally liable if the conduct is grossly negligent. On the other hand, if the harm resulted from only simple negligence, only a civil action can be supported.

It can be seen that law enforcement officers are often required to make a value judgment at the scene as to the degree of negligence involved in a particular case. If the officer, based on the guidelines above, determines that there has been a duty owed by one to another and that the standard of care has been breached, that as a proximate result a harm requiring police investigation has occurred, the officer may justifiably make an arrest. If the officer finds that the accused acted with a criminally negligent state of mind, the officer will be protected. This is so even though the prosecutor or jury later find only simple negligence.

Negligence will not support a charge for a crime requiring a specific intent. The following examples serve to consolidate the principles of negligence. Suppose that John is driving down the street and is obeying all traffic laws. John takes his eyes off the road for a second to light a cigarette and unintentionally runs through a stop sign, striking another automobile crossing the intersection and killing the driver of that car. John, of course, owed a duty to others on the road constantly to be aware of traffic conditions. He breached that standard of care by taking his eyes off the road, which proximately resulted in the harm. There is no doubt that John was negligent, but to what degree? At most, John would be civilly liable for his negligence, because the breach of the standard of care was not so severe as to constitute culpable negligence, notwithstanding the amount of injury inflicted.

On the other hand, suppose that John has been driving while intoxicated instead of merely lighting a cigarette when he ran through the stop sign. Here, John's breach of reasonable conduct was so severe that a jury could justifiably find gross negligence and hold John criminally liable.

5.5 THE MENTAL ELEMENT IN MALA PROHIBITA OFFENSES

As noted in Chapter 2, all common law crimes were considered moral wrongs or wrongs in themselves and were labeled *mala in se*. These common law crimes all required proof of intent whether general or specific. Since that time, offenses have been created that involve no moral turpitude but that are designed to protect the health, safety, or welfare of society. These offenses are not wrongs in themselves but, rather, are wrongs because the lawmakers say so. These offenses are called *mala prohibita*. A large number of these offenses require no proof of intent or other state of mind but merely proof that the act was done. A number of academics say that these offenses are not crimes. That may be so in a pure sense, but, in each *malum prohibitum* offense, a person's property or life may be restricted. The U.S. Supreme Court guarantees the right of trial by jury, and all other constitutional rights are extended to one charged with such an offense.

Among the many types of offenses considered to be *mala prohibita* are (1) illegal sales of intoxicating liquor by such conduct as selling of a prohibited beverage or selling to minors; (2) sales of impure or adulterated food or drugs; (3) sales of misbranded articles, such as selling oleomargarine as butter; or (4) shortweighting a customer by having produce or meat scales improperly balanced so that the customer is overcharged. Violations of most traffic regulations and motor vehicle laws also are considered *mala prohibita*. Running a red light, speeding, improper passing, driving without a license, and others are violations enacted for the safety, health, and well-being of the community. Most such violations are not inherently opposed to the moral concept of what is wrong. These types of offenses are wrong only because they are prohibited. The fact that the existence of a guilty state of mind at the time of commission need not be proved is only of secondary importance in distinguishing such conduct from crimes *mala in se*. It is the history and nature of the offense that controls its classification as *malum prohibitum* or *malum in se* rather than the requirements regarding proof of a guilty state of mind. It is thus improper to believe that any offense that does not require proof of intent or guilty mind, either general or specific, is, by that very fact, *malum prohibitum*. Statutory rape is a good example of this principle. Some view statutory rape as a *malum prohibitum* offense because proof of a criminal state of mind is not an element of the crime needed for conviction. The mere doing of the prohibited act is all that need be shown. Although this is true, when analyzed in terms of the nature of the offense, statutory rape takes on moral overtones that clearly reflect the opinion of society to treat that conduct as being morally unacceptable as well as legally prohibited; hence, *malum in se*. It may then be concluded that although many *mala prohibita* offenses do not require proof of intent as an element leading to conviction, this strict liability factor is not the sole criterion for distinguishing acts that are *mala prohibita* from those that are *mala in se*.

The importance of the fact that *mala prohibita* offenses are crimes was brought home in companion U.S. Supreme Court cases. In these cases the defendants were arrested for traffic misdemeanors and searched. Defense attorneys argued that one could not be searched upon an arrest for a traffic offense. The Court disagreed and said that a misdemeanor is a crime. A person lawfully arrested for a crime is subject to a full field strip search incident to the lawful arrest. It does not matter that the crime for which the person is arrested is petty or great.

Several states disagreed with the Supreme Court. Some, by court decision, interpreted their own constitutions as not allowing such searches for traffic stops unless the search is related to the reason for the stop, such as looking for whiskey or drugs when the charge is driving under the influence. Others absolutely prohibited, by legislation, custodial arrests for minor traffic offenses. Others decriminalized their traffic laws with the exception of vehicular homicide, reckless driving, and driving under the influence to avoid the potential impact of the Supreme Court decisions.

▬▬▬ DISCUSSION QUESTIONS ▬▬▬

1. Albert, a city police detective assigned to the vice detail, managed to gain the confidence of a group operating a commercialized gambling operation. One night Albert sat in on a card game for the purpose of learning the identity of participants for eventual use in prosecutions for violating the state gambling laws. Several hours after the game started, the house was raided by sheriff's deputies and state investigators. All the participants were arrested and charged with gambling, including Albert. At trial, Albert argued that his intent was only to gain the evidence he was seeking and not to gamble. The jury convicted Albert, after an instruction by the court stating, in effect, that Albert's motive was to get the evidence, but his intent was to gamble. Albert immediately appealed. What should be the decision of the appellate court?

2. A defendant ran through a stop sign while driving an automobile and convincingly showed he had done so unintentionally. Should the court have acquitted him?

3. Sam was an Olympic medalist in swimming. He secured a job as a lifeguard at Euclid Beach Park. Things were not too exciting, so Sam decided to have a few drinks while he was on duty. Sam was feeling no pain when he heard a cry for help. He saw a person in great difficulty about 100 yards offshore. Sam stumbled and finally reached the water. He was unable to get to the victim before she drowned. Under what theory, if any, would Sam be liable for the death of the drowning victim?

4. B, a single mother of two children (one eight years old, the other two and a half years) decides to go out for the evening. The mother cannot find a babysitter. The eight-year-old is given careful instructions on how to locate and call the mother, who is going to a place only three minutes away from the house. She will be gone for only two or three hours. The building burns, killing both children. How would you label the mother's mental state? Was her conduct only ordinarily negligent?

▆▆▆ GLOSSARY ▆▆▆

Intent – an evil state of mind which, when coupled with an unlawful act, will make one criminally liable.
Mens Rea – a Latin term meaning "criminal intent."
Motive – the reason one acts the way one does.
Negligence – a state of mind that accompanies an injury or harm proximately caused by a breach of a standard of care owed by one person to another. The severity of the breach distinguishes criminal from civil negligence.
Recklessness – a state of mind of one who acts in a manner, not necessarily intending harm but who acts with complete disregard for the rights and safety of others.
Specific intent – the intent to produce a specific result.
Transferred intent – when a person acts with the intent to cause a specific result and, someone or something else ends up being the target of the conduct, the law will transfer the intent to harm or injure the original target to the target actually harmed.

▆▆▆ REFERENCE CASES, STATUTES, AND WEB SITES ▆▆▆

CASES

Commonwealth v. Chapman, 433 Mass. 481 (2001).

Miller v. State, 233 Ga. App. 814 (1998).

People v. Scott, 14 Cal.4th 544 (1996).

People v. Williams, 127 Cal. 212, 59 P. 581 (1889).

Liparota v. U.S., 471 U.S. 419, 105 S.Ct. 2084, 85 L.Ed.2d 434 (1985).

Carella v. California, 491 U.S. 263, 109 S.Ct. 2419, 105 L.Ed.2d 218 (Percurr. 1989).

U.S. v. X-Citement Video, Inc., 513 U.S. 64, 115 S.Ct. 464, 130 L.Ed.2d 372 (1994).

People v. Saunders, 85 N.Y.2d 693 (1975).

Adami v Texas, 524 S.W.2d 693 (1975).

People v. Griggs, 216 Cal. App.3d 734 (1989).

People v. Birreuta, 162 Ca. App.3d 454 (1984).

State v. Fennell, 340 S.C. 266 (2000).

State v. Hinton, 227 Conn. 301 (1993).

Garrett v. City of Bossier City, La. App. 2 Cir (2001).

Atwater v. City of Lago Vista, 121 S.Ct. 1536 (Tex. 2001).

Wilkes v. State, 2001 Md. LEXIS 397 (2001).

WEB SITE

www.findlaw.com

CHAPTER 6

Matters Affecting Criminal Responsibility

■ KEY WORDS AND PHRASES ■

Alibi
Battered spouse syndrome
Common law immunity
Consent
Custom
Defenses
Diminished capacity
Duress
Durham rule
Entrapment
Fleeing felon rule
Insanity
Involuntary intoxication

Irresistible impulse
Mistake of fact
Mistake of law
M'Naghten rule
Necessity
Religious belief
Responsibility
Selective enforcement
Statute of limitations
Statutory immunity
Tennessee v. Garner
Voluntary intoxication

6.0 INTRODUCTION

A law enforcement officer often becomes frustrated when an accused walks out of the courtroom free if it appears to the officer that the defendant did in fact commit a prohibited act. Some acquittals result from improper procedures followed by officers in the areas of arrest, search, seizure, and so forth. Other "nonconvictions" occur because a court has decided that a procedure, formerly permissible, is no longer correct under modern statutory or constitutional interpretation. A third group of cases are dismissed because of some substantive rule of criminal law that constitutes a defense to the commission of a criminal act and operates to the benefit of the accused.

The **defenses** available in a criminal case are the subject of this chapter. If the rationale underlying these defenses can be understood, perhaps their existence can be better appreciated and accepted. As was observed in Chapter 1, the early history of criminal law did not take the concept of **responsibility** into account. All that was necessary to convict for a criminal offense was to show that the accused did in fact commit the forbidden act. The law during that time was not concerned with the reasons or justifications for committing the act. If this rationale held true today, an individual killing another in self-defense, police officers killing to prevent felonies, and persons forced to commit crimes against their will under threat of death or serious bodily harm, among other examples, would all be guilty of serious crimes without any opportunity to justify their acts. As a result, there is little doubt that society would soon crumble.

Our system of justice is predicated on the theory that a person acts of his or her own free will. To maintain this system, the law must allow some tolerance to those accused of crimes when they have acted against their own free will, either because they were insane or because their free will was exercised in such a way that they were compelled to commit an act to avoid more serious consequences, as in the case of self-defense or duress, or because they were compelled to act in the performance of a legal duty, as in the case of a police officer killing to prevent a felony. With this purpose in mind, let us now examine a number of these defenses.

6.1 INSANITY

Insanity is a legal defense in criminal cases. The word *legal* should be emphasized because insanity is a legal, not a medical concept. The term itself implies a sharp line distinguishing normality from abnormality. Those of the medical profession who specialize in personality disorders (psychiatrists) believe there is no such line of demarcation, that normality and abnormality are extremes of a continuum with variations in between. However, it is a fundamental principle of law that a person cannot be convicted of a crime if, at the time of committing the offense, the person was not responsible for his or her actions. A dividing line must, therefore, be drawn somewhere to separate those who are responsible from those who are not. This line is the difference between sanity and insanity.

Everyone is presumed to be sane at the time of committing an offense. This rule benefits the state because it eliminates the necessity of proving that every defendant in every criminal case is sane before the trial can proceed. However, if the defense does present some question about the defendant's sanity, the state must take the time to prove the defendant's sanity.

This traditional process was clouded by a decision of the U.S. Supreme Court. In a 1993 case, the Court held that insanity is indeed an affirmative defense and therefore states can constitutionally require a defendant to carry the burden of proving his or her insanity by a preponderance (majority) of the evidence. Most states, however, still follow the traditional requirement that mandates the defendant to introduce evidence only to rebut the presumption of sanity, at which time the burden shifts to the state to prove the defendant's sanity as a condition precedent to obtaining a conviction. The language used by different jurisdictions makes it difficult to define the exact method of rebutting the presumption of sanity. Illinois, for example, requires a defendant to raise a " reasonable doubt" of the defendant's mental state. Some states, such as Indiana, require the defendant to present "some evidence" of lack of sanity, and Washington D. C. requires only a "scintilla" of evidence indicative of the defendant's lack of sanity. Once the defendant has raised doubt about or rebutted the presumption of sanity, the burden shifting to the state to prove the defendant's sanity can also take many forms. For example, in Connecticut and Michigan, the burden is on the state to prove the defendant's sanity "beyond a reasonable doubt." There are some states where the burden does not shift to the state. In Arizona, the defendant carries the burden of proving insanity "beyond a reasonable doubt." The defendant's burden in Alabama is to prove insanity by "clear and convincing evidence." Delaware requires the defense to establish insanity by a preponderance of the evidence. Iowa's statute does not allow mental condition to be a defense but does allow evidence during trial if the defendant's state of mind is an element of the crime.

The issue of insanity may arise at any stage of the process of administering justice. The substantive criminal law is concerned with insanity at the time the act was committed, not insanity before, during, or after trial begins. These are procedural questions, and they generally have no bearing on the question of whether the accused will ultimately be liable for the prohibited conduct. For example, if an accused claims insanity at the time of trial and is adjudged incompetent at a special hearing held for that purpose, the accused cannot stand trial until he or she is declared competent at a later date. The criterion for determining competence to stand trial is whether the accused is capable of understanding the nature of the proceedings and whether the accused is able to help counsel in preparing a defense. If not, the accused is incompetent to stand trial. This does not mean the accused cannot be tried for the crime. It simply means the trial will be postponed until the accused is adjudged competent again in another hearing held for that purpose. Similarly, a person may not be executed for a capital offense while insane.

This section is concerned with insanity at the time the act is committed, for this determines liability for the commission of criminal acts. The procedural aspects are left, with the exception of the brief mention above, to the research of the individual student or to other courses.

The origin of the insanity defense is not to be found in the statute books; rather, it has developed through the case law process. In the beginning of legal systems, those who were mad were not acquitted of criminal charges. After they were convicted by a special verdict declaring their "madness," the king granted pardons. Later, the test of insanity became known as the "wild beast test," under which an accused was excused from liability if he or she were totally deprived of reason, understanding, and memory and had no more awareness of his or her own actions than a wild beast. Subsequently, a type of right-wrong test was established under which the defendant was declared insane if unable to distinguish between what was morally right and morally wrong or between good and evil.

In 1843, Daniel M'Naghten killed the secretary to Sir Robert Peel. Peel was the British Home Secretary and is considered one of the founding fathers of modern policing. M'Naghten claimed that at the time he committed the act, he was not of a sound state of mind. The decision in this case became the cornerstone of the test of insanity in modern times. The decision established the rule that if at the time of committing the act, the defendant was laboring under such a defect of reason from disease of the mind as not to know the nature and quality of the act, or if the defendant did have knowledge of the nature and quality of the act but did not know that what he or she was doing was wrong, that person was legally insane and not responsible for his or her acts. This test became popularly known as the right-wrong test of insanity. The nature and quality of the act, as referred to in this test, deals with the ability of the accused to act in a rational manner and to evaluate the circumstances at the time the act was committed.

The **M'Naghten rule** was not totally acceptable in all jurisdictions, and consequently, additional tests were developed. For example, the state of Alabama in 1886 extended the M'Naghten rule to include a test called **irresistible impulse.** In essence the test declared that, if by reason of a mental disease, the accused had so far lost the power to choose between right and wrong that he or she was unable to avoid doing the act in question, the accused was insane. Alabama was recognizing a form of temporary insanity by use of this irresistible impulse test. It should be noted, however, that since most jurisdictions have begun to recognize the many forms of "delay stress syndromes," such as battered spouse, battered child, or rape trauma, the role of irresistible impulse as a defense is now obsolete.

The M'Naghten test was severely criticized on the grounds that it was arbitrary, that it only applied to a small percentage of the people who were actually mentally ill, that it underemphasized emotional strain on an individual, that it required psychiatrists to give a yes or no answer about a person's sanity in an age when psychiatry had become much more of a science than it had been in the past, and that it took into account only one aspect instead of the whole personality of an individual. In 1954 the U.S. Court of Appeals for the District of Columbia broadened the M'Naghten test in favor of what has become known as the **Durham rule.** The Durham case held that an accused is not criminally responsible if the unlawful act was the product of mental disease or mental defect. Mental disease is defined as a condition capable of either improving or deteriorating, whereas a defect is a condition considered not capable of either improving or deteriorating. This broad test of insanity has been adopted by only a few states.

The Durham rule has also been criticized on opposite grounds from M'Naghten in that the Durham rule is too broad and places too much power in the hands of psychiatry in determining the legal issue of insanity. Critics fear that too many people will be able to escape punishment for crimes if this test is applied.

In a long line of cases following the Durham decision, the district court attempted to clarify the rule so that its application would be more understandable to juries who were required to apply it. One of the greatest problems faced by the court was in its attempt to restrict the impact of the rule on juries so that they would base their decisions as to the defendant's sanity on factual details relating to the evidence presented rather than simply on medical conclusions testified to by experts. The evidentiary distinction between the right of an expert witness to give opinions or draw conclusions on matters of a technical nature and the function of the jury to draw conclusions on factual matters became fuzzy. By early 1971, the dissatisfaction with the Durham rule in the District of Columbia courts became patently irreconcilable. As a result, the district rejected the Durham rule in 1972 in favor of the Model Penal Code test which proposes the following: The accused is legally insane if he or she "lacks substantial capacity either to appreciate the criminality of his conduct or to conform his conduct to the requirements of law."

All federal courts except the First Circuit (which includes the states of Maine, New Hampshire, Massachusetts, and Rhode Island and the commonwealth of Puerto Rico) are following the Model Penal Code test. Some jurisdictions recognize an insane delusion test under which the accused may have been partially insane in respect to the circumstances surrounding the commission of the crime, but sane as to other matters. Another suggested test goes by the name "policeman-at-the-elbow" test. The essence of this proposal is that the accused is not guilty if he or she would still have committed the crime even if there had been a police officer standing at his or her elbow. In such a case, the accused must have been insane.

All these tests rest on the premise that the accused is *non compos mentis,* that is, that the accused is unable, because of mental illness, to form the intent necessary to commit a crime. As may be observed from the conflicting tests described, there is no one test that is universally acceptable, and all are subject to valid criticisms.

The public has generally become dissatisfied with the perceived workings of the insanity defense. In the past this defense was rarely used and usually only in a few homicide cases. In recent years its use has become more commonplace and in every conceivable crime. One federal court recently said the defense could be used in a firearm possession charge case. Public awareness has grown, and displeasure reached a peak when John Hinckley was found insane in the shooting of President Reagan. Legislative reaction has been varied. Some legislatures repealed their more liberal tests and returned to M'Naghten. Others, like those of Montana, Idaho, and Utah, abolished the insanity defense. Yet others changed the form of the verdict to "guilty but insane," which allows institutionalization of the defendant for the period of the mental incapacity and imprisonment upon return to normalcy. The perception that mere antisocial conduct was being rewarded with acquittal verdicts played a large role in the legislative process of reform.

Is a state required to tolerate or provide the opportunity for a defendant to raise insanity as a defense? May a state abolish the defense? According to the U.S. Supreme Court, a state may constitutionally abolish the insanity defense since affirmative defenses such as insanity are not constitutionally required.

That the general population is dissatisfied with the insanity defense is, of course, not a new phenomenon. Even the high court of England was hauled before the English Parliament to explain their ruling in the M'-Naghten case. As already noted, a handful of states have abolished the all-or-nothing effect of a successful insanity defense by providing for the "guilty but insane" verdict. In such states the defendant does not walk away from responsibility or incarceration. The defendant is treated and when later determined to be sane, he or she is transferred from the treatment facility to prison to complete the term of punishment. Montana was one of the states to adopt this position as far back as 1979. The validity of the statute was challenged and, in a 1994 decision, was found by the U.S. Supreme Court to be constitutional.

The law does not recognize any degrees of insanity. There are no gray areas. One is considered either sane or insane. Similarly, the law is not concerned with the causes of insanity. It is felt that this is the concern of the medical profession. Even delirium tremens, resulting from excess consumption of alcohol, is treated as a type of insanity and will be a defense, notwithstanding the fact that it was probably induced voluntarily by the accused over a period of time.

With only a few exceptions, most courts agree that low intelligence or mental weakness is not the same thing as, nor will it support, a claim of insanity. A person may be an idiot, moron, or imbecile, but with sufficient capacity to know the rightness and wrongness of his or her acts, that person is legally sane. The only situation in which mental weakness may operate as a defense is one in which limited intelligence may prevent the person from having the capacity to form a specific intent where one is required. In such a case, the intent cannot be proved; hence, no conviction will ensue. This is not based on a plea of insanity, and it will not be decided on the tests of insanity.

Diminished Capacity, Diminished Responsibility, Partial Responsibility, Partial Insanity

Because the insanity defense is one of the most disliked by juries, there has been a movement since the 1970s to create a midlevel insanity defense called *diminished capacity*. It is known by different names including diminished responsibility, partial responsibility, and partial insanity. The purpose of the defense, like voluntary intoxication, is to prevent the defendant from being convicted of a specific intent crime. One federal court accepted the defense as an alternative to the insanity defense and said that the defendant could use it to show that he could not commit the specific intent crime of which he was charged.

By definition, the **diminished capacity** defense is applicable when a defendant suffers from a mental disease or defect that makes it impossible to commit certain crimes that require proof of specific intent but allows for a finding of guilt for a lesser included offense that requires proof only of a guilty state of mind or general intent. Thus the defense is not used to obtain an acquittal, but rather is used to reduce guilt to a lesser crime. It should be noted that all states do not accept this defense, and of those that do, some differences in application of the defense to specific offenses do exist. Arkansas, Arizona, District of Columbia, California, Florida, Georgia, Idaho, Iowa, Louisiana, Maryland, Massachusetts, Minnesota, Missouri, North Carolina, New York, Oregon, Pennsylvania, and West Virginia do not recognize the defense. The remaining jurisdictions do recognize a form of diminished capacity as a defense.

Related to insanity cases, but not quite constituting insanity, are the problems of persons referred to as criminal sexual psychopaths. The problems of this type of person are usually treated separately by each jurisdiction and are generally covered by statute. See Chapter 9 for a more complete discussion of this point.

6.2 ALIBI

Along with insanity, **alibi** is the most commonly known of the defenses. By claiming the defense of alibi, the defendant is saying that he or she was somewhere else when the crime was committed, and therefore, cannot be guilty. The defendant must show that he or she was at another specific place during the entire time frame involving the commission of the crime. Alibi is truly a defense of physical impossibility (Chapter 4).

Unless the evidence as to the alibi covers the time at or before the crime was committed sufficiently to render the presence of the defendant impossible or highly improbable, it proves nothing. The evidence in support of an alibi need accomplish no more than raise a reasonable doubt of the defendant's presence at the crime. But just because the defense raises an alibi does not mean it has created reasonable doubt. The evidence has to be credible.

Generally, in weighing alibi evidence, many factors are taken into account. Considerations include the interests of those who testify on the defendant's behalf—wife, husband, other relatives, business partners, and so forth—or, when the witness is a co-actor to be tried in another trial, there may be reason to doubt credibility. However, just because one is a friend or relative does not necessarily mean the witness is lying; still, it is a factor to be considered.

As indicated, the time span the alibi encompasses is important. If the testimony of the alibi witness does not cover the entire time of the crime, the defense fails. Often, the alibi witness testifies in complete contradiction to the testimony of eyewitnesses. In such cases, the jury is not obliged to believe either the alibi witnesses over the eyewitnesses, or vice versa. The jury must weigh and balance the conflicting testimony as part of its fact-finding function.

6.3   INTOXICATION

Intoxication is usually thought of in the context of alcoholic beverages. Many jurisdictions, however, also apply the same rule to intoxication resulting from use of narcotics and dangerous drugs. Intoxication as a defense may fall into either of two categories.

Voluntary Intoxication

Voluntary intoxication is ordinarily not a defense. If a person voluntarily becomes intoxicated, that person is held to be responsible for the consequences of any acts committed while in that condition. There are, however, two exceptions to this rule. In those cases in which the crime committed requires proof of a specific intent, the accused's state of voluntary intoxication may be such that he or she was incapable of forming the specific intent and, therefore, cannot be convicted (Chapter 5). Also, if the crime requires knowledge of certain facts, extreme voluntary intoxication may make the defendant incapable of having this knowledge; thus an essential element cannot be proved. This might apply in a case such as "knowingly receiving stolen property."

This principle is under attack. Indiana, for example, had a statute that limited the defense of voluntary intoxication to specific intent crimes. The Indiana Supreme Court declared that statute void. They felt that any factor that serves as a denial of the existence of *mens rea* must be considered by the trier of fact. They were not worried that otherwise guilty people would walk free. They said that it is difficult to envision a finding of not guilty by reason of intoxication when the acts committed require a significant degree of physical or intellectual skills. The court said that a person who could devise a plan, operate equipment, instruct the behavior of others, or carry out acts requiring physical skill, even though voluntarily intoxicated, was guilty.

If, however, the defendant forms the intent before becoming voluntarily intoxicated, the intent remains until the act is committed, and the intervening intoxication will not be a defense. If a legally insane person is also voluntarily intoxicated at the time of committing a prohibited act, the intoxication will not affect the defense of insanity.

In those few cases in which a person claims such extreme intoxication as to be incapable of forming a specific intent, the jury must decide the truth of the claim. There is no set point at which voluntary intoxication will serve as a defense in these cases. It is a matter of degree, for it is well known that intoxicants affect different people in different ways and to different extents. Suffice it to say that it is difficult to convince a jury to accept voluntary intoxication as a defense, and rarely is it argued successfully.

The defense of voluntary intoxication thus prevents conviction for a specific intent crime only when the intent is formed after the intoxication is achieved. The battle line is over whether the crime charged is a specific or general intent crime. One court had to decide whether assault on a federal officer was a specific or general intent crime. The court found it to be a general intent crime because the defendant did not have to know that his victim was a federal officer.

New Hampshire has held that voluntary intoxication is not a defense to their reckless murder crime. Reckless murder is extreme indifference, but the court said it is not a specific intent crime.

Involuntary Intoxication

Involuntary intoxication, on the other hand (liquor forcibly poured into someone who then commits a crime, for example), is ordinarily a defense. There is a question, however, as to whether the degree of involuntary intoxication is important. If involuntary intoxication were an absolute defense, any individual could claim that the one small drink he or she was forced to take was sufficient to excuse the commission of a subsequent crime. Most authorities would agree that the degree of intoxication is an important factor in determining criminal liability under these circumstances.

Can an alcoholic claim involuntary intoxication as a defense? Those who view alcoholism as a disease beyond the control of the individual would like the answer to be yes. Most courts, however, do not accept that analysis but say that the alcoholic does have sufficient control over his or her drinking to make the resulting intoxication self-induced.

6.4 IMMUNITY

The fact that granting immunity is now common in some criminal trials does not mean the law is new. Actually, immunity was a recognized common law doctrine.

There are basically two types of immunity that may be granted to an individual: **common law immunity** and **statutory immunity.** To convict persons accused of committing crimes, it is sometimes necessary that one or more participants be given immunity so that their testimony may be used to help convict other participants. This is the purpose of granting immunity. It is rarely done if there is sufficient evidence to convict without statements from the participants themselves. Common law immunity, sometimes called *an agreement not to prosecute* or a *contract of immunity,* exists as a matter of public policy and may prevent a person from relying on the self-incrimination clause of the Fifth Amendment to the U.S. Constitution.

The prosecutor, not the police, is the only party authorized to grant immunity. In all cases, the court in which the case is being heard should be made a party to the agreement. A few jurisdictions require the court to consent, but in all jurisdictions it is a recommended policy. The agreement may take any of three forms. It may take the form of an agreement not to prosecute the person to whom immunity is granted, or it may be an agreement to the effect that the person will be prosecuted but that no sentence will be imposed. Third, it may be agreed that the compelled testimony will not be used against the witness in any subsequent proceeding. Common law immunity is a three-party agreement involving the prosecutor, defendant, and court and can be granted only with the consent of the defendant.

Statutory immunity is not an agreement of the type existing at common law. Under statutes of the various jurisdictions, immunity from self-incrimination is involuntarily taken away from the defendant by the state and by the statutes. Statutory immunity generally takes one of two forms. Transactional immunity prevents the state from taking any further prosecutorial action against the individual concerning the event or transaction. Use immunity prevents the state from using the compelled testimony. If, however, there is other sufficient evidence aside from the compelled testimony on which to prosecute, a charge may be brought against the individual.

Generally, immunity will apply to any testimony given that has any relevance to the issue in question, even though the response indicates the defendant committed an act not thought to be relevant or connected in any way.

6.5 STATUTES OF LIMITATIONS

At common law there was no time limit on prosecuting for the commission of a crime. If the offender was not apprehended for 20 or 30 years, that person could still be prosecuted after that time, the only problem being the possibility of loss or destruction of evidence or the unavailability of witnesses.

Most jurisdictions today still place no limitation on prosecutions for capital offenses or murder in noncapital punishment states. However, in most states statutes have been enacted providing that the state has only a certain period of time after a crime is committed in which to initiate the criminal process. This time period differs from state to state and is established arbitrarily. The reason for statutes such as this is that evidence and witnesses will usually be unavailable after that period of time so that prosecutions would be fruitless.

What constitutes the initiation of the criminal process is one of the prime questions under such statutes. The criminal process begins with either the issuance of an indictment, the filing of an information, or the issuance of a valid arrest warrant. When any of these events occurs, the **statute of limitations** will stop running, and there will no longer be a time limit imposed by the statute of limitations for that crime.

The statute of limitations for any offense begins to run from the moment the crime is committed, not when it is discovered or when the defendant is actually identified, unless otherwise specified. However, if a crime is a continuing offense, the result is different. Kidnapping is a continuing offense, and the statute of limitations does not begin to run until the victim is released. Thus in one case the five-year statute of limitations did not begin to run when the children were kidnapped, but instead it began to run six years later when the children were released.

Besides issuance of an indictment, filing of an information, or issuance of a valid arrest warrant, the statute may stop running when the accused is a fugitive from justice. A fugitive from justice is one who hides. However, through a series of decisions, the U.S. Supreme Court has ruled that a fugitive is one not present in the state where the crime was committed, and thus one who leaves a state after committing an offense is a fugitive unless the state forced the offender out. A person who lives openly in another state is a fugitive as much as one who lives secretly in another state.

While a person is a fugitive, the statute of limitations clock stops running. When found, even decades later, a fugitive may be returned for trial to the state where the offense was committed. This is illustrated by a case in which a priest from Massachusetts committed certain sexual crimes against minors in the late 1970s, left the state before the statute of limitations ran out, and was found in Florida more than 20 years later. After being indicted and extradited, he lost his argument that the statute of limitations had run out and eventually pled guilty to the original charges.

6.6 MISTAKE OF FACT

One who commits a prohibited act in good faith and with a reasonable belief that certain facts exist that, if they actually did exist, would make the act innocent may base a defense on **mistake of fact.** Thus when John takes property belonging to Bill, believing it to be his own, his defense will be that he was operating under a mistake of fact. In this example, the mistake prevents John from having a criminal intent, so he cannot be charged with larceny. This is true provided that John's mistake was made in good faith and was a reasonable mistake that anyone could have made under the circumstances. If the facts actually were as he believed them to be, his act would have been innocent. The mistake must be an honest one and not caused by the defendant's own negligence or deliberation. The decision in any case as to whether the mistake was honest rests with the jury.

In three cases mistake of fact will not operate as a defense. First, in those few offenses in which intent is not a necessary element of the offense and the mere doing of the act is sufficient to convict, mistake of fact is irrelevant because there is no intent element to be negated by the mistake. A statute designed to protect consumers of goods from being shortweighted by retailers, therefore, makes the seller of goods criminally liable for overcharging a customer by shortweighing the goods sold. This is a *malum prohibitum* misdemeanor (wrong only because it is prohibited) not requiring intent. Once the act is committed, mistake of fact will be no defense, and the seller's contention of not knowing the scales were unbalanced will not be a valid defense.

The second type of case to which this defense will not apply is one in which the accused intended to commit some wrong but did not intend the actual consequences of the act. If the initial wrong was intended, the defense will not apply because of the rule that a wrongdoer is responsible for the natural and probable consequences of his acts. If Joe strikes Bob with his fist intending to injure Bob, and Bob accidentally trips over a log, strikes his head on the pavement, and dies, Joe cannot claim mistake of fact as a defense because he intentionally struck Bob and is liable for all natural and probable consequences of his act.

The third exception to the rule involves crimes committed through the culpable negligence of the accused. Because culpable negligence is a different type of mental element than intention, and results from a gross failure to exercise due care, there is no intent to be negated by the mistake and it will not be a defense. Suppose that the groundskeeper of a children's playground leaves parathion, a deadly poison, within the reach of the children. As a result, one of the children, thinking it to be face powder, applies it to her face and arms and dies within a few minutes. The groundskeeper cannot claim mistake of fact as a defense to his culpable negligence by showing that he was unaware of the presence of any children in the area.

One of the most difficult cases under a mistake of fact is in the bigamy area. When the absent spouse has been away for several years, most states allow the present spouse to reasonably and in good faith presume death of the absent spouse and remarry. But what of the situation in which a married woman tells the defendant (her husband) she is going to get a divorce? She then marries another man, takes his name, and holds the relationship out to the public as one of husband and wife. All of this is observed by the defendant, who in turn remarries. However, the wife did not get a divorce and the defendant, the "former" husband, is charged with bigamy. Should the defendant be convicted? Such were the facts in a California Supreme Court case. The court held that there was no bigamy because he held the bona fide and reasonable belief that such facts existed that left him free to marry. The court reached this conclusion by relying on the fact that there was no concurrence of the act and the intent. Not all courts would reach the same conclusion, but the case is the beginning of a trend in this area.

6.7   MISTAKE OF LAW

There is an old cliché that "ignorance of the law is no excuse." This is true in 99 percent of the cases, because everyone is presumed to know the law. It is true for aliens and citizens alike. Suppose that it is permissible in a Middle Eastern country for a man to kill another when the other has stolen his purse. Can the alien do the same thing in the United States and escape punishment? No, because the alien is held to the same standards as our citizens under our law. Less exaggerated is the example often given concerning the American tourist who travels through the United States. Suppose our tourist comes from a state permitting right-hand turns when the traffic light is red. If the tourist goes to a state that prohibits such turns, the tourist cannot use his home state law as a defense, nor can he use ignorance of the law as a defense, if the prosecution wishes to pursue the matter.

When the crime charged requires proof of a specific intent (Chapter 5), and the accused can show a mistake of law to an extent that would nullify the forming of such specific intent, the defense will be available. Admittedly this is a rare occurrence. We have been unable to discover a pure **mistake of law** case that negated a specific intent. Research reveals any number of cases that state this rule but that, in fact, involved primarily a mistake of fact and its effect on the specific intent required. For example, Sam takes a book believing it belongs to Joe, intending to steal it, when, in fact, it is Sam's own book. Here, the specific intent cannot be fulfilled because Sam did not take the property of another. Lack of knowledge of the existing law will not, of itself, establish the defense. It must be shown that because he lacked this knowledge, the accused did not actually form the specific intent required for commission of the specific offense.

There is one further way in which mistake of law may be a defense. When a legislature enacts a law, constitutional principles require that it be presumed valid until and unless a proper court holds to the contrary. Anyone who obeys the law as it is written cannot be prosecuted for following it even if a court subsequently declares the law to be invalid. The most difficult problem concerns the status of one who bases conduct on a lower court's determination that a specific statute is unconstitutional, when the decision is later reversed by a higher court and the law is held to be constitutional. The accepted view in such situations is that the statute was constitutional from the beginning and that the lower court's determination of unconstitutionality must be ignored. This rule creates a dilemma for people who conduct themselves in accordance with the lower court's determination of unconstitutionality. They are placed in a position where they must act at their own peril in deciding whether to obey the law. To illustrate, suppose state A enacts a law prohibiting the importation of alcoholic beverages. John imports alcoholic beverages and is brought to trial for violating the law. The court, at John's trial, declares the statute to be unconstitutional and releases him. Bill, hearing of this decision, begins to import alcoholic beverages, relying on the decision of the lower courts. Unknown to Bill, the lower court's decision was appealed to the supreme court of the state, which reversed the lower court and declared the statute constitutional. Even though Bill relied on the lower court's decision, he is subject to prosecution for violating the statute. It is possible, however, for the law to take a practical look at the dilemma caused, and in each case determine whether or not mistake of law will be a defense.

6.8 ENTRAPMENT

The key word in the area of **entrapment** is inducement. When an officer of the law induces an otherwise innocent person to commit a crime so that the individual will be punished, the accused can raise the defense of entrapment. Notice the two key elements of entrapment—inducement of a person by a law enforcement officer.

Inducement means that the accused person did not intend to commit a crime before being induced to do so by the law enforcement officer. If the officer placed the idea of committing the crime in the mind of the accused, the defense is available to the accused, and the case will be dismissed. Williams, an undercover officer with the Big City Police Department, manages to gain the confidence of Burke. Williams proceeds to convince Burke that he should rob Johnson. Burke had no intention of robbing Johnson until Williams talked him into it. Burke commits the robbery and is immediately arrested by Williams. If Burke raises the defense of entrapment, his case will be dismissed.

Let us distinguish this from merely presenting the accused with the opportunity to commit the crime. Paul, another undercover agent for the Big City Police Department, is walking along the street and observes Roger, a known "mugger," approaching from the other direction. Paul pretends to be a staggering drunk and falls down in Roger's path. Seizing this inviting opportunity, Roger takes Paul's wallet and turns to leave, at which point Paul places Roger under arrest. This is not entrapment. In such a case, there is no entrapment, because merely presenting the accused with an opportunity to commit a crime in no way has any bearing on his intent. If the accused, without outside assistance or pressure, forms the intent to commit a crime, the accused cannot raise the defense of entrapment. Similarly, there is no entrapment when officers, knowing a crime is about to be committed, allow the accused to commit the act before making an arrest. Entrapment is not the same thing as trapping the accused during or immediately following the commission of a crime.

The courts will not allow law enforcement officers to force, pressure, persuade, or influence a person into committing a crime that such person would not otherwise have committed, merely for the purpose of punishing that person for doing the act.

Note that entrapment is referred to here as a defense. To raise the defense, the defendant must, in fact, admit that he or she committed the crime. Thus, the defendant is guilty of the act charged, but the courts refuse to punish under such circumstances because the government that punishes is also the government that compelled this person to commit the crime in the first place, and such a situation is unacceptable in our society.

The second major element of entrapment is that it must be accomplished by an officer of the law or some other government official. That is, it does not extend to acts induced by a private citizen who is not an officer of the law. If the private citizen is acting for, or on behalf of, or by direction of an officer of the law, that individual is treated as being an officer of the law for that purpose, and any arrest resulting from the commission of crimes induced this way will be subject to the defense of entrapment.

If an officer cooperates with a suspect to influence the suspect to participate in a crime, and the officer commits the act personally without intending to commit a crime, but solely for the purpose of charging the suspect as a principal in the second degree, the suspect is not guilty of the crime because the officer, the one who actually committed the act, is not guilty of a crime.

If one merely requests another to commit a prohibited act that the other would not normally commit unless such person was ready and willing at the time of the request, no entrapment results. Thus, when Arthur approaches Bob and asks Bob to sell him some narcotics, there is no entrapment if Bob complies, because there is no inducement, only an opportunity presented to someone ready and willing to violate the law.

The New York Court of Appeals adopted a four-pronged test to determine if police conduct violated due process requirements, thus giving rise to the defense of entrapment. The four questions requiring answers in a given case are:

1. Did the police manufacture a crime, or were they merely involved in ongoing criminal activity?

2. Did the police engage in criminal or improper conduct repugnant to a sense of justice?

3. Was the defend ant reluctant to commit the crime, yet that reluctance was overcome by:
   a. Humanitarian appeals,
   b. Temptation to exorbitant gain, or
   c. Persistent solicitation?

4. Were the police solely motivated by a desire to convict with no thought of preventing further crime or protecting the public?

The court said that none of these standards, by itself, is enough to find a violation of due process, but a combination of two or more would reasonably lead to a conclusion that a defendant had been entrapped.

Not all courts accept the New York position, which is called the due process entrapment defense, but a significant number of states do use the due process analysis for entrapment issues. Although the U.S. Supreme Court has shown a reluctance to fully adopt the due process defense, the Court issued a condemnation in a case in which federal officers "wore down" a defendant and finally got him to order some material related to child pornography. The Court held: "When the Government's quest for convictions leads to the apprehension of an otherwise law-abiding citizen who, if left to his own devices, likely would never have run afoul of the law, the courts should intervene." The states following the New York case allow even criminal types to use the defense when the government's conduct is outrageous enough.

6.9 CONSENT

The victim of a crime, in most cases, can consent to the crime being committed, and the defendant will often be able to use this consent as a defense. This is true specifically where a crime, such as larceny, is directed against an individual. On the other hand, if the act is the type of offense that affects the public at large, as in the case of fighting in public or disturbing the peace, consent of the individuals involved will not be a defense, for they are all wrongdoers to a certain extent.

When lack of consent is a necessary element of the crime, as in cases of forcible rape, robbery, larceny, burglary, and so forth, lack of consent can usually be presumed. If, however, the accused raises consent as a defense, the state must then prove the defendant acted without the victim's **consent.** There are four elements to the defense of consent. First, the person giving consent must be capable of giving it. Consent would not be available as a defense to statutory rape because the class of persons protected by these statutes is incapable, by law, of giving consent. Consent by insane people or infants will not be legally recognized.

Second, the offense must be of the type for which consent may be given. Breaches of the peace, including affrays and disorderly conduct, are the offenses for which consent cannot be given, and therefore the defense is unavailable. Mayhem is a nonconsentable crime. This would also be true in many *mala prohibita* offenses, such as the sale of mislabeled goods. Similarly, murder was, at common law, and still is, a nonconsentable crime in most jurisdictions, but see the discussion on assisted suicide in Section 8.4.

The third element is that the consent not be obtained by fraud as to the nature of the act to be committed. Fraud can take one of two forms that must be distinguished, because one applies to the defense of consent, whereas the other does not. Fraud in the execution refers to the fact that the victim agreed to an act, although unaware of its nature. For example, Ann goes to Doctor Borman because of illness. Borman advises Ann that she needs an operation to which she agrees. While Ann is under anesthesia, Borman proceeds to have intercourse with her. This "consent" was obtained by fraud as to the nature of the act and, therefore, consent will not be a defense in Doctor Borman's trial for rape. On the other hand, suppose that Doctor Borman has intercourse with Ann after telling her that this will cure her ailment. Ann later learns that this was just a ruse on the part of Doctor Borman to have intercourse with her. In his subsequent trial for rape, Borman's defense of consent would be applicable. In this case, Ann did consent to the act, knowing full well its nature. However, she was induced to give her consent by fraud. Here, fraud will not negate her consent because the law is not concerned with her reasons for consenting as long as she was aware of the nature of the act to be performed. This is called fraud in the inducement.

The fourth element is that the person giving consent must have the authority to consent to the commission of the crime. Arthur may consent to have his property stolen, but his consent is no good when he agrees to allow Bill to take property belonging to Charlie.

All four of these elements must be satisfied before the accused can raise the defense of consent. If consent to perform a certain act is given, the accused may still be liable if the conduct goes beyond the bounds of the consent. If Alan gives Bob permission to take a $1 bill from the dresser in Alan's bedroom and Bob takes $10 instead, Bob has exceeded the bounds of the consent and will be liable for the theft of the extra $9.

Consent given under threat or fear of reprisal will not be a good defense if the threat or reasonable fear was so strong that agreeing was a better alternative than allowing the threat to be carried out. The reasonableness of the fear will vary according to the type and seriousness of the act. If Al says to Ben, "If you don't give me all of your money, I will never speak to you again," this is not sufficient threat or fear to negate Ben's consent when he agrees to hand over the money. If, on the other hand, Al forces Betty to consent to intercourse under threat of death, Betty's agreement will not constitute consent for the purpose of his defense.

6.10 DURESS

Duress implies that one is not acting of one's own free will. Our system of criminal law emphasizes responsibility. We look to the facts of any given case to determine whether an individual may be held accountable for his or her actions. If not, we will not punish that person's conduct. Most crimes require both the intent and the act. Duress acknowledges that a person may act wrongfully without having any criminal intent. For duress to be available as a defense, one person's will must have been substituted for that of another. The person commanding the crime forms the intent and, by imposing his or her will, forces another to commit the crime. One federal court said that duress does not preclude an intent to commit the crime but is only a justification for the conduct that would otherwise be criminal.

When one is compelled or commanded to commit a crime under fear or threats against his or her person, this will be a defense to committing the act provided certain elements are present. As soon as the threat or fear under which the accused was operating ceases, the accused must stop the wrongful conduct. Continuing such conduct, the accused alone will be held liable for the criminal acts. Suppose Art commands Bert at gunpoint and under threat of death to commit a robbery. If this threat continues up to the point at which Bert commits the robbery, Bert will not be liable, for duress will be a defense. However, if Bert discovers that Art's threat is withdrawn before he commits the act and continues because it seems like a good idea, Bert will be liable. Bert failed to stop his wrongful conduct when the threat ceased.

To be a defense, duress must involve a threat against the person, not just a threat to destroy or deprive one of one's property. Finally, the threat must be present and impending, not future. Someone who says, "If you do not help me commit this robbery, I will come back next week and kill you," is not imposing an immediate threat. A crime committed under these conditions cannot be defended on grounds of duress. Thus, in a case where a wife is riding around with her husband and another man, if the two men suggest a spontaneous armed robbery, leave the wife in the car and then demand that she drive the getaway car by saying to her that they will "whip her ass" unless she drives, then it is possible, if no death has occurred in the robbery, that the driver has a valid, arguable duress defense. The jury does not have to believe the defense, but it does test the issue of voluntary participation. The duress defense negates intent because of the immediacy and imminency of the threatened harm.

However, at least one court has decided to depart to some degree from this. A doctor raised duress as a defense to conspiring with another to obtain money under false pretenses by making out phony medical injury reports. The threat was, "Remember, you just moved into a place that has a dark entrance and you live there with your wife. . . . You and your wife are going to jump at shadows when you leave that dark entrance." The doctor said he felt he had to comply for his wife's safety and his own. The court recognized that there was a lack of present, imminent, and impending danger of harm that is normally required. The court felt that the individual is usually expected to seek police protection. However, the court ruled that said duress would be a defense to everything except murder, if the defendant became involved, because of the threat that a person of reasonable firmness would have been unable to resist.

Suppose that the threat is to loved ones who are being held at a location away from the person claiming the duress defense. Will courts allow the jury to consider the defense? Today, most courts will treat a well-proven imminent threat to others as a proper use of the defense even if those others are not physically in the presence of the coerced person.

Another development of the duress defense concerns its use as a defense in a prison escape case. Prisoners said they escaped because of bad prison conditions or brutality on the part of guards or other prisoners. Pennsylvania was one of the first states to accept such a defense. Its rule is that only in the most extreme cases is duress a defense to escape. However, Pennsylvania seems to be confusing duress with a defense of necessity. Necessity arises from the press of events and not from the imposition on the actor of the will of another person. The prisoner escaped to save his own life in the Pennsylvania case; therefore, the prisoner seems to have been acting because of the pressure of the events.

A famous case in which the defense of duress was unsuccessfully argued was the bank robbery trial of Patty Hearst. She was kidnapped by a radical group known as the Symbionese Liberation Army (SLA). Some months later, while still in "captivity," Hearst was photographed holding a weapon during the bank robbery committed by the SLA. The jury did not believe Hearst's duress claim.

Most authorities believe that duress is not a defense when it requires the taking of another's life to preserve the accused's own life. This is based on the theory that, if the choice comes down to losing one's own life or taking another's, a person is supposed to sacrifice his or her own life rather than kill. A Missouri appellate court supported this position in a 1984 decision because there was a state statute specifically prohibiting the use of duress as a defense.

6.11 NECESSITY

As indicated in Section 6.12, duress is often confused with **necessity.** Necessity is an old defense, usually adopted in cases of justifiable homicide.

However, until modern times, it had not received much treatment in other areas of the law.

Necessity comes about when the press of events causes conduct for which there is no readily apparent alternative. In an Alaska case, the defendant tried to use necessity as a defense to a charge of reckless destruction of personal property and joyriding. His truck was stuck in the mud and he was afraid it would tip over and perhaps damage its roof. Without asking permission, he appropriated construction equipment and the damage was done. The Alaska court made two points. First, the harm the defendant was seeking to prevent was no greater than the harm he caused. Second, the defendant failed to use any lawful alternatives that were available to him.

This second point is the critical one. The defendant must exhaust all lawful alternatives or show why the alternatives could not be used. This has been the critical turning point in the prison escape cases.

In one such case, the Iowa Supreme Court said:

(1) necessity is a defense to an escape charge; (2) the limited defense of necessity is available where, among other things, prisoner is faced with specific threat of death, forcible sexual attack or substantial bodily injury in the immediate future, there is no time for complaint to authorities or time to resort to courts and he immediately reports to authorities when he has obtained a position of safety, and (3) after such defense is properly raised by defendant, burden is on the State to disprove it beyond a reasonable doubt.

This was echoed by the Illinois Supreme Court, which was echoing a California decision, when it said:

In prosecution for escape, the following conditions are relevant factors to be used in assessing defendant's claim of necessity, but the existence of each condition is not necessary to establish a meritorious necessity defense: prisoner is faced with specific threat of death, forcible sexual attack or substantial bodily injury in immediate future; there is no time for complaint to authorities or there exists a history of futile complaints which make any result from such complaints illusory; there is no time or opportunity to resort to courts; there is no evidence of force or violence used towards prison personnel or other innocent persons in the escape; and prisoner immediately reports to proper authorities when he has attained a position of safety from immediate threat.

Once the defendant gets to a place of safety and is no longer faced with the specific threat, thus out of reach of the alleged danger, that person loses the case of necessity as a defense. Similarly, when the facts themselves demonstrate that the defendant acted contrary to the threat, the defendant cannot use what might have been a valid defense. For example, a defendant said the prison hospital authorities misdiagnosed his major medical problem. Yet he escaped and did not seek medical help until 13 months later. The court rejected this as a necessity.

Some have even tried to use their religion as an excuse for illegal conduct. They said it was necessary. However, most of these claims have not been given any credence.

Generally, necessity, apart from self-defense or defense of others, is not tolerated as a defense in most serious crimes. However, necessity has a legitimate function at times in the *mala prohibita* crime context. In some cases the defense might be specifically included in a *mala prohibita* statutory scheme. For example, an Oregon statute prohibiting driving on a suspended license creates the affirmative defense of necessity. The statute recognizes that there may be a real injury to a person or animal that demands that the suspended driver operate a car at that time. The problem is whether to use an objective or subjective test. The Oregon court thought the subjective belief of the driver was sufficient if he or she could point to specific facts that led the driver to that conclusion that would also lead a reasonable person to so believe. If he or she does this, the judge has to let the defense go before the jury.

The state of Connecticut made it clear that the defense of necessity is available in a criminal case despite the fact that the statutes did not create it, and they held that the defense applies to both *mala in se* and *mala prohibita* crimes because it is a legal justification defense as much as it is an intent-modifying defense. The defense is clearly a "choice of evils" defense which in fact is equally applicable to crimes with an intent as to those where no intent need be proven. In such a case, if believed, where no intent is required, it is up to the court to decide whether to treat it as a complete defense, relieving the defendant of all criminal liability, or only to use it in mitigation of the sentence to be imposed.

6.12 RELIGIOUS BELIEF

The First Amendment to the U.S. Constitution guarantees religious freedom. The right to believe anything one wants to believe is unassailable. However, can one use this right to avoid criminal liability when the practice of an individual's beliefs results in the commission of an otherwise criminal act? The answer in most instances is no. Religious practices that violate positive criminal law cannot be used to justify or excuse any criminal conduct. For example, one case held that even though church members are sincere in their belief in "peyotism," the fact that both the state and federal governments classify peyote as a controlled substance, with overriding public welfare concerns, means that the First Amendment rights must give way to the concerns about the regulation of controlled substances.

It is essential that public peace and moral order be maintained. If this were not the rule, each person could, under the First Amendment, start a religion to evade the law. In other words, by conceding the right to practice certain religious beliefs in violation of positive criminal law, society would be open to fraud and subject to the possibility of chaos.

6.13 CUSTOM

If a statute or common law rule is in force, the fact that the law has never before been enforced is no defense. If the act is prohibited, the **custom** of not enforcing the rule against it can never justify the violation. This is a universally recognized rule of law and has been upheld in case after case with no exceptions found.

Police officers are often confronted by an angry motorist who says, "I was just keeping pace with all the other cars. Why don't you give them tickets, too?" Often is heard the sound of the surprised offender who says, "That law has never been enforced before. Why now? Why me?" All these complaints have no effect on the ultimate liability of the one accused of a violation or crime, for if the police officer made a proper arrest for a valid offense, the custom of not enforcing the law or refusal in the past to prosecute for such violations will not be a defense. A violator may not successfully claim ignorance of the law because of past failure to enforce the law. Ignorance of the law is generally no defense.

6.14 VICTIM'S GUILT

The fact that the victim of a crime may have committed a criminal act is no defense for the accused. When Andy steals property from Bob, Andy cannot base his defense on the fact that Bob had previously stolen the property from Chris. In this case, both may be punished for their crimes. Even though neither Andy nor Bob had better rights to the property than did Chris, the actual owner, Bob had a better right to it than Andy did, for Bob was the first thief.

6.15 SELECTIVE ENFORCEMENT

The exercise of police discretion to determine what laws shall be actively enforced and against whom has begun to draw the attention of the courts and others. There has arisen a new defense called **selective enforcement.** The defense is recognized in Alabama, Alaska, Florida, Louisiana, Maryland, Massachusetts, New York, South Dakota, Tennessee, Virginia, and Washington, among others.

Selective enforcement of criminal laws is not in itself unconstitutional. However, when that selective enforcement is designed to prosecute a person without any intention of general enforcement, the enforcement is unconstitutional.

A defendant seeking to use selective enforcement must show that a broader class of people than those prosecuted has violated the law and that the failure to prosecute them was consistent and deliberate. Further, the defendant must show that the decision was based on some impermissible classification such as race, religion, or sex. Of course, all enforcement is presumed to be exercised in good faith. Thus the defendant has the initial burden of demonstrating the bad faith. The defendant must show intentional discrimination.

In one case the defendant demonstrated that 98 percent of those arrested for violating Virginia's liquor laws were blacks. The court held that this alone was not enough. In another case the defendant complained that he was the only one prosecuted for a prison riot. The court found that he was the ringleader. They said he failed to demonstrate a deliberate course of discrimination based upon race, religion, or other arbitrary classifications. Courts have not accepted proof of lax enforcement as meeting the standard. In another case the U.S. Supreme Court upheld the government's policy of prosecuting only those who either announced that they did not register for selective service or those who were turned in to the government for failing to register. Since two types of nonregistrants were prosecuted equally, the defendant could not prove intentional illegal discrimination. Ohio takes the position that for selective enforcement to reach the level of unconstitutional discrimination, the discrimination must be intentional or purposeful. It is not enough to allege discrimination; there must be proof that the police knew of enforcement omissions.

The area in which the defense has been most utilized is in the charge of being a habitual offender. Most states have had habitual offender statutes aimed at repeat offenders but many of the laws were not strong and had fallen into disuse. In 1993, Washington state enacted the first "three strikes" law to put habitual offenders away for good. Twenty-two other states quickly got on the bandwagon, but the majority of states have only used such laws sparingly. California has sentenced 40,000 people for second and third strikes—a quarter of the state's prison population; 4,400 of these sentences were for 25 years to life. Georgia has sentenced almost 2,000 under its law. Washington state has 120 people locked up for life without chance of parole, but the majority of other states generally have not used their statutes more than half a dozen times. Other states with such statutes are Alaska, Arkansas, Colorado, Connecticut, Florida, Indiana, Louisiana, Maryland, Montana, Nevada, New Jersey, New Mexico, North Carolina, Pennsylvania, South Carolina, Tennessee, Utah, Vermont, Virginia, and Wisconsin. The U.S. Supreme Court has held that sound discretion not affected by race, religion, or sex that is exercised is good faith in prosecuting under the habitual offender statute will not be disturbed.

A challenge was lodged against the California statute by Michael Riggs, a nine-time loser who wrote the appeal himself claiming cruel and unusual punishment. Riggs was convicted of shoplifting a bottle of vitamins from a grocery store. He previously had been convicted of four nonviolent crimes and four robberies. In 1998, the U.S. Supreme Court refused to hear the case.

There are instances in the body of criminal law in most jurisdictions where almost identical crimes containing identical elements may be found in separate sections of the statutes. The only difference is that one is punishable as a felony, the other as a misdemeanor. Defendants have attempted to get cases dismissed on a claim of selective enforcement when prosecutors choose to use the felony offense rather than the misdemeanor violation. Here again the U.S. Supreme Court has held that as long as no class discrimination is involved in the decision, a prosecutor has the absolute right to choose which form of the statute to charge.

6.16   THE USE OF FORCE AS A DEFENSE

General Comments

All too often law enforcement officers encounter situations that require them to decide whether or not to take the life of another human being. Law enforcement agencies tend to train their personnel to use firearms properly but sometimes fail to give proper instruction on the legal ramifications of the use of both deadly and nonfatal force. Without having the knowledge as to when deadly force may be used under state law and without proper policy guidance from law enforcement administrators, law enforcement officers patrol their beats often wondering what they may legally do if the unfortunate choice should ever present itself.

Many states do not specifically regulate those conditions under which a law enforcement officer may use deadly force. Some do, but without policy formulation by police executives, the statutes exist in letter only, and the interpretation of those laws made by court decisions leaves many questions unanswered. The decision of when to use a weapon is, without doubt, the most difficult decision a law enforcement officer ever has to make, and often the officer must make that decision within a few seconds.

It is both unfair and impossible to expect the law to have an answer for every conceivable situation that a law enforcement officer may encounter in which the use of force may be necessary and justified. But if general principles regarding the use of force can be instilled in their minds, law enforcement officers will obviously be more confident that their conduct is legally and morally proper when they do act.

In deciding when to use deadly force, law enforcement officers are expected to use common sense. Too often, however, instinct governs conduct, and instinct may not be compatible with common sense unless adequate training is received. Common sense must be guided by some legal interpretation and policy guidelines. What is required when the officer faces a situation in which deadly force may be operative is a reasonable belief that deadly force is necessary. This does not mean that the use of deadly force must actually be necessary. It is essential that law enforcement officers recognize that they may use deadly force when acting under circumstances that would lead an ordinary and cautious person of ordinary intelligence to believe that the only way to prevent further harm to his or her own or another's life or safety would be to use deadly force. When an officer understands this principle, common sense becomes the officer's most important asset.

The fact that the law permits an officer to use deadly force in a particular situation does not necessarily mean that this type of action is required. The words "as a last resort" must guide the law enforcement officer's use of common sense. It must not be forgotten that density of population and technical and scientific advances not only cause unfortunate situations, but also can help prevent those situations. So it is with the use of deadly force. Even if deadly force is permissible in a given situation, it does not mean that it must be used. On the contrary, when safety, duty, and common sense demand, it is preferable that means other than deadly force be used to enforce the laws. Mobility, communications, and the assistance of other personnel should govern the decision of when to use deadly force. In addition, the officer must assess the risk when endangering the lives and safety of innocent bystanders by firing a weapon.

Deadly force is that amount of force that is likely to cause, or does cause, death. In all cases, the facts of the particular case govern the amount of force a person may legally use. Most situations warrant the use of nonfatal force, whereas only a few actually justify the use of deadly force. Whenever force is justified, the law allows only a reasonable amount of force necessary to accomplish a legal objective. The reasonableness of the force used will be a matter for ultimate determination by administrative and judicial procedures. When a law enforcement officer uses deadly force, whether that officer intended to kill or only wound will never be questioned. If an officer is justified in using deadly force, it will be presumed that the officer intended it to be fatal.

If reasonable force is used, either by law enforcement officers or private citizens, to accomplish a legal purpose, any unintended death that results is excusable homicide. For instance, if Art attacks Ben and Ben, to defend himself, shoves Art aside, Ben will not be criminally liable for homicide if Art accidentally trips, hits his head on the floor, and dies. The initial force used by Ben to defend himself was reasonable and lawful under the circumstances and the result was unintended.

Warning Shots

The practice of, or the prohibition against, firing warning shots is generally one of policy only, not governed by statute. Although firing warning shots is authorized by some departmental policies, the majority of agencies prohibit this, and those courts that have treated the subject have by and large condemned the practice.

Following the definition of deadly force just given, which includes force likely to cause death, the firing of warning shots can constitute deadly force if death results unintentionally. Therefore, warning shots must never be fired unless the law enforcement officer has the right to use deadly force in that particular case and would be justified in killing.

The two most urgent problems arising from the use of warning shots are the danger imposed upon innocent bystanders and the criminal and civil liability of the officer. These two factors must govern the choice of whether to fire warning shots. A corporal injury or fatality resulting from an unreasonable use of force will subject a law enforcement officer to civil, and perhaps criminal, liability.

Homicide to Prevent the Commission of a Crime

As a general rule, one may commit a homicide to prevent the commission of a felony. This statement should not go unchallenged, because interpretations of this rule have placed some limits on its application. Before deadly force is justified in such situations, the felony must be an atrocious one involving force, surprise, danger, or death to a person. A good question to ask in these situations is whether reasonably apparent harm imminently threatens someone's life or safety. If so, deadly force would be justified in preventing the felony.

The fact that the person killed was actually committing a felony at the time of death will not be a defense if the slayer did not know this fact. In this case, the slayer would have acted with a criminal intent.

The rule is somewhat different for misdemeanor cases. Deadly force is never justified to prevent the commission of a misdemeanor. Only reasonable force short of deadly force is allowed. If, however, during the commission of the misdemeanor, the misdemeanant repels the prevention attempt with force that threatens death or serious bodily harm, deadly force may be used. It is allowed on the grounds of self-defense, not on the grounds of preventing the misdemeanor.

The following examples illustrate situations that arise within the confines of the principles discussed in this section. In these instances, the problems faced by officers are quite common. The solutions we propose should be recognized for what they are—commonsense approaches to very uncomplicated illustrations. In fact, most situations may not be this clear-cut, and additional factors must be taken into consideration. Suppose that Officer Jones comes upon a man who has thrown a woman to the ground and is apparently about to rape her forcibly. Based only on these facts, the officer's justification for using deadly force would depend on whether that officer's presence and the threat to use force did not prevent the commission of the felony. However, the officer must consider whether deadly force is necessary and to what degree he or she will endanger the victim or innocent bystanders by shooting. In addition, the officer must consider whether the situation is an actual attempt to commit rape or some other act not necessarily as serious or even criminal. By exercising restraint, the officer may be able to find out more about the facts of the situation.

In a second situation, Officer Smith observes one person robbing another. The question of Smith's right to use deadly force to prevent the commission of this crime depends upon whether the robber appears to be armed or unarmed. If it appears that the robber is armed or somehow capable of using deadly force, then the officer would be justified, as this would be an atrocious felony. An unarmed robbery would not fall into this category. As in other cases, the officer should make all reasonable attempts to prevent the felony by means other than using deadly force.

A third simplified problem often encountered involves a law enforcement officer who observes someone entering a second-story window of a house in the early hours of the morning. On the basis of just these facts, the officer definitely would not be justified in using deadly force. There are insufficient facts present. The officer must have additional knowledge concerning the circumstances before taking such drastic action. In any event, a burglary is not ordinarily classified as an atrocious felony unless the burglar is threatening to use physical force against an occupant of the dwelling. Consequently, there would be very few occasions in which an officer, observing such a scene from the outside, would be justified in shooting. Further investigation would be warranted.

The Use of Deadly Force to Effect an Arrest or Prevent Escape

A homicide committed while attempting to make an arrest or to prevent an escape can be justified in felony cases only as an absolute last resort if there are no other means of apprehending the felon. This rule covers not only the right to use deadly force to make a felony arrest, but also includes what is commonly referred to as the **fleeing felon rule.**

Until 1985, the states were not in absolute agreement about whether the right to use deadly force to effect an arrest was applicable to all felony cases. Some states held that it was, whereas others allowed deadly force to be used only in cases when the felony was an atrocious one, involving threat of death or serious bodily harm to persons. In the landmark case of *Tennessee v. Garner,* the U.S. Supreme Court resolved the issue. A Tennessee statute permitted law enforcement officers to use all necessary means to effect an arrest, after giving notice of the intent to arrest a criminal suspect. Acting under authority of this statute, a law enforcement officer shot and killed Garner's son, who, after being told to halt, fled at night over a fence in the backyard of a house he was suspected of burglarizing. The officer used deadly force even though he was reasonably sure the suspect was unarmed, youthful, and of a slight build. The Court held the statute to be unconstitutional insofar as it authorized the use of deadly force against an apparently unarmed, nondangerous fleeing felon. The Court went on to say that deadly force could be justified only when necessary to prevent the escape of a felon if the officer had probable cause to believe that the suspect posed a significant threat of death or serious bodily harm to the officer or others. Thus, the mere fact that deadly force was the only means for apprehending a suspect no longer could serve as a justification to use such force against a fleeing felon. In reaching its decision, the Court reported that it had closely examined the statutes of all states on this issue, as well as the policies of numerous law enforcement agencies, and found that the majority of states and the overwhelming policy position of law enforcement agencies already supported the position announced by the Court in the case.

Normally we equate justifiable homicide with the action of law enforcement in preventing a felony, stopping a felony in progress, or pursuing a fleeing felon (within the limits of *Tennessee v. Garner*). Can a private citizen also take advantage of the justifiable homicide defense? A California court says that the citizen is entitled to the same justification (as limited by *Tennessee v. Garner*) as the police. Michigan also has held that private citizens should be held to the same standards as police.

There is some disagreement among the states about the justification for the use of deadly force if the officer or private citizen did not know with certainty whether the person had actually committed a felony. Most states hold that a reasonable belief justifies the use of deadly force, but a few states say that the officer must know that a felony was committed. In those states, the officer acts at his or her own peril, and, if wrong in believing that a felony has been committed, the officer must be prepared to suffer any attendant legal consequences. Although the *Garner* case did not address this issue, it should be noted that the decedent in that case was only suspected of committing a burglary.

To justify a homicide while effecting an arrest, it is necessary that the arrest be lawful. A lawful arrest is not governed by the ultimate determination of the guilt or innocence of the accused; rather, it is determined by the procedures followed to bring the accused within the custody of the law, such as the issuance of a warrant, the existence of probable cause, or the commission of a crime in the presence of the arresting officer. If the arrest is lawful, the arrestee has no right to resist, and, if the arrestee does resist to the point of using deadly force, then the officer may counter with such force as is reasonably necessary to effect the arrest up to and including deadly force. If a person resists a lawful arrest with deadly force so as to warrant the use of deadly force in return to protect the life or safety of the person making the arrest, self-defense, not the arrest itself, justifies the use of such force.

If, on the other hand, the arrest is unlawful, the person being arrested has the right to use such force as is reasonably necessary to avoid the arrest, up to and including deadly force. Several states have restricted or prohibited by statute the right of one unlawfully arrested to resist with deadly force even where there is a threat to use deadly force to effect the arrest. The theory is to allow this dispute to be settled in court. However, if it reasonably appears that surrender will not stop the use of deadly force against the arrestee, the arrestee may respond with force reasonably necessary for self-defense.

The use of deadly force against an escaping prisoner, although generally guided by the rules just described, has additional ramifications in some states. A few jurisdictions classify the escape itself as either a felony or a misdemeanor, depending upon the charge for which the accused was arrested. Thus, the authority to use deadly force on an escapee in these states depends upon the charges for which the officer arrested the accused originally. If the escapee was arrested for a felony, the escape is a felony; if arrested and charged with a misdemeanor, the escape is a misdemeanor, and deadly force would not be justified. At least one state has changed the common law by making the escape itself a felony so that the inquiry as to the status of the prisoner's detention need not be involved. As a result, the escaping prisoner would be committing a felony in the presence of the person having custody, and would thus be a fleeing felon.

Again, some examples may serve as useful illustrations of the basic principles involved. A law enforcement officer sees someone crawling out of a window of a home that the officer knows is temporarily vacant. The residents are out of town. It is night, and the person is carrying something. Here again, the officer has insufficient facts to warrant the use of deadly force. Certainly there is good reason to suspect that a burglary has just been committed, but the officer may not use deadly force until in a better position to ascertain the facts.

In another case, an officer sees someone commit an armed robbery. While running from the scene, the robber throws away the gun used in the robbery. At that time the officer recognizes the robber. Obviously, because the officer can identify the suspect, an arrest may follow with a warrant at a later time. The use of deadly force is unnecessary and unwarranted in this situation. Even if the officer cannot identify and has no chance of catching the suspect, the decision in the *Garner* case would prohibit the officer from resorting to deadly force, even as a last resort, since the facts indicate that the suspect had thrown away the gun.

In another case, someone has gone berserk on a public street. Three people have already been shot when the officer arrives on the scene and sees the individual standing on the sidewalk waving a gun. In this case, if no other means appears to be reasonably adequate or safe to the officer, the officer may resort to deadly force. The situation, if allowed to continue uninterrupted, is extremely likely to endanger the lives and safety of innocent persons. The utmost care must be taken.

In a final illustration, an officer comes upon a scene where someone is running down the street followed by a shopkeeper yelling, "Stop thief!" Situations such as this often present the most frustrating and confusing circumstances in which an officer must decide whether to use deadly force. The facts as presented here do not warrant the officer shooting at the fleeing man. The officer is not sufficiently informed and does not know whether a felony has been committed, or if there is a danger to persons that warrants the use of deadly force.

Would the use of a police dog to apprehend a nonarmed, nondangerous felon that results in the death of the felon violate *Tennessee v. Garner*? One federal court said this was not the type of force prohibited by the U.S. Supreme Court in *Tennessee v. Garner*. In this case the dog was trained to go for the closest part of the body (usually the arms or legs); however, the defendant was under a car and the closest body part was his neck. The Court found this as a remote possibility and thus more in the nature of an accident.

Self-Defense

One may use as much reasonable force as is necessary to defend one's self. Deadly force is justified only when death or serious bodily harm is threatened. Before deadly force may be used, the danger to life or safety must be imminent. An assault not threatening death or great bodily harm does not justify the use of deadly force. The reasonableness of the force used is based on an objective test: How much force would a reasonable and prudent individual be justified in using under the same circumstances?

The right of self-defense implies that one has the right to defend oneself. When the threat against the person ceases, the right of self-defense ceases. If the person attacked then pursues the attacker to avenge injuries, the pursuer becomes the attacker and cannot justify the actions on the grounds of self-defense. Revenge is not synonymous with self-defense. Suppose Andrews verbally insulted Brown. In anger Brown kicked Andrews lightly, and then turned and started to walk away. Andrews's anger was aroused, and he went after Brown with the intent to injure him seriously. Brown killed Andrews to preserve his own life. Will the fact that Brown struck the first blow affect his claim of self-defense in his trial for murder? No.

This rule is qualified under one set of circumstances. When the initial attack renders the victim so insensible that the victim is unaware the attack has ceased and thereafter pursues the attacker, believing the attack is continuing, self-defense may be used as a defense. The attacker's injuries in a case such as this are caused by his or her own initial action. To illustrate, if Andrews assaults Brown and then retreats, but Brown is rendered so insensible by the attack that he believes he is still under attack, or he does not realize what is happening and pursues Andrews and assaults him, Brown will be held to have acted in self-defense.

One need not wait until one is struck by an attacker before the right of self-defense arises. Defense also implies the right to prevent. If the attack is imminent, an individual may, in fact, strike the first blow, and still justify the action on the grounds of self-defense. John, with his back against a wall, sees Bill preparing to lunge at him with a knife. John instantaneously picks up a rock and throws it at Bill, striking him in the head. If John is prosecuted for causing this injury, he may successfully claim self-defense.

At common law, one is required to retreat as far as possible before justifiably taking another's life in self-defense. This doctrine is obsolete in most jurisdictions, and the modern rule is that, when one is feloniously assaulted in one's home, office, place of business, or on property owned or lawfully occupied, that person is not bound to retreat but may stand his or her ground and defend with such reasonable force as is necessary up to and including deadly force in cases where such force is justified in self-defense. Consequently, if one is in a place where one has a lawful right to be, that person is not required to retreat as long as that right is at least equal to, if not greater than, the rights of the attacker. The exercise of a legal right is not ample provocation for the other party to attack and does not deprive the victim of the right of self-defense.

Homicides committed in self-defense are of two types: justifiable homicide in self-defense is committed when one kills to defend oneself from death or serious bodily harm resulting from a felonious attack. Excusable homicide in self-defense implies that the slayer was to some degree at fault for getting into a situation in which he or she had to kill to preserve his or her own life but, nevertheless, did kill in self-defense. The distinction between these two types of homicide was important at common law, but today it is a distinction without a difference. Some jurisdictions, however, still maintain the distinction by statute. Chapter 8 deals with the difference between justifiable and excusable homicide in more detail.

Defense of Others

Under strict common law interpretation, most authorities agree that the right to defend others was limited to the same right that one would have to defend oneself in the same situation. Thus, anyone who uses this defense "stands in the shoes of the victim" and may use the same amount of force the victim could have used in self-defense. One who comes to the defense of another, however, is held to act at one's own peril if the person assisting has no knowledge of the surrounding facts. Before defending others, particularly with deadly force, one should learn the facts. If the "good Samaritan" chooses the wrong side and aids the aggressor, he or she will be liable for his or her actions.

A few states modified this rule for cases in which the defender seeks to help a blood relative. By statute, these states recognize the emotional ties between relatives. Basically, the statutes allow defense of the relative even though the relative was the attacker, as long as it reasonably appeared to the defender that the relative needed help or that at the time of the defense it was impossible to determine which party was at fault. This view is gradually disappearing.

Other states have modified the common law by allowing the defender to act under a reasonable belief that the facts justify coming to the defense whether the defended person is the assailant or not. This is an application of the objective or reasonable person test.

The modern rules were intended to aid one who comes upon a scene and has to make a quick judgment as to which person to help while the deadly affray is in progress. When this defender chooses the apparent person in need of defense, the choice will not be condemned if a reasonable person in the same position would have come to the same conclusion.

There is one important qualification for use of the modern rules of defense of others. A person who is present at the beginning of a deadly altercation cannot use the defense if he or she aids the original deadly force aggressor by using deadly force.

Defense of Property and Habitation

Deadly force is never justified for the defense of property, even if it is one's home. Reasonable force, short of deadly force, may be used. Only when an atrocious felony, threatening life or serious bodily harm, is being committed in conjunction with the attack on property may deadly force be used as a last resort. Then that amount of force is justified as self-defense. Again, one need not retreat before one is allowed to use reasonable force to defend home or property. Consequently, the setting of a spring-loaded gun in the garage of one's home to injure intruding thieves is not a justifiable use of deadly force.

6.17 DOUBLE JEOPARDY

For the most part, double jeopardy is a procedural question. However, there is one area where the substantive criminal law and the defense meet head on. To illustrate: Suppose that A goes into a large grocery store, points a gun at cashier number one and demands the money in the till. That accomplished, A moves on to cashier two and again brandishes the gun, demanding and receiving the money from the till. Has A committed one robbery or two? If it is just one, then a second trial for the robbery of the second till would be double jeopardy. If there are two robberies, the defense would not be available.

A Case of Double Stalking

On May 24, 1995, John Francis Jones was charged with aggravated stalking which occurred on April 30, 1995. On July 21, 1995, Jones was again charged with aggravated stalking for events that occurred between May 1 and May 16, 1995. In the meantime, Jones was acquitted in the first case. Jones argued that prosecution of the second case constituted double jeopardy. It should not be allowed because aggravated stalking is a continuing crime and the offense charged should have been brought in a single prosecution. The Florida appellate court disagreed. The court clearly explained that the issue in this case concerns that aspect of double jeopardy dealing with whether a particular factual circumstance constitutes one or two or more separate and distinct factual events. In rejecting the defendant's position, the court said that the purpose of fixing the time allegations in the information was to protect the defendant from multiple prosecutions for the same offense.

The rule states that where the same act or transaction constitutes a violation of two distinct statutory offenses, if each violation requires proof of an additional fact which the other does not, two separate offenses are chargeable. In this example, the determination would rest on whether each cashier had possession or only custody of the money stolen. The crime(s) committed are directed against the rightful possessor of the property. If it was decided that the store owner was the possessor, thus the victim, only one offense could properly be charged. If, however, the cashiers had rightful possession of the money, rather than mere custody for the owner, there were two separate victims, and two charges could be made.

The example in the preceding paragraph represents the double jeopardy rule followed in most states. A few states, however, follow a more stringent and demanding double jeopardy rule known as the same transaction rule. In those states all crimes arising from the same transaction must be prosecuted at the same time unless the defendant seeks separate trials. If the state does not bring all charges at once, the state is barred by its own double jeopardy rules from initiating subsequent trials.

6.18 CRIMES INCIDENTAL TO EACH OTHER

Approximately 20 states follow a rule that allows punishment for only one of two crimes charged if the second crime is a natural or incidental part of the other crime. For example, in robbery there is usually a false imprisonment because the victim is forced or threatened to stand still. On the other hand, there may be a kidnapping incidental to the robbery if the victim is forced to move during the robbery. This is not identical to the doctrine of greater or lesser included offenses discussed in Chapter 2 because these offenses incidental to each other are not in the same chain.

Although a law enforcement officer would not be out of line to charge all offenses that may occur even if some are incidental to others, an officer should be aware that the defendant will probably only be convicted of one offense, generally the primary offense, if provable.

6.19   BATTERED SPOUSE SYNDROME AS A DEFENSE

The law makes every reasonable attempt to address the needs of society. New defenses grow out of old and continuing problems. One of those problems concerns a death caused by someone who has been the object of physical abuse by the decedent. Thus a spouse or child who was constantly being beaten finally rebels and without warning kills the abuser. In a case where no physical threat was then present, such a person could not plead self-defense. The only plausible, yet unsatisfactory, defense lay within the insanity area. No satisfactory defense existed. Could the law accommodate a defense, whether imperfect (charge reducing) or perfect (no liability) to deal with this problem?

Before any kind of defense would be recognized, misconceptions and myths about battered spouses had to be overcome. Incorrect assumptions about the ability of the victimized spouse to leave had to be neutralized. Questions like, "Isn't the battered spouse free to leave? Is the battered spouse masochistic? How bad can the beatings be if he or she does not leave?" had to be answered and explanations had to be given.

In truth, the **battered spouse syndrome** is created by a cycle of physical abuse within the relationship. There are typically three phases in the cycle. The first involves verbal abuse, minor battering, and attempts by the victimized spouse to placate the abuser. The second phase is usually where the severe battering occurs. The third phase is where the batterer becomes apologetic and loving. This gives hope to the battered spouse that her/his partner will reform. This is what keeps the victimized spouse in the relationship. Then the cycle starts again. Each time this happens, the abused spouse feels more and more helpless to improve the situation.

When the true picture was beginning to be recognized by the legal system, some states developed the battered spouse syndrome as a defense to a homicide charge. In New Jersey, which had one of the earliest cases in the country to allow the battered spouse syndrome as a defense, and other states, it is used as a division of the self defense doctrine. Expert testimony is allowed to aid the defendant and is relevant in helping the jury understand the reasonableness of the defendant's fear of imminent harm at the time of the killing.

Florida, although recognizing the battered spouse syndrome, was among a small minority of states that still imposed the common law duty to retreat (even in one's own home) before the courts would recognize the validity of a self-defense claim. This position provided the battered spouse with no more rights than anyone would have in a claim of self-defense. The duty to retreat was finally abolished in a 1999 case.

Minnesota and Texas are among a group of states that allows the battered spouse syndrome defense and the accompanying expert testimony to help buttress the self-defense claim. Minnesota will not, however, allow the expert to state whether the defendant suffers from the syndrome, but allows only general testimony about the syndrome.

Washington state also allows the defense to use an expert. It says that the general testimony about the syndrome, plus a hypothetical case that parallels the case at hand, may be addressed by the expert.

Georgia agrees that the defense can be used, especially where the defense shows the victim's reputation in the community as a violent person.

Even if the battered spouse syndrome can be used as a defense, it will not be accepted when no threat to the wife can be reasonably found but the wife attacks and kills the husband. Thus North Carolina said self-defense by reason of the syndrome cannot be used as a defense by a wife who shoots and kills her husband while he sleeps. This reversed a lower court's decision that upheld the defense. That lower court felt the sleeping period was but a "gap in the abuse" that did not destroy the defense.

Kansas agrees with the North Carolina Supreme Court that the defense does not apply when the spouse kills a sleeping mate. They too take the imminent danger approach. Thus both courts see the defense in more standard terms of an objective fear of the reasonable person rather than fear of this subjective person.

DISCUSSION QUESTIONS

1. A statute, under which John is prosecuted, forbids and punishes anyone who knowingly and intentionally takes books from a public library within this jurisdiction. John admits taking the book, knowing it belongs to the public library. At his trial, John raises the following defenses. Discuss the validity of each.

    a. John claims that he was not aware that his act was prohibited by law and therefore contends that he should be acquitted.

    b. John also claims he consulted his attorney about his right to take the book and was advised by his attorney that it was all right.

    c. Further, John says he just could not help himself, that he saw the book and had to take it. A medical examination discloses that John is a kleptomaniac.

    d. Finally, John says that when he took the book, the librarian on duty nodded his head in an up-and-down motion.

2. Assume that Officer Smith, while off duty, is just arriving home from a picnic with his family. He is carrying his off-duty revolver, as required by departmental regulations. As he approaches, Smith sees someone climb out a side window and run in the opposite direction. That indi-

vidual is carrying a sack of some kind, and it appears to be rather cumbersome. The thought also passes through Smith's mind that the person might be a teenager. Smith realizes he cannot catch the fleeing figure and knows nobody was rightfully in his home. Should Smith shoot? Would he be justified in doing so, if he killed the fleeing figure?

3. Jones verbally insulted Brown. In anger, Brown kicked Jones lightly, and then turned and started to walk away. Angry at being kicked, Jones went after Brown, intending to injure him seriously. Brown killed Jones to save his own life. Will the fact that Brown struck the first blow affect his claim of self-defense in his trial for murder?

4. The Hungry Handy convenience store is located in a less than ideal part of town. Drug deals have been known to occur in front of the store and in the parking lot on the side of the building. In an effort to reduce the problem, the owner of the store posts "No Trespassing" signs in the parking area and authorizes the police to arrest anyone who does not appear to be a customer. At approximately 6:00 p.m. on Friday, Officer Washington drives into the parking lot and sees D. Fendant standing there. As soon as Fendant sees the officer's car approaching, he runs. After a foot chase, Fendant is apprehended and charged with trespassing on posted property at the convenience store and resisting an officer without violence. Does Fendant have any defenses? Why or why not?

▬▬ GLOSSARY ▬▬

Alibi – a defense based on a claim that the defendant was elsewhere when the crime was committed.

Battered spouse syndrome – a defense used by an accused who has been the victim of domestic violence and who is accused of causing injury or death to the batterer. The defense is based on a self-defense claim.

Common law immunity – a voluntarily entered agreement not to prosecute, not to sentence, or not to use testimony against an accused, in exchange for testimony concerning a criminal case.

Consent – a defense used if the victim had the authority to and did consent to the act with which the accused is charged.

Custom – often raised but never a successful defense that a law prescribing a criminal act has not been enforced before.

Defenses – reasons why an accused should not be responsible for the commission of a criminal act.

Diminished capacity – a type of insanity defense used to establish the inability of an accused to form a specific intent, when one is an essential element of the charged crime.

Duress – a defense based upon a claim that one is compelled or commanded to commit a crime under fear or threats against his or her person.

Durham rule – one of the legal tests of insanity; an accused is not criminally liable if the unlawful act was the product of a mental disease or mental defect.

Entrapment – a defense based on a claim that law enforcement encouraged and pressured the accused into committing a crime that the accused was not already willing and able to commit.

Fleeing felon rule – allowed the use of deadly force to stop a fleeing felon; drastically changed by *Tennessee v. Garner* decision.

Insanity – A defense to a crime if the accused cannot aid in the preparation of his or her case and/or was incapable of having criminal intent at the time the act occurred.

Involuntary intoxication – generally a valid defense if the accused was forcibly intoxicated to a sufficient degree as to be incapable of having a criminal intent.

Irresistible impulse – a little used test of insanity that declares a person to be insane at the time of committing a criminal act if, by reason of a mental disease, the accused had so far lost the power to choose between right and wrong, that he or she was unable to avoid doing the act in question.

Mistake of fact – a defense used by an accused who commits a prohibited act in good faith and with a reasonable belief that certain facts exists that, if they actually did exist, would make the act innocent.

Mistake of law – ignorance of the law; generally not an excuse except when crime requires proof of a specific intent.

M'Naghten rule – one of the tests of insanity often called the right-wrong test; the defendant is insane if, at the time of committing the act, the defendant was laboring under such a defect of reason from disease of the mind as not to know the nature and quality of the act, or if the defendant did have knowledge of the nature and quality of the act, but did not know that it was wrong.

Necessity – a defense raised when an accused is faced with the press of events that causes conduct for which there is no readily apparent alternative.

Religious belief – A defense, generally unsuccessful, based on a claim that an otherwise criminal act is excused because it is an accepted or prescribed part of a religious ritual.

Responsibility – the concept that a person acts of their own free will and therefore may be held accountable for their acts, including those which society says are criminal.

Selective enforcement – a defense that the enforcement of the criminal law, with which the accused is charged, is selectively enforced based on criteria other than the violation alone.

Statute of limitations – a statute that limits the amount of time after which a crime is committed and within which prosecution of an accused must begin.

Statutory immunity – an involuntary agreement used to compel testimony by guarantying that the witness will not be prosecuted or that any compelled testimony will not be used against the witness.

Tennessee v. Garner – the U.S. Supreme Court decision effectively eliminating the use of deadly force against a fleeing felon.

Voluntary intoxication – generally not accepted as a defense unless the intoxication is so great as to render the accused from being able to form a specific intent, if the crime charged requires such proof.

▓▓▓ REFERENCE CASES, STATUTES, AND WEB SITES ▓▓▓

CASES

Medina v. California, 505 U.S. 437 (1992).

Terry v. State, 465 N.E.2d 1085 (IN 1984).

U.S. v. Jim, 865 F.2d 211 (NV 1989).

State v. Dufield, 131 N.H. 35 (NH 1988).

United States v. Garcia, 854 F.2d 340 (CA 1988).

Biddinger v. Commissioner of Police of the City of New York, 245 U.S. 128 (1917).

Commonwealth v. George, 1997 Mass, Super. 1997.

State of New York v. Isaacson, 44N.Y.2d 511 (1978).

United States v. Mosley, 965 F.2d 906 (WY 1992).

United States v. Bogart, 783 F.2d 1428 (CA 1986).

Jacobsen v. United States, 503 U.S. 540 (1991).

State of New Jersey v. Toscano, 74 N.J. 421 (NJ 1977).

People v. Harmon, 53 Mich. App. 482 (1874).

People v. Luther, 394 Mich. 619 (1975).

Commonwealth v. Stanley, 265 Pa. Super. 194 (1979).

United States v. Bailey, 444 U.S. 394 (1979).

State of Missouri v. Rumble, 680 S.W.2d 939 (MO 1984).

Nelson v. State of Alaska, 597 P.2d 977 (1979).

Powell v. Texas, 392 U.S. 514 (1967).

State v. Beeson, 569 N.W.2d 197 (IA 1997).

People v. Lovercamp, 43 Cal.App.3d 823 (1974).

Riggs v. California, 525 U.S. 1114 (1999).

People v. Unger, 66 Ill.2d 333 (1997).

State of Connecticut v. Messler, 562 A.2d 1138 (1989).

Jones v. White, 992 F.2d 1548 (Eleventh Cir. FL 1993).

Robinette v. Barnes, 854 F.2d 909 (KY 1988).

People v. Cabellos, 12 Cal.3d 470 (1974).

State of North Carolina v. Norman, 324 N.C. 253 (1989).

Smith v. State, 486 S.E.2d 819 (GA 1997).

State v. Stewart, 763 P.2d 572 (KS 1988).

People v. Goetz, 497 N.E.2d 41 (NY 1986).

McGhee v. Commonwealth, 248 S.E.2d 808 (VA 1978).

Vigil v. People, 353 P.2d 82 (CO 1960).

STATUTES

Arizona: Ariz. Rev. Stat. Ann. §. 13-502(C)

Alabama: Code of Ala. §. 13A-3-1

Nebraska: R.R.S. Neb. §. 29-2203

Delaware: 11 Del. C. §. 304

Idaho: Idaho Codes. 18-207

Oregon: O.R.S.811.180

Washington: R.C.W.9.94A.120

WEB SITE

www.findlaw.com

CHAPTER

7

Assault and Related Crimes

■ KEY WORDS AND PHRASES ■

Aggravated assault
Battery
Bias-motivated
Domestic violence

Mayhem
Simple assault
Stalking
Striking distance doctrine

7.0 INTRODUCTION

At common law there were no recognized categories of assault. All assaults were misdemeanors. Aggravated cases of assault differed only in the severity of punishment imposed by the judge. Legislation has reclassified the more severe types of assault as distinct offenses. Even though they often occur together, assault should be distinguished from battery because they are separate crimes. Assault is an attempt or offer to do harm whereas a battery is the doing of the harm. A completed battery, as in any other crime, includes an attempt (in this case an assault). Under the doctrine of merger, if the battery is completed, there is no assault, yet statutes and courts continue to refer to the completed crime as assault and battery. Properly, it should be called simply a battery.

7.1 SIMPLE ASSAULT

Simple assault is an attempt or offer, with unlawful force or violence, to do nonfatal injury to another. This definition is not self-explanatory and requires some interpretation. By definition, alternative methods of accomplishing an assault are available.

Under the first alternative, a **simple assault** may be committed by an attempt in the strict sense. In this instance, assault is identical to an attempt to commit any crime, the sole difference being that the target crime of assault must be battery. In such a case, all the elements of an attempt must be present (see Chapter 4). Suppose that Fred strikes at Bill with his fist but misses. Fred can be charged with a simple assault or with an attempted battery, because the facts satisfy the elements of either charge.

An assault may also be committed when there has been an offer to do violence. The term offer denotes the attempt of many courts to adopt the civil law definition of assault and include it in the body of criminal law as a second alternative for committing simple assault. This approach holds that assault may be committed by placing the victim in fear of immediate injury. When Fred draws back his fist threatening to strike Bill, but does no more, courts would find it difficult to support a charge of assault under the attempted battery theory. However, under the civil law theory, a chargeable simple assault would exist if Bill were reasonably apprehensive of imminent bodily harm.

The act must be done with unlawful force or violence. This means that the accused must act in some unlawful manner nonfatally to harm the victim by an act that is forceful or violent to some degree. Unlawful in this sense implies that the accused had no right or privilege to offer or attempt the violence. There are occasions when one is entitled, within reason, to use force. If such a right or privilege does exist, force is not unlawful. Thus, parents may discipline their children, and teachers may (subject to statutes) exercise discipline privileges. Individuals may defend themselves or others, and people may defend their homes.

Numerous elements must be shown to exist for the crime of simple assault. First, there must be some overt conduct on the part of the accused that would indicate the accused was about to commit battery. The overt act may be of any nature and need not go so far as would be required to charge an attempt, although assault must also go beyond mere preparation. Words alone will not satisfy this requirement. A raised fist, a pointed gun, and the like are the types of act required.

The second element deals with the *mens rea*, or state of mind, with which the accused acts. If the accused actually intends to injure and comes dangerously close to completing battery, assault is complete.

A number of states require proof of intent, including Iowa, New York, Tennessee, and Texas. It should be noted that even in these states the victim must reasonably perceive an "imminent use of force." But there is an additional manner in which this element can be satisfied. Many courts, including those of Alabama, California, Florida, Georgia, Kansas, Massachusetts, Michigan, Montana, New Hampshire, North Carolina, Vermont, and Washington recognize that an assault should be looked at from the point of view of the victim as well as the defendant. They require only apparent present intent to injure, as determined by the victim. These courts hold that if the accused acts in a way likely to create apprehension (fear of being harmed) on the part of the victim, assault has been committed regardless of the actual intent of the accused.

The intent or apparent intent must be to inflict injury at the time the threat is made. Threat to inflict harm in the future will not constitute an assault. For example, if Joe doubled up his fist and said to Bob, "If we weren't in mixed company, I'd knock your block off," there is no assault. Because they were in mixed company, Joe did not have the present intent nor could Bob be apprehensive of imminent harm. Joe's remarks clearly indicated his lack of intent to inflict harm at the present time.

A conditional threat will not negate an assault if the condition is unlawful. This refers to situations in which the accused does the necessary overt act to constitute an assault but the accused's intent shows that he or she is giving the victim an opportunity to avoid the battery by complying with a condition. If the condition imposed by the accused is one that the accused has a lawful right to make, there is no assault. If, however, the condition is unlawful, the assault is complete once the overt act is done. For example: While pointing a pistol at Ben, Alan says, "If you don't pay me the money you owe me, I will kill you." Alan is imposing a condition by which Ben could avoid the attack. However, the condition is one that Alan has no right to impose, and therefore the assault is complete. On the other hand, if Alan, after warning Ben that he is trespassing, shakes his fist and says, "If you don't get off my property, I will knock your block off," the condition is a lawful one and no assault has occurred.

The last element of simple assault pertains to the accused's ability to commit battery. At common law and in some jurisdictions today, it must be shown that the accused was actually able to commit battery when the threat was made. One who pointed an unloaded gun at another and threatened to use it could not be guilty of assault. The majority view today holds that this element is satisfied if it appears to the victim that the accused was presently able to inflict injury. In the example given, if the victim was not aware the gun was unloaded, the assault was complete. If the victim knew the gun was empty, there could be no assault, because there was no present ability to hurt nor was there an apparent present ability to inflict battery. A statute in North Dakota punishes as reckless endangerment the creation of "a substantial risk of serious bodily injury or death to another." The question that the state faced was whether pointing an unloaded gun at police constituted the crime. The North Dakota court looked not at whether actual harm could occur but rather at the issue of whether the actor's "conduct created a potential for harm." Pointing any gun, loaded or unloaded, especially at armed police, creates in their view, a potential for harm.

Another aspect of the ability problem involves what is commonly referred to as the **striking distance doctrine.** Even though the force threatened is unlawful and the means threatened will actually do harm, the accused must be close enough to actually harm the victim. Otherwise, there can be no real apprehension in the victim's mind under any rule. Suppose that Art, with an offer of unlawful force, threatens to cut Bill's throat with a bowie knife. If Art is 300 yards from Bill at the time, there can be no assault. But if the means used is a loaded M 16 rifle and the distance is 300 yards, there is an assault when Art unlawfully threatens to shoot Bill.

7.2   ASSAULT AND BATTERY (BATTERY)

The completed act, when the injury threatened by the assault actually occurs, is the crime of **battery,** or, as it is called in most jurisdictions, assault and battery.

A battery is defined as an unlawful injury, however slight, done to another person, directly or indirectly, in an angry, revengeful, rude, or insolent manner. The injury inflicted need not be serious. The least touching, done in any manner described earlier, will be enough. The force used to commit the injury need not be inflicted directly by the hand of the accused. For example, if Smith assaults Jones in a manner that makes him apprehensive enough to jump out a window to avoid the imminent battery, the injury resulting is a battery for which Smith may be held liable.

Criminal intent is a necessary element of battery, but the intent may be implied from the seriousness of the circumstances. An unlawful arrest constitutes an assault and battery, and reasonable force necessary to avoid an unlawful arrest is justified, but an excessive amount of force used to avoid unlawful arrest will also constitute a battery.

Because the offense of battery requires the accused to have acted with an intentional state of mind, injury inflicted as a result of culpable, criminal, or gross negligence will not support a charge of battery. To fill this gap, some states have enacted statutes providing punishment for injuries inflicted in a criminally negligent manner. Consent is a valid defense to an assault and battery charge, but all elements of this defense must be satisfied (Chapter 6).

7.3   STATUTORY ASSAULTS

As mentioned earlier, legislatures have divided assaults according to the degree of severity. Simple assaults remain misdemeanors. The more serious assault cases have been defined as distinct offenses.

Perhaps the most common legislative assault is **aggravated assault**—a felony in most states. An aggravated assault in general encompasses all cases of assault that are vicious in nature, including assaults with a specific intent to commit a felony and assaults with a deadly weapon. Because aggravated assaults are creatures of legislation, many states have further subdivided aggravated assaults into separate offenses called *assaults with intent to commit felonies, and assaults with deadly or dangerous weapons*. As a general rule, assaults with intent to commit felonies require specific intent. Assaults with deadly or dangerous weapons are general intent crimes requiring no proof of intent to kill.

The major difficulty with the assault with a deadly weapon statutes involves defining a deadly weapon. A deadly weapon is anything capable of causing death depending on how it is used and with what intent it is used in any particular case. The fact that a weapon is capable of causing death is not the determining factor in declaring it a deadly weapon. This determination depends on the manner and intent with which the weapon is used in addition to its natural capability. It is too easy to think of deadly weapons as being only guns, knives, baseball bats, and other things wielded in a deadly manner by a person. Most states are willing to accept that feet, legs, arms, and fists may also fall into that category. However, can one be convicted of assault with a deadly weapon by biting? A number of courts today recognize that biting could support such a conviction whether or not the defendant has some deadly transferable disease. The courts are declaring that whether a thing is a deadly weapon is a question of fact for the jury and the item used does not have to be inherently dangerous.

A Case of Biting

During a heavyweight title boxing match, Mike Tyson bit his opponent, Evander Holyfield's ear—twice. The fight was over; Tyson lost. Tyson was not charged with battery. Had he been charged, could he have been convicted? Since it was a sporting event, did not Holyfield consent to a battery on himself? (See Chapter 6.)

Good Try, But . . .

A defendant drove her car at a law enforcement officer, who, in attempting to defend himself, put his hands on the vehicle to push himself away. The defendant argued that he was the aggressor by making contact with her car first. The court didn't buy it.

7.4 DOMESTIC VIOLENCE

For years a matter of common concern has been the problem of intrafamily assaults, particularly those committed by husbands upon wives. Any experienced law enforcement officer can relate that **domestic violence** takes a considerable amount of time and presents a considerable potential danger.

The law has never offered a satisfactory solution to this social, psychological, and often economic problem. Even when the vehicle exists to prevent further abuse, the abused spouse will often fail to pursue his or her rights. Police frustration runs high, when after arresting the abusive person and thereby risking danger, the abused spouse refuses to continue the complaint and often provides bail and then withdraws the complaint. The same frustration plagues judges who prepare the warrants and do other preliminary work leading to the issuance of peace bonds only to find that the abused spouse wants to back out.

Additionally, the common law created impediments to both protective action and possible long-term solutions. Assaults by spouses without deadly weapons were at common law merely misdemeanors. This distinction carried into modern times. One of the incidents of a misdemeanor is that an arrest for a misdemeanor could only be made by an officer upon a valid warrant or if the offense was committed in the officer's presence.

Most spouse beaters do not commit their assaults in front of police officers. Unless the victim or someone else witnessing the abuse would swear out a complaint leading to the issuance of an arrest warrant, the police could not help. This was true even though the police went to the scene and observed the results of the beating.

Something needed to be done. A few legislatures reacted in an attempt to protect even those who could not or would not protect themselves. In addition to special acts creating civil relief and shelter services, many states amended their criminal codes to overcome some of the common law impediments. For example, California has both misdemeanor and felony domestic violence statutes. In a felony case, the law of probable cause and exigent circumstances comes into play that would allow an arrest without a warrant. Florida and Michigan allow arrests by police without a warrant for this one misdemeanor, even though it is not committed in their presence.

A larger problem exists. Some states still recognize the common law immunity that a husband could not commit a crime upon his wife. This immunity did not recognize the existence of two people but saw them as one unit before the bar of justice. Those states felt it necessary to create the crime of spousal assault. States such as California, Florida, Illinois, Missouri, Rhode Island, and Tennessee passed such laws, as have many other states. So, any state without such a law is now the exception, rather than the rule.

People living together without a formal marriage ceremony also seemed to demand attention as some cohabitants also refused to press charges. California, Florida, New York, Mississippi, Michigan, Ohio, Oregon, and Rhode Island included cohabitants in their spousal assault statutes.

Ohio makes the second and any later assault a felony; thus, an arrest may be made on the subsequent assault without a warrant. The only problem is discovering whether or not a particular assault is a subsequent assault.

Social Security Administration Helps

Even though states are addressing the spousal abuse problem with more effective criminal laws, an increasing number of abused spouses are seeking escape from an abusive relationship. This may involve moving and assuming a new identity. In the past, the Social Security Administration would issue a new Social Security number if the abused spouse could prove not only that he or she was abused but that the abuser had misused the victim's Social Security number. Now a new number may be issued if the victim provides written evidence of domestic violence from a local shelter, a treating physician, or a law enforcement official.

Finally, many states are recognizing that standard law enforcement training does not adequately prepare an officer to deal with the problems of domestic violence. To meet this challenge, several states have passed statutes relating to special training in such matters for law enforcement officers. These states include Michigan, Nebraska, and New Jersey.

7.5   ASSAULTS ON SPORTS OFFICIALS

Assaults and batteries on sports officials are increasing. In an effort to provide additional protections for officials while performing their sporting responsibilities, legislatures in many states have enacted laws providing criminal and, in some cases, administrative, penalties for assaulting or battering a sport official while they are doing their job as a sport official. As an example the Montana statute reads:

45-5-211

1. A person commits the offense of assault upon a sports official if, while a sports official is acting as an official at an athletic contest in any sport at any level of amateur or professional competition, the person:

   a. purposely or knowingly causes bodily injury to the sports official;

   b. negligently causes bodily injury to the sports official with a weapon;

   c. purposely or knowingly makes physical contact of an insulting or provoking nature with the sports official; or

   d. purposely or knowingly causes reasonable apprehension of bodily injury in the sports official.

2. A person convicted of assault upon a sports official shall be fined in an amount not to exceed $1000 or be imprisoned in the county jail for a term not to exceed 6 months, or both.

California's statute provides a potential penalty twice that provided for in the Montana statute and contains this definition of a sports official:

243.8 Battery against sport official

b. For purposes of this section, "sports official" means any person who serves as a referee, umpire, line judge or in any similar capacity in supervising or administering a sports event, and who is registered as a member of a local, state, regional or national organization which provides training or educational opportunities for sports officials.

Section 128C.08, of the Minnesota statute concerns assaults on sports officials in high school or other interscholastic athletic activities.

A review of the statutes seems to indicate a growing awareness of the dangers faced by the sports officials but, what about assaults committed on coaches by irate or out-of-control parents at amateur athletic events? "Why isn't my son getting more playing time?" "Why are you playing my son at this position when he's better at that other position?" "What a stupid coach you are for calling that play!" These words alone do not constitute an assault but too often when parents lose their tempers, assaults and batteries can and do occur. Is more statutory protection needed?

7.6 MAYHEM

One form of battery was given separate treatment at common law because the act deprived the victim of the ability of self-defense. This common law crime, called **mayhem,** included any act against the person that violently deprived that person of the use of any members so as to make him or her less capable of self-defense. The character of the member was the key to determining the existence of this crime and not the seriousness of the injury. Similarly, the fact that an injury disfigured the victim did not constitute mayhem unless that same injury weakened the person to a point where he or she was less capable of self-defense. Oddly enough, the common law courts held that cutting off another's ear or nose was not mayhem under the definition just given, for this only disfigured and did not weaken. Modern statutes altered the rule and include intentionally inflicted acts of disfigurement within the purview of mayhem.

The injury must be inflicted willfully and maliciously and must be of a permanent nature to constitute mayhem, although premeditation is not a required element. The commission of one of these prohibited acts in necessary self-defense is not criminal. Mayhem is among those crimes for which consent may not be given by the victim. Consent in such a case is not a valid defense to this crime (see Chapter 6). (Also see the sidebar entitled "A Case of Biting" in Section 7.3 and analyze it based on common law and statutory mayhem.)

7.7 STALKING CRIMES

Stalking has been called the crime of the nineties. Celebrities, former lovers, and former spouses, among others, have been subjected to harassment, threats, constant trailing, and other acts that over a period of time would seriously annoy any reasonable person.

Stalking is a unique form of criminal activity composed of a series of actions rather than a single act as is the case for most other crimes. It is comprised of acts that, when taken individually, might constitute legal behavior. Sending flowers, writing love notes (hard copy or e-mail), and waiting for the person outside his or her place of work are not criminal. Coupling these acts with the intent to cause fear or injury to the person who is the object of this conduct may constitute an illegal pattern of behavior. Stalking, like domestic violence, is gender-neutral. It is committed by both males and females.

Stalking first became an issue of public concern when young actress Rebecca Shaeffer was shot to death in 1989 after being stalked for two years by an obsessed fan. The initial publicity about this murder caused the frenzy in state legislative bodies to pass antistalking legislation. California was the first state to react, followed closely by Florida. By 1998, all states and the District of Columbia had antistalking legislation. Many of the initial laws, hastily drawn and passed, were flawed by ambiguities, unenforceable provisions, and confusing proof of intent requirements. At least 63 constitutional challenges have been considered in at least 24 states plus the District of Columbia. The principal arguments have been that the statutes are "void for vagueness," meaning the prohibited conduct is not specifically spelled out so that average persons would understand how to conduct themselves so as not to violate the law, or that the statutes are too broad, prohibiting too much conduct and thereby infringing upon constitutionally protected speech. Although some statutes have been successfully challenged on these bases, the courts are generally reluctant to strike down antistalking statutes for these reasons.

The issues seem to be settling down and patterns are emerging. There are three primary elements of the offense—specified conduct, threat requirements, and intent.

Conduct

Almost all jurisdictions require that the accused must have engaged in a "course of conduct," not just a single act, which, viewed collectively, constitutes a pattern of behavior. The jurisdictions differ in the types of conduct and in the number of acts required to satisfy this element but, generally, the conduct is among the following: pursuing or following, watching, lying in wait, intimidating, nonconsensual communicating, harassing, trespassing, possessing or showing a weapon, approaching, making presence known, disregarding a warning, confining or restraining, causing bodily harm, or vandalizing.

Threat

Most of the jurisdictions require the stalker to pose a threat or create a situation in which a reasonable person would feel fearful. This is similar to the definition of threat as an element of assault (as discussed in Section 7.1, "Simple Assault"). The threat does not have to be verbal or written. Some conduct, like forming the hand into a gun and pointing it at the person, if reasonably causing fear under the circumstances, can satisfy this element.

Intent

Most statutes contain words like *knowing, deliberate, willful, purposeful,* or *intentional* as the controlling determinant of intent. However, many of the jurisdictions do not require proof that the accused intended to cause fear as long as there was intention to commit the act(s) that resulted in fear. In these jurisdictions, the general intent element has been met if the victim is reasonably frightened by the accused's conduct. The Minnesota Supreme Court held that the wording of its statute requires proof of a specific intent.

The elements listed above generally describe the most serious stalking, referred to in many jurisdictions as aggravated stalking, conviction of which is a felony. Some jurisdictions recognize a misdemeanor or other less severe form of the offense where the threat is absent and the conduct might amount to misdemeanor harassment. Under other statutes the felony offense can be further aggravated where, for example, the offender waves a gun around, violates a restraining order, directs threatening conduct toward a child, or has previously committed a stalking offense.

Can there be an attempt to commit an aggravated stalking? The Supreme Court of Georgia said yes. A month after Ricky Rooks began making harassing phone calls to his ex-wife, she swore out a criminal warrant. The judge issued a bond conditioned that Rooks was to have no contact with his ex-wife or her family. Rooks made telephone calls to his ex-wife's place of business, telling co-workers things he was going to do. The state court of appeals reversed the conviction, reasoning that stalking was an assault and since an assault is itself an attempted battery, there cannot be an attempted attempt. Thus, attempted stalking is a legal impossibility (see Chapter 4). The state supreme court disagreed and reversed. The court said:

. . . a person commits stalking when with a specific intent he "follows, places under surveillance, or contacts another person." Aggravated stalking is the same behavior when done in violation of a judicial order prohibiting such conduct. Generally, none of these actions would constitute an assault, which requires a demonstration of violence and a present ability to inflict injury. The intent element of the stalking statute requires proof of intentional conduct that "causes emotional distress by placing such person in reasonable fear of death or bodily harm to himself or herself or to a member of his or her immediate family." This element differs from assault in two significant ways. The assault statute requires proof that the accused induced fear of an immediate violent injury. The stalking law contains no immediacy requirement. Secondly, assault requires proof that the victim perceived the threat of violent injury to himself, whereas stalking may be committed by inducing fear that the victim's family may be harmed. These differences in both the act and intent elements demonstrate that stalking is not "in essence a common law assault."

While assault and stalking may overlap in some circumstances, the rationale for not punishing an attempted assault does not apply to an attempted stalking. In refusing to recognize the crime of attempted assault, this court stated that to attempt an assault is "to do any act towards doing an act towards the commission of the offense" and noted the absurdity and impracticality of criminalizing such behavior. To attempt to stalk, however, is to attempt to follow, place under surveillance or contact another person. It is neither absurd nor impractical to subject to criminal sanction such actions when they are done with the requisite specific intent to cause emotional distress by inducing a reasonable fear of death or bodily injury.

7.8 HATE CRIMES

Hate crimes or **bias-motivated** crimes are offenses motivated by hatred against the victim based on the victim's race, religion, sexual orientation, national origin, ethnic background, or handicap. Hate crimes are not new. Such violations have existed throughout recorded history. The Romans persecuted Christians. The Nazis persecuted Jews. In the United States there have been and, in some cases still are, cross burnings, lynchings, assaults on homosexuals, swastikas painted on Jewish synagogues, and burning of African American churches. Muslims and Asian Americans are also increasingly becoming targeted victims.

The offenses committed in bias-motivated cases are, for the most part, already violations of the law covered by such offenses as murder, malicious destruction of property, trespass, arson, assault and battery, and so forth. But prosecution for those violations doesn't get at the core of the problem, which is that the violation was directed at a whole group because of who they are or what they believe. Such offenses can lead to group reactions and group retaliations if not "nipped in the bud."

There are no hard statistics to show that hate crimes are on the increase, but the concern over those that are occurring has led 49 states to enact some form of legislation to combat hate crimes. Most of these statutes— those of 43 states—prohibit and prescribe punishments for bias-motivated or intimidation-based violence. At last report, Wyoming was the only state that had no legislation addressing hate crimes. The states that have not enacted bias-motivated or intimidation statues are: Arkansas, Hawaii, Indiana, Kansas, New Mexico, South Carolina, and Wyoming.

Although the statutes contain the two essential components of a prohibited act against a protected or covered group, the laws vary in that three different approaches are followed to combat hate crimes—prohibiting specific intimidating conduct, prohibiting behavior motivated by bias, and increasing penalties for criminal acts motivated by bias.

California, Florida, and Ohio are among a group of states that have enacted laws prohibiting specific acts at specific places. For example, statutes punish vandalism at a place of worship or causing an intentional disturbance at a place of worship. In addition, Florida, joined by the District of Columbia, has proscribed acts such as burning a cross or placing a swastika or other offensive symbol on the property of another with the intent to intimidate.

New York is among the group of states that punishes behavior motivated by bias wherein the targeted victim is an integral part of choosing and committing the crime.

Wisconsin and others provide for enhanced penalties when the motivation for an otherwise criminal act is bias. The Wisconsin law provides that the maximum penalty for an offense is enhanced if the defendant intentionally selects the person against whom a crime is committed because of the race, religion, color, disability, sexual orientation, national origin, or ancestry of that person.

Hate crime statutes have been challenged on the grounds of violating the First Amendment of the U.S. Constitution, which provides for freedom of speech and expression. The U.S. Supreme Court has twice addressed the issue. In a 1992 case, the Court examined legislation that made particular bias an element of a crime. The defendant was accused of burning a cross on a black family's lawn. He was charged under a city bias-motivated crime ordinance that made it a misdemeanor to "place on public or private property a symbol, object, appellation, characterization, or graffiti, including, but not limited to, a burning cross or Nazi swastika, which one knows or has reasonable grounds to know arouses anger, alarm or resentment in others on the basis of race, color, creed, religion or gender."

The Supreme Court struck down the ordinance as being oriented toward particular forms of action rather than being neutral on its face and prohibiting all actions likely to provoke a violent response. In other words, the Court said the ordinance selected particular forms of conduct which were disapproved, but it did not disapprove of all forms of conduct likely to invoke that same response. Hence, by singling out specific activities to criminalize, the ordinance was invalid.

A year later the Supreme Court provided clarification when it considered the constitutionality of a Wisconsin statute that enhanced the penalty for otherwise criminal behavior when it was bias-motivated. In the case, a group of young black men, including the defendant, beat up a white boy. The defendant was the instigator of the attack after the group became hyped from discussing a scene from a movie. The defendant was convicted of aggravated battery, an offense that normally carried a penalty of two years in prison. Because a jury found that the defendant had intentionally selected the victim based on the boy's race, under the enhanced penalty statute for bias-motivated crimes, the maximum sentence increased to seven years. The defendant was sentenced to four years' imprisonment and appealed. By unanimous opinion, the Supreme Court upheld the statute as being constitutional. The Court distinguished the two cases by saying that in the earlier case, the particular bias was an element of the crime itself while in the state statute it is a factor to be considered during sentencing. Since these two decisions, other state appellate courts have examined their own statutes and, for the most part, found them to be valid including cases in Florida, Maryland, Missouri, Ohio, Oregon, Washington, and Wisconsin.

▨▨▨▨ DISCUSSION QUESTIONS ▨▨▨▨

1. A statute of the state of Excelsior allows a person to trespass on another's land for the purpose of recapturing strayed cattle. Johnson is a farmer in a small rural community in Excelsior. He makes his livelihood by raising cattle. One day he noticed that four cows had wandered out of his pasture and onto the property of his neighbor, Green. Johnson entered Green's property for the purpose of retrieving his cattle. Just as Johnson approached the property, Green rushed out of the house and ordered Johnson off his property. An argument ensued and ended in Johnson knocking Green to the ground. Johnson took his cattle and left. Shortly afterward, Green reported the incident to the police, who arrested Johnson for assault and battery. In light of the statute mentioned, would a conviction be proper?

2. Brown was to be a witness in a trial against Howard. The day before the trial, as Brown was crossing a street, he observed a vehicle approaching

at high speed. Brown hurried toward the curb, but the car veered off its path and headed straight toward him. Brown was observant enough to see and remember the license number. The following day, the police arrested a suspect who admitted driving the car. The state, however, was unable to prove any intent on the part of the suspect to murder Brown. Does the state have an alternative charge on which to proceed against the suspect?

3. The marriage of John and Lorena, husband and wife, has been in trouble for some time. She accuses him of spouse abuse and claims that he often forces her to have sex against her will. One night, in a fit of rage and while John is asleep, Lorena gets a butcher knife from the kitchen, cuts off John's penis, runs from the house, and, while running, throws the penis away but does not know where. John survives. Assume that John and Lorena live in a common law state. If Lorena was charged with mayhem, could she be properly convicted? Why or why not?

GLOSSARY

Aggravated assault – encompasses all assaults of a vicious nature including assaults with intent to commit a serious offense and assaults with deadly weapons.

Battery – an unlawful injury, however slight, done to another person, directly or indirectly, in an angry, revengeful, rude, or insolent manner.

Bias-motivated – the basis for most hate crimes.

Domestic violence – assaults and/or batteries committed by persons who live together. Primarily growing out of the abusive/battered spouse syndrome, often carries a greater penalty than other assaults and batteries.

Mayhem – any act against a person that violently deprives the person of the use of any members so as to make him or her less capable of self-defense.

Simple assault – an attempt or offer, with unlawful force or violence, to injure another.

Stalking – a crime consisting of a series of acts directed at following or contacting another person, when coupled with an intent to cause fear or injury to that person, and which conduct causes fear.

Striking distance doctrine – for an assault to occur, the person who is offering or attempting injury must be within a reasonable distance where the threat to cause harm can justify the fear of harm experienced by the victim or where the harm can be immediately implemented.

■ REFERENCE CASES, STATUTES, AND WEB SITES ■

CASES

State v. Rooks, 468 S.E.2d 354 (GA 1996).

Bryant v. State, FLW D1406B, 4th D.C.A. (FL 2001).

State v. Orsello, 554 N.W.2d 70 (MN 1996).

State v. Cole, 542 N.W.2d 43 (MN 1996).

State v. Meier, 422 N.W.2d 381 (ND 1988).

R.A.V. v. City of St. Paul, 505 U.S. 377, 112 S.Ct. 2538, 120 L.Ed.2d 305 (1992).

Wisconsin v. Mitchell, 508 U.S. 476, 113 S.Ct.2194, 124 L.Ed.2d 436 (1993).

Monhollen v. Commonwealth, 947 S.W.2d 61 (KY 1997).

State v. Plowman, 838 P.2d 11 (Or. App. 1984), rev. den., 683 P.2d 1372 (OR 1984).

State v. Talley, 858 P.2d 217 (WA 1993).

State v. Mitchell, 504 N.W.2d 610 (WS 1993).

State v. Wyant, 624 N.E.2d 722 (OH 1994).

STATUTES

California: Cal. Pen. Code §. 241.6; §. 243.8

Oklahoma: 21 Okl. St. §. 650.7

Minnesota: §. 128C.08

Montana: Mont. Code Anno., §. 45-5-211

WEB SITES

www.adl.org/99hatecrime/intro.html
www.findlaw.com

Homicide

■ KEY WORDS AND PHRASES ■

Aforethought

Assisted suicide

Culpable negligence

Excusable homicide

Felonious homicide

Felony-murder rule

Heat of passion

Homicide

Infanticide and feticide

Involuntary manslaughter

Justifiable homicide

Malice aforethought

Manslaughter

Murder

Provocation

Quickened

Voluntary manslaughter

Year and a day rule

8.0 INTRODUCTION: HOMICIDE IN GENERAL

Police officials and members of the public often incorrectly use the terms **homicide** and **murder** interchangeably. Murder is only a part of the broad criminal law category of homicide.

Homicide is the killing of a human being by another human being. This broad definition covers many forms of conduct, only some of which are criminal. It encompasses all deaths either directly or indirectly due to causes other than natural. It also includes death from natural causes resulting from the act of another human being. For example, Tom threatens Jerry with a gun and scares Jerry so badly that he has a heart attack and dies. This is a homicide. As discussed in Chapter 4, a homicide may be committed by failing to perform a legal duty—a crime by omission, or negative act.

At common law, if the victim did not die within a year and a day after the injury was inflicted, but died after that time, the perpetrator could not be charged with homicide. The reason for this rule was that there could have been too many other factors contributing to death during that intervening period, making the cause-and-effect relationship too remote to satisfy the legal requirement of causation (Chapter 4). The fact that there were no other intervening causes was immaterial because this was a rule of law, not a determination of fact. The **year-and-a-day rule** is part of the common law of the United States and, unless changed by statute, will remain so.

At least one court in Massachusetts decided that since the year-and-a-day rule was judge-created, it could be undone by judges. The court reasoned that the original rule rested in the lack of scientific knowledge to determine the critical issue of whether the original injury truly caused the death or whether some independent, intervening, superseding cause had, in fact, caused it. The court decided that since science is now advanced enough to make those determinations, the reason for the rule no longer exists and abandoned it. Iowa, among others, has also reached the same conclusion.

When is a person dead? This has not always been legislatively defined in the past. Ordinarily, death has been determined according to the traditional criteria of irreversible cardiorespiratory repose. However, we can now maintain heartbeat and breathing mechanically. We now know medically that breathing and heartbeat are not independent indications of life but are part of systems normally controlled by the brain. The brain is therefore dominant. Since 1970 the law has been catching up with science, and some states now define death as the cessation of brain functions evidenced by a "flat" reading of the electroencephalograph. Why does all of this make a difference? Suppose that A shoots B, and B is put on a life-sustaining device. Later, requisite readings are "flat" and B, a self-designated organ donor, has his heart and kidneys removed. The machine is turned off and B is buried. Who killed B? A? The doctors? In a state with brain death laws, A is the killer.

In Court

Q. Doctor, before you performed the autopsy, did you check for a pulse?

A. No.

Q. Did you check for blood pressure?

A. No.

Q. Did you check for breathing?

A. No.

Q. So, it's possible the patient was alive when you started the autopsy?

A. No.

In Court—*continued*

Q. How can you be so sure, doctor?
A. Because his brain was sitting on my desk in a jar.
Q. But could the patient still be alive nevertheless?
A. It's possible he could have been alive and practicing law.

In Another Court

Q. Do you recall the time that you examined the body?
A. The autopsy started around 8:30 p.m.
Q. And Mr. Dennington was dead at the time?
A. No, he was sitting on the table wondering why I was doing an autopsy.

In Still Another Court

Q. Doctor, how many autopsies have you performed on dead people?
A. All my autopsies are performed on dead people.

8.1 NONCRIMINAL HOMICIDES

Not all homicides are criminal. Some homicides are specifically deemed to be noncriminal in the eyes of the law. These homicides fall into two categories: justifiable homicide and excusable homicide. At common law, the distinction between justifiable and excusable homicide was important for more than purposes of definition. Although noncriminal in nature, excusable homicide was not entirely free from penalty. A person who committed excusable homicide forfeited certain lands or goods according to the circumstances surrounding the homicide. This is no longer true, but statutes in the various jurisdictions continue to distinguish these noncriminal homicides.

Justifiable Homicide

The common law defined **justifiable homicide** as the necessary killing of another in the performance of a legal duty or the exercise of a legal right where the slayer was not at fault. This classification includes execution of convicted capital offenders, homicides by police officers in the performance of a legal duty, and so forth. Also included are slayings in self-defense when the slayer is feloniously attacked and has to kill to preserve his or her own life, provided that the slayer is not in any way at fault for the attack on his or her person. Justifiable homicide carries with it no penalty because the slayer was not at fault.

Excusable Homicide

Excusable homicide differs from justifiable homicide in that one who commits an excusable homicide is to some degree at fault but the degree of fault is not enough to constitute a criminal homicide. Excusable homicide covers two fundamental situations resulting in homicide. First is when death results from misadventure. This is similar to what may be termed "accidental" death at the hands of another. Misadventure is death occurring during commission of a lawful act, or a *malum prohibitum* unlawful act, committed without any intent to hurt and without criminal negligence. Examples of misadventure include the death of a child who is being lawfully punished, or the death of a person who runs in front of a moving automobile when the driver is unable to avoid the collision. The second type of excusable homicide involves self-defense when the slayer is not totally without fault. For example, someone gets involved in a sudden brawl and has to kill to preserve his or her own life. Note that self-defense may be considered either justifiable or excusable homicide. However, it is more properly treated as a matter affecting criminal responsibility, and for that reason it was discussed in Chapter 6.

8.2   CRIMINAL HOMICIDES

Felonious homicides are those treated and punished as crimes. They fall into two basic categories: murder and manslaughter.

Murder

The common law defined murder as the felonious killing of any human being by another with malice aforethought. Murder was a general intent crime. Specific intent to kill any particular person was not required, although it would satisfy this requirement. The difficulties in proving a case of murder at common law basically involved the interpretations given the words **malice** and **aforethought** and the interpretation given both words together as a phrase.

The word *malice* in legal usage connotes something different from what it does in the popular sense. *Malice* in the popular sense is often used as a synonym for hate, ill will, bad feelings, and the like. For purposes of the law of homicide, *malice* means the intentional doing of a wrongful act in such a way and under such circumstances that the death of a human being may result.

Malice may be either express or implied. To prove this element satisfactorily, it was necessary to show that the perpetrator either actually intended to kill (express malice) or killed while committing a deliberate and cruel act likely to cause death (implied malice). There are basically four situations in which the law would imply malice at common law. These instances involve, in effect, unintentional killings. The first was intentional infliction of great bodily harm on someone, unintentionally resulting in death. This would warrant a murder charge at common law. John hits Bill on the head with a tree branch, intending to injure Bill seriously but not intending to kill him. If Bill dies from the blows, malice will be implied.

The second situation in which the law would imply malice involved no actual intent to kill on the part of the perpetrator. Instead, it involved a deliberate act or omission, of such a nature that it tended to cause death or serious bodily harm. If Sam deliberately drove his automobile in excess of one hundred miles an hour through a crowded city intersection to see how fast it would go, and killed somebody as a result, the law would infer malice. Third, the law would also imply malice when death resulted during the commission of a felony under the felony–murder rule discussed in detail in the next section.

Finally, when death was caused by one resisting a lawful arrest, the common law would imply malice. The law requires a person to submit to a lawful arrest but permits the person to resist an unlawful arrest with necessary reasonable force. If the jury finds deadly force was reasonably necessary to resist an unlawful arrest, the defendant will be found to have committed excusable homicide.

In a more recent development, a number of states have begun to rule that some conduct may be so "indifferent" as to constitute murder. The impetus for this has come from deaths caused by illicit drugs. Tennessee, for example, has held that the sale of heroin resulting in death may constitute murder. The specific case involved a dealer who was told not to sell the drug until it was cut. He ignored this warning and a death resulted. The court said the prosecution might be able to demonstrate that his actions amounted to such conscious disregard of the consequences that malice could be implied. Can a person be tried for a homicide that requires as its mental element "extreme indifference to the value of human life" when his pit bull kills a neighbor's two-year-old child even when the child wandered into the man's yard and the dog was chained up? Yes, said a California court. They felt that if a dog was bred and trained to be a vicious fighting dog, its owner could not escape a trial in that state.

The term **aforethought** conveys the meaning of planning ahead. It means substantially the same thing as do the words *premeditated design* commonly found in modern murder statutes. Although the time lapse between planning and doing the act that causes death is immaterial, there must be a deliberate design. This design may be formed seconds, minutes, or days before the act is performed and may be inferred by the courts and juries from the circumstances surrounding the homicide, but it must be proved. Premeditated design or aforethought was the prime element that distinguished murder from manslaughter or "accidental" homicide at common law. It distinguishes the various degrees of murder under present statutory law in many jurisdictions.

The phrase **malice aforethought** can thus be defined as the intentional killing of one human being by another without legal justification or excuse under circumstances insufficient to reduce the crime to manslaughter or a lesser degree of murder.

The malice and the aforethought must exist simultaneously with each other and with the act. For example, Smith says on Tuesday that he will kill Jones. On Wednesday, by accident and misfortune, Smith kills Jones. In this case the forethought may still have existed, but, under the circumstances, there was no malice. Thus, murder would be an improper charge.

The Felony-Murder Rule

The common law courts held that the **felony–murder rule** applied to deaths occurring during the commission of, or an attempt to commit, arson, burglary, larceny, rape, or robbery. In essence, under this rule any death resulting from commission of any of those felonies was murder. For a killing to be murder, there had to be malice aforethought. This element was satisfied by treating such a case as one involving implied malice. Whoever committed a felony acted in a deliberate and cruel manner, and that person was held responsible for the natural and probable consequences of the act. One of the probable consequences of the commission of a felony is that someone might get killed. This is true even though the killing is unintentional and accidental.

The common law felony-murder rule applied only to certain enumerated felonies. Some modern statutes have expanded this and allow the rule to operate in cases of homicides occurring during the course of committing, or while attempting to commit, *any* felony. In most of the states allowing felony-murder to operate under any felony and in those states having only a general statute on the subject, the courts have been reluctant to extend the coverage beyond felonies that are not in themselves inherently dangerous to life. Some states allow a broader interpretation and say that the felony must be either inherently dangerous or committed under circumstances that are inherently dangerous. Others continue to specify only certain felonies to which the rule applies, but some of the felonies may be different from those enumerated at common law. Even in states that follow the latter course, killings in the course of other felonies, not enumerated in the felony-murder rule, are generally covered by a statute making them lesser degrees of murder.

Even modern statutes will not answer the practical questions that arise in applying this rule to the facts of a given case. Some important questions arise: Does the rule cover all homicides during commission of, or while attempting to commit, one of these felonies, or does it have to be a homicide actually committed by one of the perpetrators of the felony? What does "during the commission of a felony" mean? Does it include fleeing from the scene? If so, is the time and distance factor important? What types of causation factors apply to this rule? These and several other questions can be answered only by referring to the cases that have been decided by the appellate courts.

As to whether the rule applies to all homicides committed in the perpetration of, or in attempting to perpetrate, the enumerated felonies, it has been said that there are at least 16 possible combinations to which the rule may apply. Included in these combinations are (1) victim killing co-felon, (2) co-felon killing victim, (3) co-felon killing a bystander or police officer, (4) victim killing a bystander or police officer, (5) police officer killing victim, bystander, other police officer, or co-felon, and so on. Among the courts that have struggled with the problem of whether the rule should apply to all these possible combinations is the Supreme Court of Pennsylvania. In the 1940s and 1950s this court was faced with a series of four cases involving the felony-murder rule. The court changed its mind several times before deciding on the scope of felony-murder under Pennsylvania law.

In *Commonwealth v. Almeida* (1949), a police officer was killed during an exchange of shots between police and robbers. The defendant was convicted of first-degree murder committed in the course of the robbery.

The Supreme Court of Pennsylvania, in affirming the conviction, said:

> Their acts were "the cause of the cause" of the murder. They "set in motion the physical power" which resulted in Ingling's death and they are criminally responsible for that result. Whether the fatal bullet was fired by one of the bandits or by one of the policemen who were performing their duty in repelling the bandits' assault and defending themselves and endeavoring to prevent the escape of the felons is immaterial.

Six years later, the same court ruled on another felony-murder case. In *Commonwealth v. Thomas* (1955), the facts disclosed that Thomas and a confederate committed a robbery, and, while fleeing from the scene, the confederate was shot and killed by the store owner. Thomas was indicted for the murder of his co-felon. The trial court granted the defendant's motion to dismiss and the state appealed. The Pennsylvania Supreme Court reversed the judgment. In answer to the question, "Can a co-felon be found guilty of murder where the victim of an armed robbery justifiably kills the other felon as they flee from the scene of the crime?" the court said, "The felon's robbery set in motion a chain of events which were or should have been within his contemplation when the motion was initiated. He therefore should be held responsible for any death which by direct and almost inevitable sequence results from the initial act." It is interesting to note that after this decision, the district attorney *nolle prossed* the murder charge against Thomas.

In 1958, the Pennsylvania Supreme Court repudiated its earlier decision in the Thomas case. In *Commonwealth v. Redline*, (1958), involving facts similar to the Thomas case, the court said:

> In adjudging a felony-murder, it must be remembered at all times that the thing which is imputed to a felon for killing incidental to a felony is malice and not the act of killing. The mere coincidence of a homicide and a felony is not enough to satisfy the requirements of the felony-murder doctrine. It is necessary . . . to show that the conduct causing death was done in furtherance of the design to commit the felony. Death must be a consequence of the felony . . . and not merely coincidence.

The court also stated that its research found no cases either in this country or under the English common law in accord with the Thomas decision.

Oddly enough, on the same day that the Redline case was decided, the court also handed down a decision in *Commonwealth v. Bolish*. Bolish and an accomplice, Flynn, planned an arson. In carrying out the plan, Flynn was fatally injured by an explosion that occurred when he placed a jar of gasoline on an electric hot plate. Bolish was convicted of first-degree murder. The judgment was affirmed on appeal. The court held that the defendant:

> was actively participating in the felony which resulted in death. The element of malice, present in the design of the defendant, necessarily must be imputed to the resulting killing, and made him responsible for the death. . . . The fact that the victim was an accomplice does not alter the situation, since his own act which caused the death is in furtherance of the felony.

A review of these cases reveals the confusion that exists as to the scope of the felony-murder rule. Pennsylvania's initial attempt to apply a blanket rule to homicides occurring during the commission of a felony was later rejected because of lack of support from any other American jurisdiction or from English law. Applying the felony-murder rule to all homicides, regardless of by whom committed, fails to take the rules of proximate causation into account (Chapter 4). By following the rules of causation, courts would find it difficult to hold a felon liable for murder in the highest degree when, unrelated to the actual conduct of the felon, one innocent bystander happens to kill another innocent bystander while attempting to help apprehend the felon. Consequently, courts in most jurisdictions where this problem has arisen have extended the felony-murder rule only as far as the rules of proximate causation can justify.

New York was one of those jurisdictions that did not follow the reasoning of proximate causation and limited felony-murder liability to those persons who delivered the death blow while perpetrating the felony. In reaction to that position of the courts, the New York legislature rewrote the state's felony-murder statute to include punishment for those who were in the act of committing or attempting to commit a felony, no matter who dealt the death blow. Faced with this statutory language, the high court of New York upheld a conviction where a police officer killed a fellow officer who was attempting to prevent the escape of the defendant. The court said that the statute requires a focus on whether the acts of a third person in such circumstances are reasonably foreseeable, thereby justifying proximate causation as a means of finding liability. If the causal act (the shooting) is part of the felonious conduct and not merely coincidental, there is ample reason for holding the nonshooting defendant liable.

Confusion and differences still exist among the states regarding the application of the felony-murder rule. The major differences occur in situations involving killings of or by persons resisting the felony such as victims, police officers, or bystanders. Many courts are holding that the killing must be by the defendant or an accomplice for the rule to apply. Some courts hold that, if a co-felon is killed by one who is resisting the felony, the rule cannot apply because the killing would have been lawful; therefore, there is no killing for which the defendant can be held liable. Those courts that follow this view seem to distinguish between whether the killing of the co-felon was justifiable or excusable. If excusable, the defendant can be held liable.

Courts have almost uniformly found liability under the felony-murder rule in "shield" cases in which the perpetrator uses another person as a shield to assist in an escape. Courts apply the rule even if the killing of the shield is by one who is resisting the felony, but it is very unclear as to whether liability is being based on the felony-murder rule or on some other theory of murder.

Those courts that tend toward limiting the application of the felony-murder rule seem to be concerned with three major issues: the facts of the case, the exact wording of the statute, and a desire not to extend application of the rule beyond the rational limits that it was designed to serve. Some statutes provide that any death resulting during the commission of a felony is covered. Others are worded so as to apply in cases when the killing is by a person engaged in the commission of a felony.

The remaining unanswered questions involve an overall understanding of when the commission of a crime begins and when it ends. Because of the requirement that the homicide occur during commission of or while attempting to commit the felony, time, distance, and escape all become important matters.

Most courts have agreed that the time and distance separating the killing and the felony are not the sole determining factors in deciding when commission of the felony ends. Each case must be judged on its own facts. It is agreed that escape from a crime scene is to be considered as part of the commission of a felony for purposes of the felony-murder rule. The rationale for this rule is simply that one does not intend to commit a crime and be caught.

The problem is best summed up by the Florida Supreme Court, which said:

It is a sound principle of law which inheres in common reason that where two or more persons engage in a conspiracy to commit robbery and an officer or citizen is murdered while in immediate pursuit of one of their number who is fleeing from the scene of the crime with the fruits thereof in his possession, or in the possession of a co-conspirator, the crime is not complete in the purview of the law, inasmuch as said conspirators have not won their way even momentarily to a place of temporary safety and the possession of the plunder is nothing more than a scrambling possession. In such a case the continuation of the use of arms which was necessary to aid the felon in reducing the property to possession is necessary to protect him in his possession and in making good his escape.

Although the court indicates that it is laying down a rule for robbery cases, it is safe to assume that any homicide committed during an escape or flight from the scene of one of the enumerated felonies will fall within the rule if the felons have not yet reached a point of reasonable safety and freedom from immediate pursuit by law enforcement officers. The time separating actual perpetration of the felony and the commission of the homicide, as well as the distance from the scene of the felony where the homicide occurs, are factors to be considered in each case to decide the appropriateness of the felony-murder rule, but, in themselves, they are not the sole determining factors.

The felony and the homicide do not merge, and it is proper to charge the perpetrators with both the felony and the murder either in separate counts of a single indictment or with separate indictments. The risk run by separate trials is that of a possible double jeopardy problem when a felony-murder is tried and there is an acquittal. The subsequent trial on the underlying felony would be double jeopardy, since it must be proven in the felony-murder trial. Since an acquittal resulted, the defendant will have already been tried on the entire issue of the underlying felony (see Chapter 6).

The confusion and dissatisfaction with the felony-murder rule have led at least one state to abolish it. A Michigan court held that since they had no statutory felony-murder rule, they would exercise their right to abrogate the common law felony-murder rule. The court went on to say that to prove murder it must be shown that the defendant acted with intent to kill or the natural tendency of his behavior was to cause death or great bodily harm.

Courts continue to wrestle with the felony-murder rule. Most have a tendency to restrict its application only to those listed crimes to which it may apply. Thus the underlying felony must be the exact one designated by the legislature. Some courts also require that the mere fact of an underlying felony's being present is not enough unless the death truly occurred as a result of that underlying felony. For example, in a Virginia case the underlying felony was drug distribution. The plane the marijuana was being carried in crashed and killed a co-felon. The state failed to prove the plane crashed because of the underlying felony; therefore, the conviction could not stand.

Although Kansas, on the other hand, generally takes a less restrictive view and says the state need prove only the underlying felony, a court held that child abuse cannot serve as the underlying felony, because child abuse is an assault that merges into the homicide as a lesser included offense. A California court agrees with Kansas on this interpretation.

In summary, felony-murder was created when there were a very limited number of felonies. The major felonies originally made applicable to the felony-murder rule were arson, burglary, rape, robbery, and larceny. All of these felonies carried the death penalty and all, except larceny (a crime of stealth), invited armed defense. As legislatures dropped the death penalty from most felony convictions, and as they created new felonies not known at common law, courts had to decide which felonies should be subject to the felony-murder rule. Primarily, courts look to see if the felony is inherently dangerous. Kidnapping, a common law misdemeanor is, today, recognized as an inherently dangerous felony.

The nonapplicability of the felony-murder rule to crimes that are not inherently dangerous does not mean that there can be no murder prosecution. It simply means that the prosecutor must follow the standard processes and the rules of proximate causation to obtain a conviction.

Degrees of Murder.   Unlike modern law, the common law did not recognize degrees of murder. If a homicide failed to contain the necessary elements of malice aforethought, the proper charge was some other crime and not murder. Modern statutes have modified this rule in most jurisdictions. With the exception of Alabama, Georgia, Kentucky, Montana, New Jersey, and North Dakota, all other states and the District of Columbia divide murder into degrees. Only three of those states, however, have more than two degrees of murder. They are Florida, Minnesota, and Wisconsin. The aim of this legislative action was to limit the use of capital punishment without reducing the seriousness of the crime.

The most common division of murder used by the states is based on the seriousness of the conduct. Thus a planned murder is always murder in the first degree, as is felony-murder. The reasoning is that a person who is already engaged in some inherently dangerous conduct should pay the price for deaths that he or she causes directly or indirectly, while engaged in that conduct.

Murder in the second degree is often succinctly defined as follows: Any murder that is not murder in the first degree is murder in the second degree. This may be succinct, but it is not necessarily crystal clear: To interpret such a statute properly, one must still understand the basic common law definition of murder as a malicious killing.

Manslaughter

Manslaughter is the second major category of criminal homicide. Manslaughter is charged when a homicide is committed under circumstances not severe enough to constitute murder, yet not mild enough to be either justifiable or excusable homicide. At common law, manslaughter was divided into voluntary and involuntary manslaughter, each containing certain elements to be proved before a conviction could be obtained.

Voluntary Manslaughter

Voluntary manslaughter is an intentional killing but does not contain the elements of malice or premeditated design. Its essential elements include a legally adequate **provocation** resulting in a killing done in the heat of passion, before cooling. The law recognizes the weakness of people and regards this type of killing as less severe than killing in cold blood. This is the only justification for the difference between murder and voluntary manslaughter.

Adequate Provocation. Incidents that might provoke a person are not necessarily legally adequate to reduce the crime from murder to manslaughter. The types of provocation and the degree to which they must exist to be legally recognized are questions that are left to the jury to decide. In making this determination, however, the jury must be guided by certain legal principles. The jury must decide not whether the facts of the particular case provoked the defendant but, rather, whether those same facts under similar circumstances would have provoked a reasonable and prudent person to kill. This is known as the objective test. As a general rule, a simple assault or a mere technical battery will not be sufficient to constitute provocation adequate to reduce a homicide charge from murder to manslaughter. Words alone, unaccompanied by any conduct, are never sufficient to constitute provocation. If certain types of conduct accompany the words, adequate provocation may exist. Insulting gestures (not identical to assaults) are not adequate. With these guidelines to work by, the jury determines the adequacy of the provocation. For example, juries have found the following to be legally adequate provocation: seduction of a man's wife, knowledge of the rape of a man's wife acquired within a few minutes after the rape, and adultery.

All of the foregoing examples illustrate a revenge motive as adequate provocation. Is passion limited to a revenge motive? Passion need not be limited only to rage or anger. One court turned to the dictionary and found that passion could mean any violent, intense, highly wrought, or enthusiastic emotion. The court then declared that the duty of the jury is to judge whether the defendant's reason was, at the time of his act, disturbed or obscured by some passion to such an extent as would render the ordinary person liable to act rashly and without due deliberation and reflection. In this case the deceased had threatened suicide and unendingly provoked the defendant until out of wild desperation he shot her as she had requested.

Heat of Passion. **Heat of passion** is not the same thing as insanity, but mere anger is not enough. All that must be shown is that the adequate provocation just described was of such a nature that the defendant's mind became so inflamed to the point of not knowing what he or she was doing. No cold-blooded killing can ever be mitigated to manslaughter. The jury must be convinced that the provocation prevented thought and reflection and the formation of a deliberate purpose. Perhaps the most common case arises when the defendant discovers that someone is having, or has had, unlawful sexual intercourse with a female relative. Illinois supports this position.

Cooling. At the time the homicide occurs, the accused must still be acting under heat of passion. If the accused has cooled to the point of knowing what he or she is doing, the homicide is in cold blood and will be considered murder. Just as in determining legally adequate provocation, the jury must apply an objective test to determine whether a reasonable person would have cooled under the facts of the particular case. This is not governed by whether the accused had actually cooled or not. All the circumstances must be taken into account. Although a lapse of time between the provocation and the actual killing is important in determining whether a reasonable person would have cooled, this is not the sole determining factor. There is no definite rule for determining when cooling has taken place, but, generally, the greater the passion, the longer the cooling period. On January 1, Mr. Smith had shot and killed the son of Mr. Jones. Smith was tried for that homicide and was acquitted in March. Nothing further happened until November 29. On that date, while walking down the street Mr. Jones observed Smith standing on the corner talking to Parker. Jones walked up to Smith and said, "Hello, Dan," and without further warning fired two shots into Smith in quick succession. As Smith was falling to the ground, Jones fired two more shots into Smith, and then turned and walked away. After going some distance, Jones turned and came back. He put his pistol close to the head of Smith, pulled the trigger two more times (the gun was empty by this time), and said, "You damn son of a bitch, I told you I would kill you; you killed my boy." Murder would appear to be the proper charge as the time lapse would ordinarily be sufficient for the reasonable person to cool off. In addition, the deliberate act of repeatedly firing at the victim evinced a determination to commit cold-blooded killing.

Causal Connection. Even where the three prime elements of voluntary manslaughter exist, it must be shown that there is a causal connection between all three elements. In other words, the provocation must cause the heat of passion, which causes the homicide (Chapter 4).

Involuntary Manslaughter

Involuntary manslaughter is an unintentional killing. Death resulting from the commission of a *malum in se* unlawful act, or from culpable negligence, is involuntary manslaughter. This definition of involuntary manslaughter contains several essential elements discussed in detail in the following paragraphs.

Unlawful Act. The death must result during the commission of an unlawful act. However, if the unlawful act is a felony covered by the felony-murder rule, murder will generally be charged, as it is more serious than involuntary manslaughter. In those jurisdictions in which all homicides occurring during commission of a felony are, in some degree, classified as murder, involuntary manslaughter can only apply to misdemeanor cases.

The Nature of the Act Must Be **Malum in Se.** The unlawful act must be *malum in se*; that is, the act must be one that is wrong in itself as opposed to *malum prohibitum*, wrong merely because it is prohibited. For example, a motor vehicle operator who unintentionally strikes a pedestrian while exceeding the speed limit by five miles per hour would probably not be charged with involuntary manslaughter if the pedestrian dies. But a motorist who strikes a pedestrian while driving under the influence of intoxicants might very well be charged with involuntary manslaughter if the pedestrian dies, as driving while intoxicated is an act that is wrong in itself.

There Must Be a Proximate Causal Connection between the Unlawful Act and the Homicide. To charge involuntary manslaughter properly, it must be shown that the unlawful act was the direct or proximate cause of the homicide. For example, if a child runs out in front of a car and is unavoidably struck and killed, it cannot be said that the intoxicated condition of the driver was the cause of death when even a sober driver could not have avoided striking the child. In such a case, despite the fact that a *malum in se* unlawful act occurred, there was no proximate causal connection between the act and the homicide, so involuntary manslaughter would be an improper charge.

Culpable Negligence. Involuntary manslaughter may occur either through commission of an unlawful act or through **culpable negligence.** Simple or ordinary negligence is not sufficient to justify a charge of involuntary manslaughter. The negligence must reach such a degree of blameworthiness that a jury could say it was culpable or criminal in nature. Only then will it suffice to support a charge of involuntary manslaughter.

If a parent avoids medical treatment and under religious scruples "treats" the child only with prayer, can the parent be convicted of involuntary manslaughter if the child dies? Yes, said the California Supreme Court. The state's paramount interest in a child's welfare does not unreasonably burden freedom of religion, said the Court. Thus religiously motivated conduct is subject to regulation.

Failure to help a guest in one's home who overdoses on drugs can leave one responsible for involuntary manslaughter, said a California court. This was especially true when the guest was invited to the defendant's home for the purpose of using drugs and the defendant knowingly supplied a spoon and her bathroom for drug injection purposes. This conduct raised the duty of care, and therefore her neglect amounted to criminal negligence.

A few jurisdictions have a negligent homicide, or fourth-degree manslaughter statute, on which a conviction can be obtained for a death resulting from ordinary negligence. One may then ask the question, "If I investigate a traffic fatality in which there occurred an unlawful act but it was neither *malum in se* nor culpable negligence, what do I charge?" The answer, of course, is that you can charge for the unlawful act itself. The death will be ruled an excusable homicide. This is a common occurrence, for it is well understood that not all traffic fatalities result in someone being charged with manslaughter. When manslaughter is not charged, the officer, in effect, and possibly without knowing it, is saying, "There is not enough evidence here, or there is no indication of manslaughter." By not charging manslaughter, the officer is ruling that the incident was an excusable homicide.

Vehicular homicide is discussed in Chapter 18. However, the entire issue of drunken and reckless driving and homicides resulting therefrom has put considerable pressure on the criminal justice system to provide harsher penalties. Since 1983, prosecutors throughout the United States have begun to prosecute such conduct as an offense greater than a general vehicular homicide, and they have been proceeding under the general murder/manslaughter statutes. How have they fared? Consider the following cases.

The basic issue in these cases is: Can the malice needed for murder be inferred from a drunk driver's recklessness? Several courts have said yes to this issue. For example, the U.S. Court of Appeals for the Fourth Circuit upheld a murder conviction where the driver's conduct indicated a "depraved disregard for human life" and was enough for malice aforethought. The driver had driven at excessive speeds, crossed the median at rush hour, and scattered the oncoming cars. His blood alcohol level was measured at 0.315 percent.

The California Supreme Court came to the same conclusion. Alaska also felt that a murder conviction was warranted where a defendant ignored a friend's warning that the defendant was too drunk to drive and then drove, ultimately killing another. The court felt this manifested an extreme indifference to human life. Several other states have agreed with this position.

However, not all states and not all courts agree. A Virginia court said that a drunk driver is one who is grossly negligent only and he or she could be convicted of nothing more than manslaughter. The court felt that intoxication is relevant only as an aggravating factor that increases the degree of negligence and the severity of the sentence. A California Court of Appeals felt that its own Supreme Court was wrong in allowing a murder charge to flow from such conduct and felt it had overreacted to public outcry.

Statutory Extension of Manslaughter

Statutes throughout the jurisdictions in the United States label many types of prohibited conduct as manslaughter. Many of these offenses do not fit into the definitions of either voluntary or involuntary manslaughter as described here. Legislatures have created new crimes whose elements do not conform to the traditional categorization of manslaughter, but which are not serious enough to be punished as murder. This factor has produced a third category of manslaughter violations not known to the common law. The most common example of such legislation labels the killing of an unborn child as manslaughter.

Infanticide and Feticide

The ever-continuing debate over when an infant falls under the protection of the criminal law began early in the common law. Many arguments, moral, legal, ethical, and religious, have all contributed somewhat to the confusion. There are those that insist that, from a legal point of view, the law cannot protect an unborn child for it is not yet a human being. Others argue that an infant is a human being from the moment of conception and therefore should be protected by the criminal law from that point forward. Many others propound theories that hold that an infant becomes a human being at some point between conception and birth. In the political debates over abortion rights, "anti-abortion" activists and "pro-choice" activists use all or none or a combination of these arguments to substantiate their respective positions.

The courts generally agree that killing an infant before its mother starts giving birth is not a common law homicide. These same courts do not agree, however, as to the point at which protection does start once delivery has begun. Some contend the infant is a human being at the beginning of the delivery process. A second group holds that the child must be fully born before its death can be a common law homicide. A third group of courts holds that the child must first breathe before its death can be **infanticide.** Still others say that not until the umbilical cord is severed is the criminal law applicable. A large segment of the courts seems to follow the more widely accepted view that death at the hands of another person is infanticide only after the child has established an independent circulation, separate and apart from its mother.

Whichever of these views is followed by the courts, the required event must occur before killing of the infant can be treated as either common law murder or manslaughter depending on the punishment provided in the various jurisdictions for infanticide. This is not to say that the injury that produces death must occur after delivery. The injury may be inflicted while the child is still in the womb, but, if it is born alive and later dies, infanticide has occurred. If, on the other hand, the child was born dead, the common law courts did not recognize this as a common law homicide.

The need for clarification in this area was recognized by most legislatures. They have enacted statutes making it criminal homicide to kill an unborn child. These statutes have taken a variety of forms, but they all have one thing in common. They all have accomplished the goal of placing the unborn child within the protection of the criminal law. These statutes use different names but the crime may be commonly called **feticide,** the killing of a fetus.

Although under these statutes killing an unborn child is almost a universally punishable crime, the statutes disagree as to when in the development of the unborn child its destruction is criminal. Nonmedical legislatures have used medical terms that must be interpreted by the courts. The statutes fall into these categories: (1) killing an unborn child "in a pregnant woman," (2) killing an unborn child in a "woman with a quick child," (3) killing an unborn child in a "woman with child," and (4) killing an unborn "vitalized embryo or fetus."

Medically, there are three stages of intrauterine development. The first two weeks are known as the ovum stage. The second through eighth weeks constitute the embryonal stage. From that point to birth the unborn child is said to be in the fetal stage. The difficulty that arises with the first two types of statutes is in deciding at what point killing the unborn child is criminal— at the ovum, embryonal, or fetal stages.

Georgia, Illinois, New Jersey, and Wisconsin, which recognize feticide with regard to killing an unborn child in a "pregnant woman," disagree. Some say the unborn child must be **quickened,** that is, the motion of the unborn child must have become perceptible to the mother by kicking or other independent movement. Others indicate that any stage of pregnancy is sufficient for criminal liability to arise.

The "pregnant with a quick child" states, including Florida and Missouri, attempt to avoid the problem by using "quickness" as the turning point. As long as the definition of "quickness" is uniform, there will be no problem in knowing what that turning point will be. However, each child quickens at a different time, which, for the prosecutor, presents problems of proof. Kansas uses this standard in civil wrongful death cases.

The third state position requires a "woman with child" and by decision, merely requires proof of pregnancy at any stage. Arkansas follows this path.

Nebraska alone has adopted the fourth possibility and, by spelling out embryonal and fetal stages, has eliminated the ovum stage.

If one refers to local statutes it may be difficult to find the crime labeled feticide. States that have extended criminal law protection to the unborn child through these statutes have attached varying labels to them. In one state you may find it specifically punishable as manslaughter. In another it might be found in the abortion laws. In still another state it may be intermingled with the homicide statutes, but not assigned a specific traditional homicide label.

Three Charged with Capital Murder in Death of Fetus

By Kristen Evertt—Associated Press—September 3, 1999

Little Rock, Arkansas – A man hired three youths to kick his girlfriend in the abdomen, killing her nine-month fetus, a prosecutor said . . . in bringing charges under Arkansas' new Fetal Protection Law. . . . A young man and his two teen-age brothers were charged with capital murder and police issued an arrest warrant for the boyfriend. . . . Once he is arrested, Bullock [the boyfriend] will be charged with capital murder too. . . . The 23-year-old woman, a junior psychology major at the University of Arkansas at Little Rock, told police. . . He told her before that he did not want any children and was not going to have anything to do with this one. . . . He was afraid. He didn't want his parents to know about the pregnancy . . . [the victim] remained hospitalized. . . . She has a broken wrist and bruised face and had to have her spleen removed.

Massachusetts took a relatively bold step in this area with its decision. Cass, a drunk driver, struck the auto of an eight-and-a-half-month pregnant woman. The fetus died in the womb and Cass was charged with vehicular homicide. The court was asked to determine whether the fetus was a "person" within the meaning of that statute. Feeling that the legislature had not distinguished between born and preborn, the court held that when it can be proven that the defendant's act caused the death of a viable fetus, that death could be charged as a homicide. The court did not believe, as other courts, that this was a legislative matter. Not long after this decision the South Carolina Supreme Court joined Massachusetts in holding that a viable fetus was entitled to homicide protection and that manslaughter could be charged.

Although, in most instances, a lawful abortion of a child *in utero* (within the uterus) will not constitute homicide, does this mean that all *in utero* injuries will escape penal sanction? No, it does not. A growing number of states are beginning to recognize degrees of punishable homicides in this context. Maryland, for example, has held that criminally inflicted injuries upon a pregnant woman that result in the death of her child shortly after its live birth are punishable as homicide. The fact of a live birth was a critical factor.

But not being "born alive" was important to the Kansas Supreme Court. A viable fetus was killed due to a vehicular accident. The court felt that since the legislature did not define a viable fetus as a "human being," the defendant could not be charged with its death under the vehicular homicide law.

North Carolina said the willful killing of a viable but unborn child is not murder within the meaning of its state murder statute. The court ruled that extending the coverage of the statute required legislative action and not court interpretation.

8.3 SUICIDE

Suicide was a common law felony for which forfeiture of lands and goods could be imposed if the suicide was successful. Suicide is, strictly speaking, outside the scope of homicide because it is not death caused by another human being but it is often discussed as a related subject.

Under modern legislation, suicide has been maintained as a felony in some states, has been completely abrogated by statutes in others, and remains as a holdover from common law as a misdemeanor in a third group of states. Nobody has yet found a means of directly punishing the successful suicide, but attempted suicides are punishable. It would seem that applying the criminal law to attempted suicide would have little significance in a day when medical treatment would have much more lasting value. However, collateral problems surrounding a successful or attempted suicide do involve significant legal questions.

An often-cited situation involves the "suicide compact": two people agreeing to commit suicide. If, in the attempt, one of the participants dies and the other survives, the survivor can be charged with murder unless he or she abandoned the compact and urged the other also to abandon the compact.

What happens if, while a person is attempting suicide, he or she kills another? Is there a chargeable homicide? Yes. The degree of liability will be judged by the traditional elements of the various homicide laws including the ever-essential intent element.

8.4 ASSISTED SUICIDE

At common law and before the words **assisted suicide** were so popularly used, and before Dr. Jack Kevorkian became a household name, one who helped another commit self-murder or suicide would generally be guilty of the suicide offense as an accessory before the fact or as a principal in the second degree. The stupidity of that can only make sense if the reader remembers that suicide required forfeiture of goods and lands at common law. In states that abolished the common law, the principal and accessory distinction no longer exists. In either case, however, if one who aids the suicide does so in such a direct manner as to be the proximate cause of the death, such person could and can, of course, be charged with murder or manslaughter, as appropriate in a given situation.

The activities of Dr. Jack Kevorkian, a physician who has actively assisted people in his home state of Michigan and in other jurisdictions to commit suicide, have been a hot media topic for several years and have caused a flurry of legislation to be passed in states desiring to prohibit physician-assisted or any other kind of assisted suicide. Although 45 states have existing prohibitions about helping another to commit suicide, 35 states jumped on the bandwagon to enact statutes explicitly criminalizing assisted suicide: Alaska, Arizona, Arkansas, California, Colorado, Connecticut, Delaware, Florida, Georgia, Hawaii, Illinois, Indiana, Iowa, Kansas, Kentucky, Louisiana, Maine, Minnesota, Mississippi, Missouri, Montana, Nebraska, New Hampshire, New Jersey, New Mexico, New York, North Dakota, Oklahoma, Pennsylvania, Rhode Island, South Dakota, Tennessee, Texas, Washington, and Wisconsin.

Nine states still criminalize assisted suicide through common law: Alabama, Idaho, Maryland, Massachusetts, Michigan, Nevada, South Carolina, Vermont, and West Virginia.

North Carolina, Utah, and Wyoming have abolished the common law and have no statutory prohibition against assisted suicide. Virginia does not criminalize assisted suicide but has a statute that imposes civil sanctions on persons who assist suicide. Ohio's highest court has ruled that assisted suicide is not a crime.

Only the state of Oregon statutorily permits physician-assisted suicide.

The United States Supreme Court has addressed the issue in two civil cases presenting constitutional challenges. In both cases—one arising in New York, the other in Washington state—the petitioners were groups of physicians who would have assisted suicides if it were not prohibited by the respective state statutes. The Court found that statutes prohibiting assisted suicide do not violate either due process or equal protection under the Fourteenth Amendment.

Meanwhile, back to Dr. Jack Kevorkian, who was tried three times for assisted suicide in the state of Michigan, acquitted each time, and had two murder charges dismissed. But see the accompanying box on page 149–150.

The "60 Minutes" Tape

On November 22, 1998, Dr. Jack Kevorkian appeared on the TV show "60 Minutes." He brought with him videotape, which appeared to show him administering a lethal injection to Thomas Youk, a 52-year-old man who suffered from advanced amyotrophic lateral sclerosis, or Lou Gehrig's disease. The 70-year-old physician, who boasted that he had assisted in the suicides of more than 130 people with debilitating

---

The "60 Minutes" Tape—*continued*

diseases since 1990, indicated he participated in the euthanasia and challenged Michigan authorities to charge him. They did. He was charged with first degree murder, which carries a mandatory life penalty; assisted suicide, with a maximum penalty of five years' imprisonment; and a controlled substance charge with a seven-year maximum penalty. The videotape showed Youk consenting, but the prosecutor pointed out that consent is not a defense to homicide in the state of Michigan (see Chapter 6). Kevorkian turned himself in to police after it was announced that the state would prosecute.

Dr. Kevorkian represented himself in his trial. The suicide charge was dropped, but he was convicted of second-degree murder and on the controlled substance violation.

DISCUSSION QUESTIONS

1. Dan was the proprietor of a small retail grocery store. One day while Dan was alone in the store, John, a stranger to Dan, entered the store, suddenly drew a revolver, pointed it at Dan, and ordered him to the rear of the store to open the safe. Dan, believing John's revolver to be loaded, made as if to obey. After taking several paces, Dan whirled suddenly, drew a revolver from under his coat, and fired twice in rapid succession at John. One of the bullets killed John and the other, missing John, killed Xavier, a passerby who was lawfully proceeding along the street in front of the store. In fact, Dan was carrying his revolver in violation of a state statute making it a misdemeanor for any person to carry a concealed weapon without a permit. A subsequent examination of John's revolver revealed it was unloaded. On the foregoing facts, Dan is separately indicted for the murder of John and the murder of Xavier. Is Dan guilty of the crimes charged, or of either of them? Why?

2. Gamblers were threatening Don about debts he owed. Don decided to resort to robbery to get money to pay the gamblers. He concealed himself at about 11:30 P.M. in a dark alley and sprang on Alan, manager of a theater, on his way to a night bank depository with the evening receipts. Don struck Alan on the head with a section of lead pipe, seized the bag containing the money, and fled. Alan died about two hours later as a result of the attack. Don arrived home about 30 minutes after committing the robbery. He heard a man's voice in the bedroom he shared with his wife. Don peeped through the keyhole and saw Fred, partially disrobed, in conversation with Don's wife. Don waited until Fred left and then discussed the situation with his wife. A bitter argument ensued. About three hours later, Don, armed with a gun, proceeded to Fred's home. Don shot and killed Fred without saying a word when Fred opened the door. What liability, if any, does Don have for the deaths of Alan and Fred? Why?

GLOSSARY

Aforethought – in murder, conveys the meaning of planning ahead.

Assisted suicide – an offense in most states involving assisting another to commit suicide

Culpable neglige1.ce – negligence with a breach of a standard of care so great as to rise to the level of being a crime.

Excusable homicide – a homicide committed by accident and misfortune or during the commission of an act *mala prohibita.*

Felonious homicide – the killing of one human being by another under circumstances that would make it criminal.

Felony-murder rule – malice will be implied to make the offense a capital murder if committed during the commission of a specified felony.

Heat of passion – a condition caused by sufficient provocation that leads to manslaughter.

Homicide – the killing of one human being by another.

Infanticide and feticide – the killing of a newly born or an unborn child.

Involuntary manslaughter – an unintentional killing during the commission of an act *malum in se* or from culpable negligence.

Justifiable homicide – necessary killing in the performance of a legal duty or exercise of a legal right where the slayer is not at fault.

Malice aforethought – premeditated design; planned ahead with an evil purpose in mind.

Manslaughter – a criminal homicide not severe enough to be murder.

Murder – the felonious killing of one human being by another with malice aforethought.

Provocation – an element of voluntary manslaughter based on legally sufficient grounds to excite and anger a person enough that the person kills in the heat of passion.

Quickened – when movement or other independent action of an unborn child is felt by the mother.

Voluntary manslaughter – a killing, less than murder, done in the heat of passion.

Year and a day rule – at common law, if death from a criminal homicide did not occur within this time period, the offender could not be held liable for the death.

REFERENCE CASES, STATUTES, AND WEB SITES

CASES

Washington et al. v. Harold Glucksberg, 521 U.S. 702, 117 S.Ct. 2258, 138 L. Ed. 2d 772 (1997).

Vacco v. Quill, 521 U.S. 793, 117 S.Ct. 2293, 138 L.Ed.2d 772 (1997).

Commonwealth v. Almeida, 362 Pa. 596, 68 A.2d 595 (Penn. 1949).

Commonwealth v. Thomas, 382 Pa. 639, 117A.2d 204 (Penn. 1955).

Commonwealth v. Redline, 391 Pa.486, 137 A.2d 472 (Penn. 1958).

Commonwealth v. Bolish, 381 Pa. 500, 113 A.2d 464 (Penn. 1955).

Commonwealth v. Lewis, 409 N.E.2d 771 (MA 1980).

Iowa v. Ruesga, 619 N.W.2d 377 (2000).

Tennessee v. Randolph, 676 S.W.2d 943 (1984).

Berry v. Santa Clara County, 208 Cal. App.3d 783 (1989).

People v. Pavlic, 199 N.W. 373 (MI 1924).

Enmund v. Florida, 458 U.S. 782 (1982).

Hornbeck v. Florida, 77 So.2d 876 (1955).

Kansas v. Hupp, 248 Kan. 644 (1991).

People v. Smith, 678 P.2d 886 (CA 1984).

Illinois v. McCarthy, 547 N.E.2d 459 (1989).

Walker v. Sacramento County, 763 P.2d 852 (CA 1988).

People v. Oliver, 210 Cal. App.3d 138 (1989).

USA v. Tan, 254F.3d1204 (10th Cir. CO 2001).

People v. Whitfield, 868 P.2d 272 (CA 1994).

Pusich v. Alaska, 907 P.2d 29 (1995).

Essex v. Commonwealth, 322 S.E.2d 216 (VA 1984).

People v. Watson, 637 P.2d 279 (CA 1981).

Commonwealth v. Cass, 467 N.E.2d 1324 (MA 1984).

State v. Horne, 319 S.E.2d 703 (SC 1984).

Williams v. State, 561 A.2d 216 (MD 1989).

State v. Trudell, 755 P.2d 511 (KS 1988).

State v. Beale, 376 S.E.2d 1 (NC 1989).

STATUTES

California: Ca. Penal Code Ann. §§.189, 190.2(a)(17) (West Supp. 1982)

Florida: Florida Statutes 782.04(1)(a), 775.082(1),921.141(5)(d)(1981)

Georgia: Ga. Code §§.26-1101(b),(c); 27-2534.1(b)(2) (1978)

Mississippi: Miss. Code Ann. §§.97-3-19(2)(e); 99-19-101(5)(d) (Supp. 1981)

Nevada: Nev. Rev. Stat. §§. 200.030 (1)(b); 200.030(4); 200.033(4) (1981)

South Carolina: S.C. Code §§. 16-3-10; 16-3-20 (C)(a)(1) (1976 and Supp. 1981)

Tennessee:Tenn. Code Ann. §§. 39-2402(a); 39-2404(I)(7) (Supp.1981)

Wyoming:Wyo. Stat. §§ 6-4-101; 6-4-102(h)(iv) (1977)

WEB SITE

www.findlaw.com

CHAPTER 9

Sex Offenses and Offenses to the Family Relationship

▓ KEY WORDS AND PHRASES ▓

Adultery
Bawdy house
Bigamy
Carnal knowledge
Fabrication exception
Forcible rape
Fornication
Incest
Obscenity
Pandering or pimping

Polygamy
Rape shield law
Rape trauma syndrome
Rebuttable presumption
Roe v. Wade
Seduction
Viability
White Slave Traffic Act (Mann Act)

9.0   INTRODUCTION

One need only glance through a daily newspaper to see how much sex is in the limelight in contemporary societies throughout the world. Nevertheless, certain segments of many societies still hold the view that sex is not a proper subject for public display or discussion. It is not our purpose to debate the pros and cons of such arguments, nor is it our function to comment on the effect that the changing attitudes of society will ultimately have on the laws regarding sex offenses and related areas. Suffice it to say that the law must change to conform to the desires of society. Any major laws that are contrary to the desires of society will not, as our experience shows, exist for long. This point was classically illustrated in the early part of this century with prohibition of liquor.

In the public spotlight today are the laws concerning narcotics and dangerous drugs. A movement is afoot to legalize the use of certain previously prohibited drugs. Whether such steps should or should not be taken is a subject we will not discuss now, but this trend illustrates the effect of society's attitudes on the laws of the day.

Closer to the subject matter in this chapter is the ever-growing debate over liberalization of abortion laws. The past and present status of abortion laws will be fully discussed later in this chapter. What will happen in the future depends on the ability of the various factions in society to convince the lawmakers of the rightness of their separate views on the issue.

There is one further controversial issue in the forefront of today's news. Are the X-rated movie films, so prevalent in this country, affecting the sex crime statistics in any way? To our knowledge, no statistical proof has been presented to show either an increase or a decrease in sex crimes as a result of the more liberal attitude of society toward films such as these. Statistics from Sweden, where pornography laws have been liberalized or eliminated except in the case of juveniles, show some variations in the figures regarding sex offenses, but officials caution against assuming any direct correlation between these statistics and the laws. The status of pornography laws will also be discussed in this chapter.

9.1 FORCIBLE RAPE

Forcible rape is defined as having unlawful carnal knowledge of a female by force and against her will. In the following paragraphs we discuss each of the elements involved. This crime was recognized at common law.

Unlawful

The unlawfulness of the carnal knowledge requires no great amount of legal discussion. *Unlawful* implies that intercourse takes place without the consent of the victim and against her will. The significance of the word *unlawful* as used in defining this crime, however, does raise one interesting question. Can a man rape his wife? Sexual intercourse is an integral and essential part of the marital relationship. Such conduct is not only expected but is wholeheartedly sanctioned by all segments of society. As a consequence, it was until recently a universally accepted rule that a man cannot rape his wife even though he may, on occasion, need to resort to physical force to accomplish his purpose. Subject to this exception, any female may be the victim of forcible rape. In connection with this point, we have often heard the comment that a prostitute cannot be raped. This statement, of course, is completely without merit. If the elements of the crime of rape are present in a particular case, it makes no difference what the victim's background happens to be.

Since 1976, legislative and decisional activity have raised questions about the soundness of the rule that a man cannot rape his wife. A much publicized case in Oregon kept the nation's attention as the state prosecuted a husband for the rape of his wife.

The most significant case, however, was *State v. Smith* because of the scholarly manner in which New Jersey approached the matter. The New Jersey court made several points. The early basis for finding that a man could not rape his wife was the contract of marriage. The wife gave up refusal rights for the protection that the marriage offered.

But the court recognized that rape is an act of violence, while sexual communion is an act of love and respect. As a result, rape tends to leave emotional scars, the court noted. Coupling this with an enlightened age that today should recognize the right of women to govern their own sexual desire, the court found that the contract theory is too mechanically postured against modern realities.

This led the court to find as follows:

> It is small comfort to a married woman whose husband has forcibly ravished her against her will to know that she may resort to the matrimonial courts to recapture or retrieve her right to sexual privacy. If she chooses not to seek such formal (and sometimes, formidable) judicial relief, how can one logically defend the result—that a husband has an unbridled right, protected by law, to force himself sexually upon her at any time he chooses, no matter how far the marriage relationship has deteriorated between them? Truly, society should not today suffer such a situation to continue upon so callous a basis as applications of contract law and the doctrine of consent. It has always been recognized that "marriage contracts cannot be placed on a par with ordinary contractual obligations" because "(I)n every marriage contract, the state is an interested party. . . ."

The court then went on to say that current rape laws discriminate against a married woman. However, the court asked, is it the job of the court to rewrite the law? The answer to their own question was that such changes will have to come from the legislature. They held:

Thus viewed, it is more properly a legislative, rather than a judicial function, to determine or redetermine the type of conduct which will constitute the substantive crime of rape, especially when, as here, serious societal objectives, philosophical evaluations and moral judgments are involved. . . .

If there is to be change, this court says, that change must come from the legislatures and not the courts. What have the state lawmakers done?

Kentucky, Maryland, Montana, and South Carolina have modified the earlier rule by saying that if the parties are separated by judicial decree, then the husband loses his exemption; the marital consent no longer exists. In Minnesota, the consent disappears if the parties are living apart and one has filed for divorce or separate maintenance.

In New Mexico, the exemption disappears if the husband and wife are living apart or if one has filed for divorce.

Arkansas, Georgia, Hawaii, and Iowa allow the charge to be prosecuted when the victim did not consent to the act upon which the specific charge is based.

In summation, the presumption of consent that goes with marriage is no longer indulged where the parties are living apart under a formal separation agreement or where formal divorce proceedings have begun. However, states are split where the parties are living apart, but no agreement has been entered into and no divorce action has been filed. Those with a sexual assault statute avoid the rape issue by focusing on and severely punishing the assault. Those without such a statute will either continue the presumption of consent and charge only a standard assault or will come to grips with the facts of a *de facto* dissolution of the marriage and find that, when the parties have lived apart for a significant period of time, there is no longer a marriage upon which to base presumed consent. This revocation must be demonstrated by a manifest intent to terminate the marital relationship. Each case stands on its own facts.

A man may be found guilty of the crime of rape. As was pointed out in Chapter 4, aiding and abetting another to commit a crime makes one liable as a principal in the second degree. This rule applies to the husband and wife situation. If a man aids and abets in the rape of his wife, he is liable for rape as a principal in the second degree or simply as a principal in states that have abolished the degrees of principals.

Carnal Knowledge

Carnal knowledge is a legal term for sexual intercourse. In a rape case, it must be shown that there has been sexual intercourse. The law defines sexual intercourse as penetration of the vagina by the penis. Consequently, this requirement can be satisfied if the slightest penetration can be shown. It is not essential for the completion of this crime that there be any emission of sperm.

Of a Female

There is little need for discussion on this point except to raise a few questions. Can a man be the victim of a forcible rape? Many cases have arisen in which a man has been sexually assaulted. To our knowledge there is no case on record in which such facts have supported a forcible rape charge. All the elements of the crime clearly indicate that it would be inconsistent to call such acts forcible rapes.

Nevertheless, it should be remembered that, even though a female cannot actually rape a man, a female can be just as easily convicted of rape as can a male. If a female acts either as an accessory before the fact or as a principal in the second degree, her liability is no less than that of the man she procured or aided in committing the rape (Chapter 4).

An interesting question in this area of the law is raised by the modern medical phenomenon of sex change. Will a person who has had a sex alteration from male to female be considered a female for purposes of the law of forcible rape? We make no attempt even to guess at the possible answer to this perplexing question. To our knowledge the case has never yet come before the courts. Where, however, as in Louisiana, the state permits a change in the original birth certificate designation from male to female, it could be reasoned that such official recognition of gender is tantamount to total acknowledgment of the person as a female.

Finally, to be a female one must be alive. Odd as it may seem, this concept is not always clearly understood. In 1989 a California court had to bring this point home to the prosecution. Rape can only be committed against a living person, because rape is a crime against the will of a person and only a live person has a will. The California court said a dead body has no feelings of outrage. The court thus rejected postmortem rape as a transaction for which there could be a conviction.

By Force

The law requires that force be used to establish the crime of forcible rape. How much and what type of force will satisfy this requirement is discussed in a number of court opinions. Today, it is rarely held that actual physical violence must be shown. Of course, if there is physical violence, the element is satisfied. Most modern courts also recognize threats of force or violence as a sufficient alternative. If a woman were threatened to the point where she justifiably feared immediate death or great personal injury to herself or to an immediate member of her family who was present, the requirement would be met. Notice that the threat has to involve immediate personal injury rather than future injury. Threat to property will not supply this element.

No discussion of force can be complete without an inquiry into the amount of resistance the victim must offer. If the victim offers no resistance, it cannot properly be said that intercourse was forced. Formerly, the majority rule was that the victim must have resisted to the utmost of her ability. Today, this view has become fairly obsolete, and the majority of courts now require only that the victim resist as much as her age, strength, and all the other circumstances surrounding the event allow. If the victim has a knife at her throat and death is threatened, there is little point in physically resisting. If the victim does resist, as required by the law, and is still overcome, there is no further requirement that she continue to resist. Once there has been slightest penetration, the crime is complete.

Against Her Will

The element requiring that carnal knowledge occur against the will of the victim also conveys the idea that it be without her consent. This may seem like a distinction without a difference, but we will see that there are situations in which these two phrases may, if construed literally, convey separate and inconsistent meanings. There is some overlapping between this element and the one discussed previously. If, in fact, physical violence does accompany the attack, lack of consent will be implied.

As a general rule, lack of consent will be implied in forcible rape cases. If, however, the defendant raises consent as a defense to the charge, the state must prove that there was no consent and that intercourse was against the will of the victim. This is true because lack of consent, being an essential element of the crime, is a part of the burden of proof that the state must carry in order to convict. Therefore, let us reexamine the elements of the defense of consent (Chapter 6) and determine how they apply to the law of forcible rape.

First, rape is the type of crime for which consent can be given. Second, the victim is giving consent for herself and not for a third party, so she would have the authority. The last two elements, however, present some problems. Even though the victim has the authority to give consent, she may not be legally capable of giving consent. This lack of capacity may arise either as a matter of law or as a matter of fact. Let us examine a few problems in this area. The most obvious situation that comes to mind concerns the young female. In all states, a child under a certain age is, by law, treated as incapable of giving consent. Thus her "consent," if she did give it, would not be recognized and would not be a valid defense. Because of the nature and frequency of such occurrences, state legislatures have enacted separate statutes dealing with these cases. The penalties for this offense are usually less severe than for forcible rape because the victim was, in fact, a willing partner. If force is used to have intercourse with a child under the statutory age, however, forcible rape is a proper charge.

There is an exception to this exception. A child under the age of ten years was, and still is, considered totally incapable of consenting or even being a willing partner to sexual intercourse. Consequently, intercourse with a female child under 10 years is forcible rape, regardless of any surrounding circumstances.

Capacity to give consent also brings up the question of intercourse with an unconscious woman. It is here that the semantic difference between the phrases "against her will" and "without her consent" could be disastrous were it not for the interchangeable interpretations given these phrases by the courts. It is difficult to say that intercourse with an unconscious woman is against her will for, in fact, her will is not being exercised at the time. However, there is little difficulty in holding that the act would be without her consent if she is unconscious. This is the interpretation given in cases such as this, and courts uniformly hold that this constitutes forcible rape. The problem is similar when the victim, although conscious, is so drunk that she does not understand the nature of the act to which she is "consenting." Here, again, courts hold that, if the victim is so drunk, a forcible rape charge will stand. The consequences are the same when the victim has been given the so-called date rape drug.

A similar question is raised when a man has intercourse with a female idiot. Many legislatures have enacted statutes specifically covering this situation. If the victim's mentality is so low that agreement to the act of intercourse is really beyond her ability to comprehend, her agreement will not be recognized as legal consent.

The last problem with consent involves consent obtained by fraud. Fraud in consent may take either of two forms, only one of which will invalidate the defense. If the fraud concerns the very nature of the act as where the victim consents to have an operation and the doctor has intercourse with her while she is anesthetized, there is no consent. The victim did not consent to intercourse. On the other hand, if the victim agrees to intercourse, knowing full well the nature of the act, but is fraudulently induced to consent for whatever reason, her consent is valid and will be a defense to a charge of forcible rape.

As one can see, the area of consent is laced with a myriad of rules. Unfortunately, and for too many years, victims of rape, unless severely bloodied and beaten, were not given sympathy or credence when reporting their plight. Jury attitudes were often no better. Other than child abuse and incest, rape was the leading unreported crime. Over the past few years police departments through training have learned how better to deal with these problems. The public on the other hand has not made as much progress and it is from them that the jury is selected. Prosecutors have needed a new tool with which to work when meeting a consent defense.

That new tool has come from the psychiatric profession. It has been labeled the **rape trauma syndrome,** gaining in acceptance within the psychiatric community and the few states that have ruled such expert opinion testimony is admissible. The thrust of the syndrome is that a woman who has been raped will demonstrate certain unusual behavior patterns—a series of posttraumatic stress disorders. The symptoms are fear of offender retaliation, fear of being raped again, fear of being home alone, fear of men in general, fear of being out alone, sleep disturbance, change in eating habits, and a sense of shame. When these are actually present, it is not hard to understand why victims delay reporting, are reluctant to identify, and are reluctant to testify.

Considerable issues concerning the use of the rape trauma syndrome have arisen. Courts have had to decide whether and under what circumstances the expert will be allowed to testify and whether either side can introduce such testimony.

Many states take the position that the expert cannot be used for rape trauma syndrome when its purpose is to enhance the credibility of the victim such as to explain why the victim could not identify the attacker. Courts feel this would invade the province of the jury. Some courts, like New York, do not agree, however, and say that an expert can help where some forms of behavior may not be commonly understood by the average juror; New York feels that failure to identify is one of those areas.

The key to the state's use of rape trauma syndrome is the centrality of the issue of consent. Many states are coming to the conclusion, as did West Virginia, that such evidence is relevant and admissible. Most states, however, limit the testimony and will not, as West Virginia held, allow the expert to say whether or not the specific victim in the case at hand was raped.

Rape trauma syndrome is not a government exclusive. The defense may use it also. Indiana had such a case and said that even a defendant can introduce evidence of the syndrome to show that postrape behavior would not include the kind of conduct the victim engaged in on the days following the alleged rape. In this case the victim went back to the bar from which she said she was abducted and was seen dancing and drinking the very next day.

At common law, a male under the age of 14 was held incapable of committing forcible rape. This followed the arbitrary common law classification of infants and their capacity to form the necessary mental element to commit crimes. A child under the age of 7 years was conclusively presumed incapable of criminal intent and a child between 7 and 14 years was rebuttably presumed incapable. In the case of rape, this **rebuttable presumption** was made conclusive. Some modern legislatures have recognized the impracticality of this common law rule and have either modified it so as to make it a rebuttable presumption again or eliminated the rule completely from their statutes.

If all the elements of forcible rape are present, the victim's subsequent forgiveness of the act will not bar prosecution for the crime. Upon satisfaction of all the elements, the crime is complete. Similarly, a subsequent marriage between the accused and the victim is no defense. Admittedly, there are some practical problems created by either of these situations, especially if the victim refuses to testify, but legally such circumstances do not bar prosecution.

The modern trend is to view rape not so much as a sex crime as one of violence. This trend comes as a result of the emphasis placed on the violent nature of the act as seen in Model Penal Code Section 213.1. Sexual intercourse still has to be proven, but the Code clearly defines the conduct which constitutes rape. Threat of imminent death, use of force that causes serious bodily injury or extreme pain or kidnapping are those violent acts which will not be condoned. The Code also prohibits the use of drugs or alcohol without the victim's knowledge to secure intercourse.

The Model Penal Code also creates the felony of gross sexual imposition. In this crime the sexual intercourse comes about because of threats that would prevent resistance of any ordinary woman, or because the woman suffers from a mental disease or defect that renders her incapable of resistance.

Taking their lead from this recommended change, several states have enacted statutes emphasizing the assault aspects of the crime. In a like vein, several states either absolutely prohibit the use of prior sexual behavior of the victim as evidence or will not allow it unless the prior conduct was with the alleged attacker.

These character-limited statutes are called **rape shield laws.** Although often strictly enforced when first enacted, there have been interpretations easing the impact of such statutes. For example, if the victim takes the stand and lies about previous chastity, courts, like those in Illinois, hold that evidence of previous sexual conduct may come in despite a rape shield law. In this case the victim lied about prior sexuality. The court said the victim made this an issue, thus making her past sexual conduct relevant to the issue of her credibility.

Michigan held the rape shield law should not be used to keep a defendant from introducing evidence that his alleged victim was and is a prostitute. The court felt that in this case the evidence of prostitution went to the base issue of consent and whether the defendant and victim entered into a financial arrangement.

A rapidly developing phase of this area concerns the creation of a **fabrication exception.** Several states, including Arizona, Kansas, Nevada, Pennsylvania, and Virginia, recognize such an exception. These states say that a defendant should be able to introduce evidence of the fact that the victim has fabricated similar charges against others. In the Virginia case the victim had twice falsely claimed she was pregnant. Virginia reasoned that such statements are not conduct nor evidence of previous unchaste character and thus not really covered by the rape shield law.

9.2   SEXUAL ABUSE, SEXUAL ASSAULT, SEXUAL BATTERY

In an effort to ensure punishment for sexual conduct that includes most but not all the elements of rape, (i.e. it is done without force or violence), legislatures have statutorily created a class of offenses alternately called *sexual abuse, sexual assault,* or *sexual battery* depending on the jurisdiction and the content of the statutes. These were not common law offenses but were created by legislation to overcome some of the problems associated with common law forcible rape. Only males could be offenders in statutory rape and only females could be victims. In addition, rape required carnal knowledge, but there is other sexual contact.

If the statutes punish a person who touches another in a sexual way even if there is no penetration, the statute may be called sexual assault or sexual abuse. More severe acts in which penetration occurs may be called sexual battery in the statutes. The Florida statute by way of definition, for example, provides,

> Sexual battery means oral, anal or vaginal penetration by, or union with, the sexual organ of another or the anal or vaginal penetration of another by any other object; however, sexual battery does not include an act done for a bona fide medical purpose.

Besides creating a class of violations that are much easier to prove because the element of force or violence is no longer applicable, the penalties, although generally severe, are not as severe as for forcible rape, which is often a capital offense or would result in a life term of imprisonment in states that have abolished capital punishment. At the same time, the penalties are much more severe than would be the case if the convictions fell under the assault and battery statutes of most states. The harsher penalties are much easier to justify by calling these violations sex offenses rather than assault and battery offenses.

9.3 STATUTORY RAPE

The reader will note that throughout Section 9.1, the crime of rape was always referred to as forcible rape. This was done deliberately to avoid using a broad term to cover all types of rapes. *Statutory rape* is a label given to a type of crime that was unknown at common law and that was created by statute. However, even at common law, a female under the age of ten years was considered incapable of consenting to sexual intercourse. This was true even if she knew the nature of the act. Her consent in such a case was no defense to a charge of forcible rape. The key difference between forcible and statutory rape is that, in the latter, force and consent are immaterial. At common law, it was not rape to have intercourse with a female over ten years of age if she consented. Legislatures took the view, however, that a child under a certain age limit was not mature enough to understand the full implications of consenting to sexual intercourse. As a result, the statutory rape statutes were enacted. Under these statutes the adult partner is held liable for the underage partner's lack of responsibility.

An interesting result of the women's rights movement is the reexamination of all laws in which there is a sexual bias. Sex crimes, particularly rape of an underage female, are no exception. Male defendants have challenged the constitutionality of such statutes because there can be (in most states) only a female victim and only a male perpetrator (as a true principal in the first degree).

The U.S. Court of Appeals for the First Circuit held the New Hampshire statutory rape law unconstitutional because it punished only male perpetrators and protected only female victims. They said it was a violation of the equal protection clause of the U.S. Constitution.

The Maine Supreme Court, fully aware of the New Hampshire decision, had a very similar statute and arrived at a different conclusion. Citing the theory of "an important governmental objective" and "substantially related to achieving that objective," the Maine court upheld its own law. The court said there was strong evidence that because females could get pregnant or have their sexual organs injured due to penetration, this was a proper objective of such discrimination. Thus Maine joined Arizona, California, Colorado, Illinois, Iowa, Maryland, Texas, West Virginia, and Wisconsin in upholding such statutes.

As mentioned in Section 9.1, consent by the underage partner is not a valid defense to the crime of statutory rape. Similarly, absence of force in these cases is immaterial. The mere doing of the act is all that is necessary to convict. However, if the victim does not consent to intercourse and if force or the threat of force is used, all the elements of forcible rape have been satisfied and the age of the victim is immaterial. The prosecution may very well choose to prosecute for forcible rape, which carries a much more severe penalty than statutory rape.

Some jurisdictions may not recognize statutory rape by that name. There is some misconception fostered by this name because the elimination of force and consent really takes this offense out of the rape classification. But however the equivalent statute reads, it is still designed for the purposes set forth in the preceding paragraphs.

Some further qualifying remarks are needed to cover this offense fully. Many of these statutes require that a victim of statutory rape be unmarried. Further, many states require that the victim be of previous chaste character. This requirement is designed to prevent fraud on the courts by young persons of promiscuous habits. In a statutory rape case, if it can be shown that the victim was of previous unchaste character, the statutory rape charge will not stand. If, however, it can be shown that, whenever the victim has had intercourse it has been with the defendant, the courts will not allow his prior misdeeds to bar his conviction.

A defense often raised in statutory rape cases is mistake of fact. The defendant claims that his underage female partner looked older than her actual age, misrepresented her actual age (sometimes to the point of producing falsified identification), and was physically endowed beyond the norm for her actual age. Because statutory rape is a crime requiring no intent, there is nothing for the defense of mistake of fact to operate on, and these attempts at justification will usually be to no avail. The defendant is held to act at his peril. Because most courts have some concerns about holding people absolutely liable for strict liability crimes, some states, like Idaho, agree that a mistake of fact as to age can be used as a mitigating factor in sentencing.

California agrees and feels that since the crimes of adultery and fornication have been repealed, there is no underlying criminal conduct that supports the denial of the mistake-of-fact defense for this crime. The courts of that state were concerned with strict liability being imposed for such a seriously punishable offense.

Massachusetts, on the other hand, finds that whether or not adultery or fornication are punishable, they are immoral acts and therefore, because they were punishable at common law, there is sufficient *mens rea* (general evil state of mind) to go along with the act and thus to justify the denial of the defense.

The reasonable mistake-of-age defense is not, however, a dead issue. Utah, for example, says that since the crime is not a strict liability offense, the critical issue is what mental state must exist as to the victim's age. The Utah court further declared that there must be proof that the defendant was at least criminally negligent as to the age of the partner. This position is unique to that single jurisdiction. A claim that the defendant is going to or has married the victim after committing the offense is not a defense.

9.4 SEDUCTION

Seduction was not a recognizable common law offense but has developed by statute. Definitions and elements of this offense vary from jurisdiction to jurisdiction, but several basic requirements are common. The gist of the offense is enticing or luring an unmarried female of previous chaste character to engage in sexual intercourse by fraudulently promising to marry her or by some other false promise. Reliance must be placed on the promise as the reason for submitting to intercourse. If the female consents solely for the purpose of engaging in the sexual activity without relying on the promise, the crime is not complete. Many statutes that make seduction a criminal offense provide that if, in fact, the promise is kept—when the parties subsequently marry, for example—the crime may not be charged.

9.5 FORNICATION AND ADULTERY

Although fornication and adultery are found in the statute books of most jurisdictions, these laws are rarely enforced. The state of Florida, for example, has prosecuted only one case in the last hundred years for violation of the adultery statute. Perhaps the reason for this lack of enforcement is the fact that normally these violations occur under conditions not generally public (despite what happens in the White House) in the sense that they are injurious to the people of the state. Notwithstanding the enforcement practicalities, both offenses are criminal, if declared so by statute, and deserve mention at least by way of definition. Yet even the definitions of these offenses are not settled, and they may vary from state to state.

Fornication is sexual intercourse between unmarried persons. **Adultery** is generally sexual intercourse between a male and female, at least one of whom is married to someone else. In the definition of adultery, there is considerable variation. Some states require both parties to be married to other people before the crime is satisfied. Other states direct that only one of the parties need be married but disagree as to which one. In many of the latter jurisdictions, the law may still hold the unmarried partner liable for fornication. There is no universally accepted rule as to what constitutes these crimes, and local statutes must be consulted.

9.6 INCEST

Incest is a crime in all states. It involves sexual intercourse between persons who are related to each other by blood. The closeness of the relationship varies from state to state and is spelled out in the statutes. Sexual relationships between father and daughter, or mother and son, or brother and sister are uniformly prohibited throughout this country. Sexual intercourse is an essential element of this crime. In some states, marriage between close blood relatives is included in the criminal statute if accompanied by proof of sexual intercourse. Marriage alone between close blood relatives will, of course, be grounds for divorce in a civil action but will not, of itself, necessarily constitute the crime of incest.

With the rise in concern about child abuse has come the unfortunate revelation that much child abuse involves incest. Now that those who suffered directly or indirectly (as with the mother who feared her husband) have protective shelters to which they can go, an increasing number of incest prosecutions have been made.

The trouble with some incest statutes is that the statute may prevent only incestuous marriages and does not necessarily speak to incestuous relationships where no marriage is involved. Some states have sought to remedy this problem. Delaware, Georgia, Indiana, Iowa, Kentucky, Minnesota, North and South Carolina, Texas, Utah, West Virginia, and Wyoming seek to prevent sex between certain relatives.

The dual function statutes that prohibit marriage or sex are found in Alabama, Arizona, Arkansas, California, District of Columbia, Florida, Hawaii, Idaho, Louisiana, Maryland, Massachusetts, Mississippi, Missouri, Montana, Nebraska, Nevada, New Mexico, New York, North Dakota, Oklahoma, Pennsylvania, South Dakota, Virginia, and Wisconsin. This is the position taken by the Model Penal Code.

Some states, perhaps recognizing that their incest statute is too broad, make it a more serious crime to marry or have sex with certain close relatives. Those states are Colorado and Kansas.

Maine punishes incest only if one of the participants is 18 years of age or older and then punishes only the person or persons who are aged 18 or older. Michigan punishes when the victim is under 13 years of age and is a member of the same household or is a person who is under the authority of the perpetrator. Thus a foster child or stepchild would be protected.

Oregon makes such intercourse a rape if it is with a female under the age of 16 and is the male's sister, or half-sister, or daughter, or his wife's daughter. Ohio and Michigan did away with the crime of incestuous marriage, but both states punish incestuous relationships with minor children by prohibiting sex with the child, adoptive child, or child who is the ward or responsibility of the perpetrator.

Local statutes should be consulted to determine if the incest statute applies to any sexual conduct, intercourse only (allowing the other sex crimes to cover the fondling–sodomy crimes), or nonintercourse conduct (allowing the general rape statute to cover the intercourse crimes).

The only problem with allowing the rape statute to cover intercourse is that, if the child is a male, it is not always a crime or if female, the child is over the statutory rape age and force may not be present, as when the child does it to please "Daddy."

9.7 ABORTION

Abortion was a recognized common law offense classified as a misdemeanor. It was defined as causing the miscarriage of a woman quick with child, with or without her consent, when it was not necessary to do so to preserve the life of the mother. The common law did not protect an unborn child. As to the death of the fetus, no crime was committed. Legislatures have changed this situation, but what of the death of the mother during an abortion? The common law courts held that, if the mother died during commission of an abortion, it was criminal homicide. The degree of the homicide depended on the manner in which the abortion was performed. If the abortion was performed by means likely to inflict serious injury or death, the crime was murder if the mother died. On the other hand, if done in a manner not likely to cause death or serious injury to the mother, an abortion that did cause the death of the mother would result only in a charge of manslaughter.

The common law definition of abortion required that the woman be quick with child. Quickness is defined as the stage in development when the fetus first starts to move in the mother's womb. This occurs around the middle of the term of pregnancy—four and one-half months after conception.

Despite legislation concerning feticide (Chapter 8), the homicide laws still left a gap in the criminal law when a miscarriage occurred before the fetus had quickened. It was basically for this reason that the abortion laws were established. In many jurisdictions, the scope of abortion laws was extended to include the aborting of any pregnant woman, eliminating the necessity that the fetus be developed to the quickened stage. In still other jurisdictions the crime was complete if the accused committed some overt act toward abortion, such as using an instrument or administering a drug intending to commit an abortion. In these states, the statute did not require that the victim even be pregnant. Thus criminal intent became the crucial factor in determining whether the crime had been committed. Usually, in these states the act must have gone further than what was required to charge an attempt.

Everything changed in 1973 when the U.S. Supreme Court rendered two opinions regarding the constitutionality of abortion statutes. In the first, *Roe v. Wade,* an unmarried, pregnant woman challenged the Texas statute that allowed abortions only when the life of the mother was threatened. The Court said that the state has a legitimate interest in seeing to it that safe and proper medical procedures are followed. But the Court went on to say that the right of personal privacy includes the abortion decision. This right is not unqualified, however, and must be considered against important state interests. The state's important and legitimate interest in the health of the mother, the "compelling" point in the light of present medical knowledge, is at approximately the end of the first trimester. Prior to this the physician in consultation with the patient is free to determine, without regulation by the state, that the pregnancy should be terminated. After the first trimester, or 90 days, the state may regulate the abortion procedure. After viability, the Court concluded the state may promote the potentiality of human life and prevent abortion except where necessary to preserve the life or health of the mother. The courts define **viability** as the capability of the fetus to have a meaningful life outside the mother's womb, and after viability, the state's interest has both logical and biological justification.

The second case challenged the Georgia statute that provided for a committee to approve all abortions. The Court said this was not constitutionally justifiable. Similarly, the Court declared the practice of requiring the patient to be a resident of Georgia invalid. Thus the Court paved the way for the liberalization of abortion practices. However, the state can still require some medical expertise by the person performing the abortion.

In 1972 the American Bar Association gave its approval to the Uniform Abortion Act. Under this act the decision is left to the doctor and the patient through the first 19 weeks. After that, the abortion must be justified on the well-being of the mother. If her life would be endangered, or if her health or mental condition would be gravely impaired, or if the child would be born with a grave physical or mental defect, or if the pregnancy were the result of rape or illicit sex, an abortion could be performed after 20 weeks. This statute requires an abortion by a licensed physician, performed in a physician's office, a hospital, or clinic. This act is based on the New York Abortion Act.

A number of states had revised their abortion statutes along the lines of the Model Penal Code. In that code the abortion must be justified, such as to protect the mother's health, but the code is more liberal than older statutes.

States adopting this law, known generally as "therapeutic abortions," are Arkansas, California, Colorado, Delaware, Georgia, Kansas, Maryland, Mississippi, North Carolina, Oregon, South Carolina, and Virginia. Other states have considered this type legislation. It has become quite controversial.

Many states have repealed criminal penalties for abortions performed in early pregnancy by a licensed physician subject to certain procedural and health requirements. Local statutes should be consulted to see if the Uniform Act or Model Act version has been adopted.

9.8 SODOMY

Sodomy was a felony at common law and included such sexual acts as intercourse between a human and an animal or anal intercourse between human beings. The acts prohibited by this law are still violations in most jurisdictions today and are often entitled "crimes against nature" or the like. Homosexuality is usually encompassed under these laws, and in no state does homosexuality appear in the statute books as a separate offense.

Until the so-called sexual revolution of the 1960s, there were few challenges to the sodomy statutes on constitutional grounds. There were also few prosecutions, because these acts took place in private. Though the husband–wife team was not much involved, homosexuals felt that such statutes were unconstitutional as applied to consenting adults in this victimless crime. Of course, there is a victim when the sodomous act is forced on a nonconsenting person or upon a person incapable of giving consent.

The U.S. Supreme Court upheld a Virginia statute even as to consenting adult homosexuals. It did so without much ado or argument and by a very brief summary disposition. The status of an earlier case that recognized that married persons had a right of sexual privacy was thus thrown in doubt. Several courts had held after a birth control case that criminal penalties could not be placed on married adults who consented to different sexual behavior.

Many legislatures enacted statutes that punished sodomy of the forced or public type but not as between consenting adults. A few states have repealed those consensual sodomy statutes, but most have kept them. Since a Supreme Court decision upheld the right of states to prohibit sodomy even as between consenting adults, Maryland has decided to uphold its crime punishing private consensual acts of sodomy involving unmarried heterosexual adults.

This entire controversy should not be considered a tempest in a teapot. The penalty for sodomy often carries a maximum of 10 years. Thus there is a substantial interest in defending oneself against such a charge.

Louisiana has added an aggravated sodomy statute. It provides for a more severe penalty when an "unnatural carnal copulation" takes place upon a victim of less than 17 years of age by a person more than three years older than the victim.

9.9 INDECENT EXPOSURE

The act of exposing oneself in a public place or in a place in which the public could view such conduct was a common law misdemeanor. This offense has been included in the statutes of most jurisdictions and remains a misdemeanor.

The most significant decisional activity in this area involves what constitutes a public place. Can a person, while standing within his own property, be convicted of this crime? The answer is generally yes. In one case the defendant stood in his brightly lit dining room and allowed himself to be viewed by passersby on the sidewalk. This met the "public place" or "to public view" element of the crime. In most instances those who "moon" the public have been found guilty. However, there are exceptions to the "mooning" cases as when the statute requires exposure of the genitals and the "mooners" did not show these.

What about a male who has to urinate badly and goes near a tree because there is no bathroom around? Is this indecent exposure? Does it matter: where the tree is located?; the time of day, evening, late night?; how close to a public place?; how much public traffic? Answers to these questions are important. Since the offender does not appear to have the intention of exposing himself to public view, many jurisdictions treat this conduct as a lessor misdemeanor such as a breach of the peace or disorderly conduct.

9.10 OBSCENITY

In 1957 the U.S. Supreme Court was confronted with an important constitutional test. In *Roth v. United States,* the defendant, who published and sold books, was charged with violation of the federal obscenity statute. In a companion case, *Alberts v. California,* the defendant conducted a mail-order business in the state of California. He was charged with violation of that state's obscenity laws for keeping obscene and indecent books for sale. The question presented to the Court in both these cases was whether or not the statutes under which these defendants were charged and convicted (the federal statute in the *Roth* case and the California statute in the *Alberts* case) were constitutional. The decision turned on whether these obscenity laws violated the First Amendment's guarantee of freedom of speech and the due process clause of the Fourteenth Amendment. The defendants argued that the wording of the respective statutes did not provide reasonably definite standards of guilt so that people could understand and conform their conduct to the requirements of the laws. The Court held that the statutes were constitutional; that not every form of speech or expression is protected by the Constitution; that obscenity laws, designed to protect the morals of the people, do not infringe on the area of protected speech or expression under the First Amendment; and that the wording of the statutes gave adequate warning of the types of conduct prohibited. The court went on to say that the test of obscenity is whether or not, to the average person applying contemporary community standards, the material appeals to prurient interest.

Thus the Court established the broad rule that obscenity was not constitutionally protected. This view has been reaffirmed in case after case since 1957. The problem then becomes one of defining the term **obscenity**. The test propounded in the *Roth* and *Alberts* cases appears rather subjective, as it takes a court to determine whether or not the material at issue in a particular case is, in fact, obscene.

Since 1957 several state statutes prohibiting the use, sale, or possession of obscene material have been ruled unconstitutional by the courts. In general, these decisions have not held that obscenity is lawful and fully protected by the First Amendment and the Fourteenth Amendment. What they have said is that, although obscenity is still not protected, the statutes that were attacked did not adequately define obscenity or provide reasonably ascertainable standards of guilt. The statutes were too vague to be understood by the people of the state who must conform their conduct to the requirements of the law. If these states were to draft new laws prohibiting obscenity and word them in such a way as to define clearly the prohibited conduct, the statutes would be constitutional.

From 1959 to 1973 the courts utilized a test for obscenity based upon the social value of the work. If a book, film, or other material had some redeeming social value, it was not obscene. It was up to the courts to determine whether a particular item had redeeming social importance. This test was applied along with the test propounded in the *Roth* and *Alberts* cases, under which the courts had to look to contemporary community standards.

But what community? The nation? The state? The city? Much debate arose as to the scope of the community. A good many people said that a national standard should be applied.

In 1973 the U.S. Supreme Court defined the community and redefined obscenity. In a series of cases the Court made major revisions concerning the right of the community to protect itself from allegedly obscene materials.

The first and most important case dealing with the definition of obscenity was *Miller v. California*. The Court here defined obscenity and the community, pointing out that obscene material is not protected by the First Amendment. The newer definition, as the Court saw it, is that any work that depicts or describes sexual conduct, that taken as a whole appeals to the prurient interest in sex and portrays sexual conduct in a patently offensive way, and that taken as a whole does not have serious literary, artistic, political, or scientific value is obscene. The court rejected the utterly without redeeming social value test.

The Court defined what patently offensive means in two parts. First, *patently offensive* means representations or descriptions of ultimate sexual acts, normal or perverted, actual or simulated. Second, *patently offensive* means representations or descriptions of masturbation, excretory functions, and lewd exhibition of the genitals.

As to what community should be looked to with regard to the standard of offensiveness, the court said, "It is neither realistic nor constitutionally sound to read the First Amendment as requiring that the people of Maine or Mississippi accept public depiction of conduct found tolerable in Las Vegas or New York City." The Court recognized the differences in people and was unwilling to impose uniformity. Therefore, a state, through individual courts and by the decision of individual juries, is permitted to determine whether any work is patently offensive.

In a companion case, *Paris Adult Theater v. Slaton*, the Court said that even though obscene materials are shown only to consenting adults, the state has a right to challenge such materials and, if they are found to be offensive, to prevent their sale and distribution.

In yet another case decided on the same day, *U.S. v. Orito*, the Court held that Congress has the power to prevent obscene material from entering the stream of commerce. The Court therefore held that the decision in an earlier case, that protected the right of people to keep obscene materials in the privacy of their homes, did not extend beyond the home. Therefore, although a person can have such material at home, it appears that he or she could have difficulty transporting it there.

The final case dealing with the definition of obscenity is *Kaplan v. California*. The Court here simply said that books without pictures can be obscene.

With these definitions in mind, the law enforcement officer will naturally ask, "How can I determine if the crime of selling or distributing obscene materials has been committed?" Because a book, film, or other material is not obscene until so declared in the proper forum and because the law enforcement officer must guess whether or not the community standards are offended, the Supreme Court is reluctant to allow seizure of such materials without some kind of neutral magistrate's intervening.

At the same time that the definitional decisions came down, the Court rendered two opinions dealing with the seizure of films. In the case of *Roaden v. Kentucky*, the court found these facts: A sheriff saw the film at a local drive-in theater. At the conclusion of the film, without a warrant, he arrested the theater manager and seized a copy of the film. The Court said this was unreasonable and violated the Fourth and Fourteenth Amendments. The Supreme Court seemed to prefer the method followed in the companion case of *Heller v. New York*. In the *Heller* case police went to the theater and viewed the film. After that, accompanied by a state prosecutor and a judge, the police went back and again viewed the film. They had already prepared affidavits and warrants that needed only the judge's signature. The judge agreed that it was obscene and signed the arrest and search warrants. The Court upheld this action.

What it comes down to is simply this: Without a prior determination of obscenity and without absolute knowledge of what a community will think is obscene, there is no way for law enforcement to know if a crime has been or is being committed. Unlike other crimes, as discussed throughout this book, the crime involving obscenity does not have clear-cut elements about which, when observed, one can say, "I just saw a crime."

Despite the fact that the U.S. Supreme Court in *Miller* severely limited what could be found to be hard-core pornography, some prosecutors and police still brought borderline cases to the courts. The Court reemphasized that only hard-core materials are obscene. The Court said, "It would be a serious misreading of *Miller* to conclude that juries have unbridled discretion in determining what is 'patently offensive.'" Occasional scenes of nudity are not enough to make material legally obscene under the *Miller* standards. Only the public portrayal of hard-core sexual conduct for its own sake and for ensuing commercial gain is punishable.

The use of children in pornography is not an entirely new problem. Its growth has been explosive. One study revealed that in 1979 over 260 magazines showed children engaged in various sex acts. Some of the children were as young as three years of age, according to congressional hearings. No one knows exactly how many children have been involved in the production of child pornography films, but one estimate indicates the figure to be no fewer than 30,000 children.

There is no doubt that the greatest potential for harm involves those children used to produce the material. That harm can be both physical and psychological. Thus the problem can be seen as a child abuse problem. Unfortunately, there is not sufficient evidence to prove abuse to children outside the industry caused by the distribution of the material. At least one witness did show that, in several cases of child molestation in Los Angeles, pornography was involved in every case. This report covered an eight-month period.

Every state prohibits child pornography and other sexual exploitation of children. This protection may be found in statutes on child abuse, obscenity, prostitution/pornography, or a statute called sexual exploitation of children. States in which this latter title is used include Arizona, Colorado, Idaho, Indiana, Maine, Montana, Oregon, Texas, Utah, and Vermont.

9.11 PROSTITUTION AND RELATED OFFENSES

Prostitution is defined as the offering by a female of her body for intercourse with men for monetary or other gain. Prostitution is an offense in most states but is not criminal in England anymore. Most American jurisdictions follow the common law definition and require that the sexual activity be for hire, implying some gain, usually money. A small number of states have eliminated this element of the offense and define the crime as any indiscriminate sexual activity whereby a female offers her body to a number of men for sexual intercourse whether for gain or not. Some states have, by statute, either included males or have eliminated the distinction, so that members of either gender may commit the offense.

The crime of **pandering,** also generally prohibited in the prostitution statutes, is the offense commonly called **pimping.** This offense, usually carrying even more severe penalties than prostitution because of its social degradation, consists of procuring a female to work in a house of prostitution or procuring a house of prostitution in which the female may work.

Bawdy houses, also called houses of ill fame and houses of prostitution, were prohibited at common law as nuisances. Today, of course, it is still unlawful to maintain such a place. However, proof of this offense generally requires more than an isolated incident or two. It must be shown that the "house" is regularly used for prostitution. Similarly, one who leases premises for the purposes of, or knowing that it will be used for, prostitution is guilty of an offense.

The word *house* as used to describe a house of prostitution is not confined to its meaning as a residence. The courts have treated such other structures as a tent, boat, or even an automobile as coming within the definition of a house of prostitution, provided all other elements of the offense are satisfied.

There is a federal statute on this topic that deserves mention. The **White Slave Traffic Act,** or **Mann Act** as it is called, was enacted by Congress in the exercise of its powers over interstate commerce. The federal law prohibits interstate transportation of a female for the purpose of prostitution or for other immoral purposes. Prosecutions for violation of this statute are held in federal court. Although the legislative history of this statute indicates that the intent of Congress was to attempt to close an avenue of revenue for organized crime, time after time courts have held that the statute applies to isolated cases in no way connected with organized crime. To prove this offense, interstate travel must be shown to bring the case under federal jurisdiction as well as the accused's intention of traveling for immoral purposes, such as prostitution.

9.12 BIGAMY AND POLYGAMY

Bigamy was not a common law offense but was first created by statute in England. All U.S. jurisdictions have enacted statutes to prohibit marriage between two people when one of them is legally married to someone else. Bigamy is not a specific intent crime. The elements are satisfied by showing that the accused married again, did so intentionally, and knew, at the time, that he or she was already legally married to another. Such defenses as mistake of fact and mistake of law (Chapter 6) will be taken into consideration in a bigamy trial.

Polygamy is the act of being married to several spouses at the same time. This, too, is prohibited by the laws of the United States. In one respect, however, the outlawing of polygamy is a rather recent innovation. The religious beliefs of certain groups allowed and sanctioned the practice of polygamy in the United States. This led to a court case in which the Supreme Court declared polygamy to be unlawful. The decision hinged on the First and Fourteenth Amendments to the Constitution. The argument for allowing polygamy as a religious practice was based on the First Amendment guarantee of separation between church and state, prohibiting the state from intervening or interfering with religious practices. The argument for outlawing polygamy was that it would be a denial of due process and equal protection to allow one portion of society to practice polygamy and subject someone outside the particular religious group to criminal penalties for doing the same thing. The second argument won out, and polygamy is now uniformly prohibited in this country.

9.13   CRIMINAL SEXUAL PSYCHOPATH LAWS

In many states, legislatures have enacted laws to provide for institutionalizing and treating persons with a background or history of deviant sexual behavior. These statutes are called *criminal sexual psychopath laws,* or *mentally disordered sex offender laws.* The statutes usually identify the types of individuals to whom the law applies and the procedure by which the state may commit and treat these individuals. These laws do not define another substantive criminal offense. They are merely tools by which the state seeks to prevent further deviant sexual behavior.

Another tool used in many jurisdictions today requires known sexual deviates who have been released from prison following conviction of a sexually deviant act to register with local law enforcement. In some cases, statutes require the law enforcement agency to ensure publication of such a person's identity, thus putting the citizenry, including potential employers, schools, parents, day care centers, and so on, on notice that the convicted sexual deviate is in the community.

▬▬ DISCUSSION QUESTION ▬▬

1. Smith has intercourse with Alice, a prostitute, who says she is 19 years old. Actually, she is only 17 years, 10 months. Assume that the state law sets 18 years as the age of consent. What is Smith's legal position?

▬▬ GLOSSARY ▬▬

Adultery – sexual intercourse between a male and a female, at least one of whom is married to someone else.
Bawdy house – a house of prostitution.
Bigamy – marriage between two people when one of them is legally married to someone else.

Carnal knowledge – the legal term for sexual intercourse.

Fabrication exception – exception to the Rape Shield Law if it can be shown that the victim fabricated similar charges against others.

Forcible rape – unlawful carnal knowledge of a female by force and against her will.

Fornication – sexual intercourse between unmarried persons.

Incest – sexual intercourse between persons who are close blood relatives.

Obscenity – sexual materials appealing to prurient interests and in violation of community standards.

Pandering or pimping – procuring a female to commit prostitution.

Polygamy – the act of being married to several spouses at the same time.

Rape shield law – refusal to allow the prior sexual behavior of the victim in to evidence.

Rape trauma syndrome – a series of symptoms indicative of posttraumatic stress disorder occurring from a rape.

Rebuttable presumption – a presumption the law requires to be made unless it can be overcome by contrary evidence.

Roe v. Wade – Supreme Court decision allowing mothers the choice to have an abortion.

Seduction – enticing or luring an unmarried female of previous chaste character to engage in sexual intercourse by a fraudulent promise to marry her or by some other fraudulent promise.

Viability – capability of a fetus to have a meaningful life outside the mother's womb.

White Slave Traffic Act (Mann Act) – a federal law making it a crime to transport a female across state lines for prostitution or other immoral purposes.

▓ REFERENCE CASES, STATUTES, AND WEB SITES ▓

CASES

New York v. Taylor, 552 N.E.2d 131 (1990).

West Virginia v. Jackson, 383 S.E.2d 79 (1989).

Henson v. Indiana, 553 N.E.2d 1189 (1989).

Kansas v. Blue, 592 P.2d 897 (1979).

Winfield v. Virginia, 301 S.E.2d 15 (1983).

Virginia v. Beverly, 52 Va.Cir. 255 (2000).

Meloon v. Helgemoe, 564 F.2d 602 (NH 1977).

State v. Rundlett, 391A.2d 815 (ME 1978).

State v. Weatherby, 43 Me. 258; 1857 Me. LEXIS 23

Pratt v. Pratt, 32 N.E. 747 (1892).

Buchanan v. State, 55 Ala. 154 (1876).

People v. Rouse, 2 Mich. 209 (1871).

Roe v. Wade, 410 U.S. 113, 93 S.Ct. 705, 35 L.Ed.2d 147 (1973).

Doe v. Bolton, 410 U.S. 179, 93 S.Ct. 739, 35 L.ed.2d 201 (1973).

Bowers v. Hardwick, 478 U.S. 186, 106 S.Ct. 2841, 92 L.Ed.2d 140 (1986).

Romer v. Evans, 517 U.S. 620 (1996).

Louisiana v. Smith, 661 So.2d 442 (1995).

Roth v. United States, 354 U.S. 476 (1957).

Alberts v. California, 354 U.S. 476, 77S.Ct. 1304, 1 L.Ed.2d 1498 (1957).

Miller v. California, 418 U.S. 915, 94 S.Ct. 3206, 41 L.Ed.2d 1158 (1974).

Paris Adult Theater v. Slayton, 413 U.S. 49, 93 S.Ct. 2628, 37 L.Ed.2d 446 (1973).

United States v. Orito, 413 U.S. 139, 93 S.Ct. 2674, 37 L.Ed.2d 513 (1973).

Kaplan v. California, 419 U.S. 915, 95 S.Ct. 194, 42 L.Ed.2d 154 (1974).

Roaden v. Kentucky, 413 U.S. 496, 93 S.Ct. 2796, 37 L.Ed.2d 757 (1973).

Heller v. New York, 413 U.S. 483, 93 S.Ct. 2789, 37 L.Ed.2d 745 (1973).

WEB SITE

www. findlaw.com

CHAPTER 10

Theft

■ KEY WORDS AND PHRASES ■

Abandoned property
Animus furandi
Asportation
Bailee
Bailment
Bailor
Concealing stolen property
Confidence game
Control

Conversion
Custody
Dominion and control
Embezzlement
False pretenses
Identity theft
Insurance fraud
Intangible
Larceny

Larceny by trick
Lost property
Mislaid property
Odometer fraud
Ownership
Personal property
Possession

Receiving stolen property
Shoplifting
Swindle
Tangible
Theft
Title fraud
Trespassory taking

10.0 INTRODUCTION: THEFT IN GENERAL

The word **theft** in its broadest sense describes many forms of criminal conduct. In this chapter we discuss four basic types of theft offenses: larceny, embezzlement, obtaining property by false pretenses, and receiving, concealing, and possessing stolen property. In fact, this list does not cover all the possibilities. The crime of robbery discussed in the following chapter is a theft offense of sorts because it involves the stealing of property. Because of the force or fear element in robbery, it is usually classified as a crime against the person rather than a theft offense. Also, forgery and uttering and publishing worthless checks are theft offenses. Because of their frequency and importance, however, they will be treated separately, as they are in most statutes. Many other crimes also involve theft, for example, a majority of burglary cases. Of the more serious offenses, theft is probably the most frequent. Finally, there are a group of offenses, most of which contain elements of theft, that are statutorily created and grouped together as white collar crimes. These are described in Chapter 22.

Before discussing specific theft offenses, certain terms should be defined. Throughout the chapter, reference is made to possession, custody and control, and ownership. The presence or absence of any or all of these factors will largely control the type of offense committed, if any.

Possession is a word thrown around loosely in everyday life. The phrase "Possession is nine-tenths of the law," has become a cliché. Although few laypersons really understand the meaning of this expression, from a legal standpoint, it contains more truth than falsity.

A person has **possession** of property when that person may exercise discretion in the use and handling of that property. The scope of this discretion may or may not be limited by other factors. Possession must thus be distinguished from mere custody and control over an item. One may be physically in control of an article without necessarily having possession of it. If a person has no right to exercise discretion as to its use or handling, there exists only **custody** and **control**, not possession. If Sam says to Bill, "Hold my coat while I change this flat tire," Bill has custody and control of the coat but not possession. If Sam says to Bill, "You may borrow my coat for your date tonight," Bill has discretionary use of the coat during that time, so he has possession. Similarly, a shopper in a store who picks up an item from the counter to examine it has only custody and control. If the shopper takes the item home on approval with permission of the store before purchasing it, the shopper has possession of the item.

It is not very difficult in the average case to distinguish between possession and custody and control. When an employee steals from an employer, however, it becomes necessary to determine whether the goods stolen were in the possession of the employee or whether the employee had only custody and control. Determination of this issue will govern the charge to be placed against the thief. As a general rule, employees have only custody of the property of their employers, and possession remains with the employer. If the property is entrusted to the employee in a position of trust and confidence, however, that employee is considered to be in possession of the goods. The classic illustration of an employee in such a capacity is the bank teller. Although discretion is limited, the teller is permitted to distribute monies given by one person to another.

Possession may be obtained either lawfully or unlawfully. If other elements of a particular theft offense are satisfied, the fact that one gained possession lawfully does not mean the person has committed no crime. The lawfulness or unlawfulness of possession governs only the type of crime for which one may be held liable.

Ownership confers complete and unlimited discretion. Legal title is the essence of ownership. An owner is always entitled to possession, but the owner may give up possessory rights in property, either temporarily or permanently, and still retain ownership. A leased automobile is owned by the leasing company but is possessed by the lessee. When property is owned and possessed by the same person, few problems will arise, but the distinction is important because larceny and embezzlement are crimes against the possessor of property, not necessarily the owner of the property. Legal title will be important to any discussion of the crime of obtaining property by false pretenses.

I. LARCENY

10.1 LARCENY IN GENERAL

The legal definition of common law **larceny** contains five elements: (1) the taking, (2) and carrying away, (3) the personal property, (4) of another, (5) with the intent to permanently deprive. Each of these elements is subject to various interpretations. For this reason, we must look to the decisions of courts for their true meaning. It would be unwise to rely solely on the wording of the larceny statutes for an understanding of this offense.

Taking

To accomplish a larceny, property must first be taken. The taker must have no right to possession of the property. This is called **trespassory taking**.

One cannot commit a larceny against property he or she lawfully possesses, because larceny is a crime against possession. All that is necessary to satisfy this element is to show that the taker has gotten possession of the property and that the possession is wrongful. This can be established by showing that the taker exercised discretionary control over the object without being entitled to do so.

The trespassory taking usually occurs in a manner that presents little difficulty. For instance, Fred picks up an item of merchandise from a store counter and walks out with it concealed under his coat. This example obviously satisfies the element of trespassory taking. There may be variations on this theme, however, such as the cases covered by the rule of continuing trespass. Every crime generally requires an act and a simultaneous intent. This rule presents some difficulty in the law of larceny. Suppose, for example, that Al wrongfully and without Bill's consent takes some property belonging to Bill. When he takes the property, Al does not intend to steal it but he later forms that intent. This case presents a dilemma because the property was taken at one time and the criminal intent was formed at a different time. Act and intent are not simultaneous. This problem is solved through a legal fiction—the widely accepted rule called continuous trespass. In essence, this rule provides that, if the initial taking involves trespass, the trespass continues until the criminal intent required for larceny is formed. At this point, the act and intent become simultaneous, and, if all other elements are satisfied, the crime of larceny occurs.

The distinction between possession and custody and control discussed previously plays an important part in determining whether the trespassory taking element is satisfied. If someone possesses but does not own property, later conversion of that property in a way that permanently deprives the owner of the rights to that property will not be larceny. This situation does not fulfill the requisite taking element of larceny. If one merely has custody and control of an article, that person's **conversion** of that article deprives someone else of his or her possessory rights and does constitute a trespassory taking.

Suppose that Sam asks Bill to keep a sealed carton for him. Bill agrees. Bill then has possession of that carton. Bill does not, however, have possession of the contents of the carton. He has only custody and control. Under the doctrine of breaking bulk, if Bill steals the carton, it is not larceny. If he breaks open the carton and steals the contents, he has committed larceny.

A further complication of the taking element involves delivery by mistake. Suppose that property is given to the accused by mistake and the accused is not aware of the error. The accused later discovers the mistake and decides to keep the property. This is not larceny, because the accused was in possession of the property at the time the intent was formed. If, however, the accused knows about the mistake at the time of the delivery and decides to keep the property, his or her acceptance of delivery will be considered a trespassory taking. For example, John goes to a bank to cash a check for $1.27. By mistake, the new teller reads the date line instead of the amount. The date is 1/27/99. She pays John $127.99. If John is unaware of the mistake and takes the money but later learns of the error and decides to keep the extra money, he has not committed larceny. On the other hand, if at the time of the transaction, John knows about the mistake and decides to keep the money, he has committed a trespassory taking and may be liable for larceny.

Consent may be a defense to larceny if it is given by someone who has authority to give it. But consent is no defense if the amount taken is more than the amount consented to. If John allows Bill to take a $1 bill out of his wallet and Bill takes $10 instead, the taking of the extra $9 is trespassory. All the elements of consent must be present to support this defense (Chapter 6).

Sometimes It Doesn't Pay to Take

Washington: When a man attempted to siphon gasoline from a motor home parked on a Seattle street, he got much more than he bargained for. Police arrived at the scene to find an ill man curled up next to a motor home near spilled sewage. A police spokesman said that the man admitted trying to steal gasoline but plugged his hose into the motor home's sewage tank by mistake. The owner of the vehicle declined to press charges, saying that it was the best laugh he'd ever had.

Carrying Away (Asportation)

Once the property has been taken, it must be carried away. The Latin-derived term for this element is **asportation**. The general rule of asportation is that the article taken must be entirely removed from the place it formerly occupied. The article does not have to be carried any great distance as long as it is entirely removed from the place it occupied just before being taken. Although this element usually presents no significant difficulty, in some cases it may become an issue. For example, an accused pickpocket has been caught in the act. The wallet that the pickpocket was removing from the victim's back pocket was only partially removed when he was caught. Because there is not much room in a pocket, the wallet would have to be completely removed before asportation is complete. If it is not completely removed, there is no larceny, only an attempt.

A number of stores selling expensive items of clothing such as coats have tried to deter thefts by using metal cables attached to racks running through sleeves of the garments. A customer cannot try on a garment until a salesperson unlocks the cable device. A thief who tries to carry away such a garment will obviously have considerable difficulty. Asportation has been thwarted. As a result, many courts today would treat this as an attempt rather than a completed larceny because of the impossibility of total removal from the store.

Other concerns regarding retail stores are to be considered. Does a person have to be leaving the store to be charged appropriately? Not necessarily. The type of store, the location, the circumstances of the apprehension, and the defendant's conduct while in the store can all contribute to whether the element of asportation has been satisfied. Apprehending a person in a self-service store before he or she has reached the checkout area raises some possibility of the "I forgot I put it in my pocket" defense. On the other hand, in the same self-service store, the outcome may be different, as in the following example. The defendant picked up a large box that contained a $19 basket. He took the basket out and filled the box with more expensive items. The checkout cashier noticed the unusual weight of the box, and the contents were discovered. The jury convicted the defendant after disagreeing with his claim that there was no asportation because he had not left the store. On appeal, his conviction was upheld because his conduct was consistent with proof of a completed intent to steal.

The means of accomplishing asportation are immaterial. The act may be accomplished by the hand of the accused or through an innocent human agent, an inanimate object, or an animate nonhuman agent such as a monkey or dog (see Chapter 4).

He Forgot to Carry It All Away

Tennessee: A man successfully broke into a bank after hours and stole the bank's video camera. While it was recording. Remotely. (That is, the videotape recorder was located elsewhere in the bank, so he didn't get the videotape of himself stealing the camera.)

Personal Property

Personal property is anything capable of ownership, except land or things permanently affixed to land. Things like cars, animals, furniture, and money are all classed as personal property. Trees and houses are generally considered real property, but once they are detached from the land, as when a tree is cut down or a house is torn down, they take on the characteristics of personal property and may be the subject of larceny. Courts have held that minerals, once they are mined, become personal property.

Personal property may or may not be **tangible**. In a number of cases, it has been held that services, such as water, gas, electricity, and oil, may be subject to larceny. This is frequently done by tapping electric or telephone lines or tapping water or oil lines. The identity of the victim may depend on whether the tap is made before or after the electricity or water has gone through a householder's meter. A theft may involve both tangible and **intangible** property as where the offender steals an electric meter from one residence and installs it at another building that has previously had its electricity turned off and its meter removed. The meter and the electricity used without payment are both thefts.

The advent of the computer caused some dilemmas regarding unauthorized use of personal property. Is such use, in the absence of a specific statute, a theft? Some courts, despite their position that electrical services can be the object of theft, held that unauthorized use of computer time is not chargeable. Others came to a contrary position while expressing a desire for legislative refinement in this area.

An example of such legislation is found in the Comprehensive Crime Control Act of 1984. Discussed in Chapter 22, the Counterfeit Access Device and Computer Fraud and Abuse Act punishes knowingly accessing a computer without authorization, using a computer for nonauthorized attempts to gain government secrets or financial records of private institutions and persons, or using a computer to modify a program or block authorized access.

New York enacted a statute prohibiting the use of business, commercial, or industrial equipment or facilities of another person, without entitlement, to derive a commercial or other substantial benefit. In one case interpreting this statute, the prosecution charged a government employee for using his agency's computer to conduct personal business. The court said this statute did not apply in this case because the government is not a commercial entity. More refinement of the legislation was suggested.

The Model Penal Code continues and expands upon the intangible theory. Section 223.7 concerns the theft of services. Services as a "takable" property include labor, professional services, telephone or other public services, accommodations, admission to exhibitions, and the use of vehicles or other movable property. It also indicates that the refusal to pay or absconding without paying gives rise to a presumption that the service was obtained by deception as to intention to pay.

Virtually all jurisdictions now have legislation covering computer-specific offenses, but there is little agreement among them as to what constitutes the crime. Local statutes and case decisions should be consulted for more in-depth study.

Of Another

Because larceny is a crime against possession, one cannot commit larceny against goods already possessed by such person. The identity of the person who has possession is important, not after the theft has occurred, but immediately before the theft. It is obvious that the thief has possession afterward. Theft must be committed by one who does not have possession against one who does have possession to constitute larceny. A person who has only custody and control of goods may commit larceny of those goods, because that person does not have possession immediately before the theft. An owner may commit larceny against goods he or she owns if someone else has possession. The classic example of this involves the innkeeper who keeps the luggage of a guest who has refused to pay for the previous night's lodging until the bill is paid. If the owner of the luggage takes the property back without paying the bill, the luggage owner is guilty of larceny. The law permits the innkeeper to retain possession of the luggage under these circumstances even though the innkeeper does not own it.

At common law, crimes of property could not exist between husband and wife, because the law treated them as one person. Can a man be convicted of theft of personal property from his wife today in light of modern statutes? Yes. Throughout the 50 states, legislatures have enacted married women's property acts that recognize the power of women to deal with their own property without interference from the husband. That being so, the nonconsensual, stealthy taking of property with the required intent is prosecutable even when a husband steals from a wife.

Whoops! Whose Property Got Taken?

Kentucky: Two men tried to pull the front off a cash machine by running a chain from the machine to the bumper of their pickup truck. Instead of pulling the front panel off the machine, though, they pulled the bumper off their truck. Scared, they left the scene and drove home. With the chain still attached to the machine. With their bumper still attached to the chain. With their vehicle's license plate still attached to the bumper.

With the Intent to Permanently Deprive (*Animus furandi*)

Larceny is a specific intent crime. The thief must intend to permanently deprive the rightful or previous possessor of discretionary use of the property. This intent in larceny is referred to as **animus furandi**. The intent must be to permanently, rather than temporarily, deprive. This intent may be shown from the facts and circumstances surrounding the incident. It is provable just as the intent for any crime is provable—by what is said and done. In many cases in which it is difficult to prove intent to permanently deprive, legislatures have created separate offenses. Even temporary deprivation, if it is done wrongfully, should not go unpunished. Many jurisdictions have resolved the problem by altering the common law limitation and provide proof of intent to be of permanent or temporary deprivation. If an automobile is stolen and the intent cannot be shown to support an auto theft or larceny charge, some modern statutes still allow the larceny to be charged. Other legislatures have created statutory offenses, such as driving away an automobile without the owner's consent. Statutes of this kind do not require that intent to permanently deprive be shown.

These temporary deprivation statutes when related to automobile takings are often called "joyriding" statutes. The tradition of these statutes has been carried on by the Model Penal Code and its Section 223.9, entitled "Unauthorized use of automobiles and other vehicles." It is a misdemeanor to operate any motor-propelled vehicle without the consent of the owner. The statute provides an affirmative defense, however. It is the defense of the actor who reasonably believed the owner would have consented to the operation of the vehicle had the owner known of it.

At common law, if property was taken with the intent to permanently deprive, it was immaterial whether or not it was subsequently returned even if the thief did so voluntarily. This may have affected the severity of the sentence, but the crime was complete when the property was taken. If the thief held property until the rightful possessor agreed to pay a reward for its return, the thief was considered to have intended permanent deprivation. It was not necessary that the thief kept the property or gained any benefit from it as long as the intent was that the rightful possessor be deprived of the property permanently.

He Must Have Forgotten His Intent Was to Permanently Deprive!

New Jersey: A woman was reporting her car as stolen, and mentioned that there was a car phone in it. The policeman taking the report called the phone, and told the guy who answered that he had read the ad in the newspaper and wanted to buy the car. They arranged to meet, and the thief was arrested.

10.2 DEGREES OF LARCENY

The common law knew no degrees of larceny. All such crimes were felonies. Most modern statutes divide larceny into two degrees: grand larceny and petit larceny. The distinction hinges on the value of the article stolen. Each state establishes its own arbitrary line between the two degrees. In some states it is $50. Others define the theft of property valued at $100 or more as grand larceny. The important thing for law enforcement officers to remember is that the value of the property must be provable in court. Because law enforcement generally investigates on behalf of the prosecutor, it is imperative in investigating larceny cases to determine the value of the property stolen. Determining value may present problems. There are many different measures of value: for example, fair market value, cost value, replacement value, reasonable value, sentimental value, and so forth.

As a general rule, courts base their decisions on fair market value if one is available. Otherwise cost value as opposed to resale value or replacement value is preferred. Sentimental value is not a consideration in the criminal law, because this might be subject to exaggeration. Local requirements should be studied to determine the rules for evaluating stolen property.

No matter which rule is followed in a particular jurisdiction, there is usually little problem in determining the degree of larceny committed when the theft involves a single transaction. There may be a problem when a series of small thefts are committed by an individual over a period of time. If the circumstances are right, two alternatives may be open to the prosecution. Several petit larceny offenses may be charged, or all the transactions may be combined into one grand larceny charge. To have this option, the prosecution must show that the accused committed a series of separate small thefts with a single felonious intent according to a common scheme or plan to accomplish a single overall objective. Suppose that grand larceny in state A is the theft of property worth $50 or more. John, a resident of state A, finds out his wife is going to have a baby some time in July. John begins to worry about how he will pay the doctor and hospital bills after the baby is born. John recalls that there is a certain parking meter downtown that has a faulty lock and that the meter can be opened by simply tapping it in the right place. Nightly, John empties the meter of its contents. At no time does he take more than $3 or $4. By July, John has accumulated over $400 from this single meter. What degree of larceny has John committed? Of course, he could be charged with each of the separate petit larceny offenses. But in this instance, it can be shown that John acted with a single felonious intent to get enough money to pay the doctor and hospital bills, taking each time from the same victim under an overall general plan. This would give the prosecution the choice of combining all the separate offenses into one grand larceny charge.

Would the result be the same if John committed several small thefts against different people under different circumstances to achieve the same objective? The answer is no. The prosecution would be permitted to charge only the separate counts of petit larceny (hoping the sentences would run consecutively if John was convicted). Although John's motive was the same, his intent was to steal from different victims by different means.

10.3 SHOPLIFTING

All states have created an additional larceny offense for retail establishments generally called **shoplifting**. The requirement that the theft be from a retail store is the only additional element of this offense. Penalties are provided independent of the value of the goods taken.

The most significant issue involving shoplifting is whether the various legislatures intended to restrict prosecutions to shoplifting if the theft occurs in a retail establishment or whether prosecutors would retain the option to charge larceny. The problem is illustrated as follows. State X has petit larceny, grand larceny, and shoplifting statutes. Each offense has a separate penalty, with a shoplifting conviction carrying the least amount of possible jail time and the lowest possible fine. The grand larceny may be charged in this state when the goods taken have a value of $200 or more. Fred walks into the store and takes goods valued at $201. He is charged with grand larceny, a felony. Fred seeks dismissal of the charge, claiming that the shoplifting violation, a misdemeanor, was intended to cover his crime. Fred will not win the argument. The states have determined that unless provided to the contrary, prosecutors retain the option as to which type of theft violation to charge.

10.4 LOST, MISLAID, AND ABANDONED PROPERTY

When lost, mislaid, or abandoned property is found, some interesting problems in the law of theft are presented. Property is said to be **lost** when the possessor unintentionally parts with it under such circumstances as to be unaware it is missing. For example, John gets on a bus, and, unknown to him, his wallet falls out of his pocket.

Property is **mislaid** when the possessor intentionally places the property in a certain location but forgets to retrieve it. For example, John takes out his wallet to pay his dinner bill and sets it on the cashier's counter. After receiving his change, he fails to pick up the wallet and walks off, leaving it there. In each case, the possessor did not give up possession. The original possessors retain possession until the property is discovered and taken by the finder. Whether the taking is legal or illegal depends upon the intent of the finder and other circumstances at the time of the taking.

Abandoned property, unlike mislaid or lost property, cannot be subject to larceny. A person abandons property when he or she intentionally decides to give up possessory and ownership rights. Only an owner can abandon property. The intent of the owner determines whether or not the property is abandoned, not the circumstances under which the property is found. If John takes possession of an apparently abandoned automobile, which has been left unattended for a period of many months, he acts at his peril. If, in fact, the true owner did not, and does not, intend to disclaim ownership, John may be guilty of theft. He should make every effort to determine whether the car is truly abandoned or not.

The finder of lost property takes possession either rightfully or wrongfully depending on the nature of the article found and the place and circumstances under which it is found. If, because of the nature of the property, it would be impossible to locate the former possessor, the finder takes rightful possession. If Sam finds a $10 bill in a crowded parking lot without seeing who dropped it, he may take it. His possession is rightful. Because of the nature of the article and the place where it was found, it is highly improbable that the original possessor will be identified or will return for the money. The fact that Sam intended to convert the money to his own use at the time he found it would not make his conduct criminal because his possession is legal.

Suppose Sam finds a wallet in that same crowded parking lot under circumstances that indicate the wallet is obviously lost. Can Sam claim rightful possession when he takes it intending to convert it to his own use? No. Sam has found an article that normally contains items that would identify either the true owner or possessor or would give hints as to the identity. Therefore, the law says it is wrong to take the wallet except for the purpose of returning it to the rightful possessor. Sam's act of taking the wallet, combined with his unlawful intention of keeping it, constitute larceny. If, however, Sam took the wallet intending to return it to its true possessor, but some time later decided to keep it and its contents for himself, there would be no larceny, because his original taking was lawful.

Thus, lost property may be the subject of larceny if, at the time it is found, there is a reasonable apparent possibility of identifying the true possessor and yet the finder instead intends to keep the found property permanently. The criminal nature of the act depends on the nature of the property, the place where it is found, and the intent of the finder.

Mislaid property may also be subject to larceny. If someone discovers a box of new clothes on the seat of a bus, the property may not be taken unless the finder intends to return it to the rightful possessor or owner. Because of the nature of mislaid property and because of the probability that the true possessor will remember where it was left and will return for it, any other intent is wrongful. The fact that the owner cannot be identified by examining the mislaid article is immaterial to a charge of larceny, provided the finder takes it intending to permanently deprive the rightful possessor of the property.

Abandoned property can never be subject to larceny. The first finder who takes possession becomes the owner of the property. This is true, however, only when the property has been completely abandoned by the previous owner. The intent of the finder when taking the property is unimportant. It is up to the finder when taking the property to determine the previous owner's intention. The intent of the taker is unimportant, because takers of apparently abandoned property usually take such property for their own use and benefit.

II. EMBEZZLEMENT

10.5 INTRODUCTORY COMMENTS

Although embezzlement was not one of the common law crimes, it was defined as a crime in England early in the history of Parliament. It is a statutory crime throughout the United States today. Embezzlement as a statutory crime was created to fill the gaps in the common law left by the law of larceny.

Conversion of property completely depriving the true owner, committed by one who has rightful possession, is not larceny. Here the distinctions between possession and custody and control become important. If a trespassory taking could not be shown, there was no crime at common law. To alleviate this situation, the crime of embezzlement was created.

10.6 EMBEZZLEMENT DEFINED

Even though all statutes are not exactly alike, embezzlement can be generally found in the laws of all states. It may be called larceny by a **bailee**, larceny by embezzlement, larceny after trust, or by some other name. Essentially, however, the offense is composed of certain standard elements.

Embezzlement is a crime against ownership. The essence of the crime is conversion of property by someone to whom it has been entrusted. Possession is obtained rightfully, and the thief later decides to steal the property. The key distinction between larceny and embezzlement is that embezzlement involves no trespassory taking. In all other respects, the elements of embezzlement and larceny are identical.

What is meant by entrusting property to another? Entrustment involves possession given by an employer to an employee, by a master to a servant, or by a principal to an agent. An agent is someone who represents the interests of another (the principal) and who can act for and bind the principal to contracts and the like. Entrustment is also characteristic of the **bailor**–bailee relationship created by leases, or by leaving property with someone to be repaired or cleaned. A **bailment** can also be created in a number of ways when property is given to another for safekeeping.

Under some circumstances, rightful possession may be obtained without entrustment. This may happen when lost property is found. If the finder later decides to permanently deprive the previous possessor of the property, this is not generally considered embezzlement because no entrustment is involved. Usually, there is no crime in a situation such as this except in a few jurisdictions that have stretched their embezzlement statutes.

Many embezzlement statutes tend to restrict the offense to those who hold special positions of trust, such as public officials, stockbrokers, bank personnel, and others in similar positions. A thorough reading and proper interpretation of these statutes will generally show that they cover most situations not covered by larceny laws.

III. THEFT BY FRAUD

10.7 INTRODUCTORY COMMENTS

Next we discuss three additional theft-related offenses created by legislation. Because obtaining property by false pretenses, larceny by trick, and confidence games involve fraudulent means used to obtain ownership, these crimes are set apart from larceny and embezzlement. Each of these offenses involves obtaining ownership by some fraudulent means. But the fraudulent method used in each differs so that the *corpus delicti* of each crime requires different types of proof. In each of these crimes, the objective is to gain **dominion and control** over the property for the converter's sole use.

10.8 FALSE PRETENSES

The crime of obtaining property by **false pretenses** always involves misrepresentation by the accused of a past or present fact. The accused must be aware that the statement is false, whereas the victim believes it to be true. The property must be delivered in reliance on the misrepresentation to the detriment of the victim and the benefit of the accused. Each of these elements will be discussed separately.

The accused must misrepresent a past or present fact. The accused must misrepresent fact—not opinion. Opinion may play a part in the confidence game but not in this crime. The fact misrepresented must be a present or past fact and not conjecture about the future. This problem becomes complicated when the accused says that the "representations were merely dealers' talk or puffing," as it is known in the law.

The misrepresentation must have been false when it was made. If it was true at that time, the accused did not obtain property under false pretenses even if it later turns out the accused did not know it was true at the time. This means it must be proven that the accused knew the representation was false. It is also essential that the victim not know the claim was false. If the victim knew it was false, the accused cannot be convicted, for no one was deceived. The victim, however, is not under any obligation to verify the accused's statements. It is no defense that the victim was negligent or even stupid not to check.

The victim must part with the property in reliance on a belief in the misstatement of fact. If the victim gives property to the accused for some other reason, there is no crime. However, if the victim parts with the property partially in reliance on the misrepresentation and for other reasons as well, the crime will still have been committed. The misrepresentation need not be the sole factor compelling delivery.

Because this is a statutory, not a common law, crime, the property obtained must be of the kind and value described by the statute. In general, modern statutes recognize the same subject matter in the offense of obtaining property by false pretenses as in larceny.

The accused must intend to permanently deprive the true owner of the property. The misrepresentation can be communicated by any method. It can be written or oral. As a matter of fact, it can be communicated by conduct indicating that some fact is true. Failure to disclose facts, however, even when directly asked, will not make the accused liable.

Let us illustrate this crime by examining the following situations. Bill wishes to sell his zircon ring for a considerable sum to pay off a gambling debt. He knows that Mary, a wealthy widow, always has ready cash and likes rings and other precious baubles. Bill calls Mary and tells her that he has a diamond ring for sale. She tells him to bring it over. Bill shows her the ring and says it is a two-carat diamond worth $20,000, which he will sell for $5000. Mary buys the ring, believing it to be a diamond. She has it appraised some weeks later and learns for the first time that it is worth only $25. Bill's statement to Mary about the ring's worth will not support a criminal charge. A statement of worth is merely an expression of opinion. But as to the nature of the article, Bill misrepresented a known fact and Mary relied on this statement to her detriment and to Bill's benefit. Therefore, the charge of obtaining money or other property under false pretense would be proper.

Suppose, however, that Smith went to a car dealer and bargained over the purchase of a new car. Smith and the dealer settled on the price of a new convertible. By doing some fancy talking, Smith convinced the dealer to sign the title at that time, agreeing to return within the hour with the full cash purchase price. Smith left with the car, fully intending to return to pay for it. Smith later changed his mind and ran off with the automobile to which he had title. Because Smith did not misrepresent any fact at the time the car was delivered, his failure to return and pay the purchase price would not support a charge of false pretenses. In fact, under these peculiar circumstances, it is difficult to identify any chargeable criminal conduct under present law. Would the result be any different if Smith had not intended to pay for the car when he obtained the title? Was the promise to return, which the dealer relied on, a misrepresentation of a present fact or a misrepresentation of a future fact? At first glance it would seem to be a misrepresentation of a future fact—a promise to pay later—which would not constitute a criminal false pretense. Some courts follow this logic. Others would say that the promise itself was a present fact that was misrepresented and that would satisfy this element of the crime.

10.9 CONFIDENCE GAMES: THE SWINDLE

Related to the crime of false pretenses is a series of statutory offenses commonly known as obtaining property by means of **confidence games**. The gist of these crimes differs from that of false pretenses in one important aspect. False pretenses involves reliance by the victim on a misrepresented fact. The personality of the accused has nothing to do with the crime. Under the confidence game statutes, reliance is placed on the accused more than on the representations. The con man sells himself to the victim. The victim is often out to get something for nothing. The facts represented may even be true. But when the accused has no intention of applying the property for the reason obtained, the crime is complete and chargeable.

10.10 LARCENY BY TRICK

One further offense related to obtaining property by false pretenses is the offense of **larceny by trick**. Basically the same as false pretenses, this crime differs only in one respect. Under false pretenses, title is intentionally passed to the accused by the victim in reliance on the misrepresented fact. In larceny by trick, only possession is surrendered because of some artifice, trick, deception, or fraud. The victim expects to get the property back. For instance, Bud, who intends to convert a rental car to cash, goes into a car rental office and says he will use the car for a day. He pays the deposit required and signs the contract with a false name and address. When he gets the car, he sells it to the underworld for cash as he had intended. The car rental agency merely intended to give up its possessory interest for a short period of time.

10.11 OTHER FRAUDS

There are many other offenses that carry the word *fraud* in their title. Like most of the other fraud offenses previously discussed they all are closely related to or are examples of obtaining property by false pretenses or larceny by trick. Let's examine a few:

Odometer Fraud

When a person purchases a used car, he or she normally places a great deal of reliance on the mileage to ascertain the value of the vehicle, how much it should cost, an estimate of how long it will last, and how much maintenance it will require. On a late model, high–mileage car, the price can rise several hundred dollars for each ten thousand miles taken off the odometer. The offense is usually a serious felony in most states as well as under federal law, and it is a crime against ownership, not possession.

Title Fraud

Also an offense against ownership, **title fraud** occurs when an individual knowingly submits falsified paperwork to a state motor vehicle agency which would alter the ownership records of the state. This may be done in an effort to establish new ownership for a stolen vehicle, to modify the odometer reading on the paperwork to cover an odometer rollback, or for a variety of other reasons.

Insurance Fraud

Insurance fraud is big business. Insurance companies spend many millions of dollars in fraudulent claims each year and spend as much or more on investigative efforts to reduce fraud. In the long run, each of us, as consumers of insurance companies, pays higher premiums because of fraud. The insurer becomes the victim and the ownership of the insurer's money, obtained from our premiums, is the object of gain. Some examples are:

Staged Accidents

Some people make a living getting themselves "hurt" in automobile accidents just to file insurance claims.

Paper Vehicles

Insuring a vehicle that does not exist just to be able to report it stolen shortly after a policy is issued certainly is an attempt to defraud.

Inflated Theft Loss

When an insurance loss does occur, whether through a fire, an auto accident, or a theft, inflating the value of lost items is a fraud and if provable, can be prosecuted.

IV. RECEIVING OR CONCEALING STOLEN PROPERTY

10.12   GENERAL COMMENTS

The reader may, at one time or another, have heard the words *possession of stolen property* used to describe a criminal offense. This is a popular misnomer. The crime referred to by this phrase is accurately known as **receiving or concealing stolen property**. Statutes rarely, if ever, punish a person for merely possessing stolen property. Perhaps the reason for this, as discussed in Chapter 4, is that possession is a weak act and cannot be made criminal unless accompanied by a readily provable intent. The terms *receiving* and *concealing* include, by implication, possession with knowledge. Receiving and concealing also imply some overt physical activity on the part of the accused to satisfy the criminal law requirement that every crime consist of an act.

   Although some jurisdictions label this crime receiving and concealing stolen property and others call it receiving or concealing stolen property, the distinction is without a difference. There is no legal significance to using the different conjunctions. By almost unanimous decisions of the courts, the words *receiving* and *concealing* are treated as separate and distinct types of conduct, either of which may apply to any given case. This does not mean that in a particular case the accused may not both receive and conceal stolen property. It means only that one or the other is essential to every case.

   There is some evidence that the crime of receiving or concealing stolen property did not exist at early common law—at least not by this name. At early common law, persons who committed acts such as these were usually charged with either compounding a felony, a misdemeanor committed by agreeing not to report a felony or to withhold evidence, or misprision of felony, a misdemeanor committed by failure to report a felony without agreeing to do so (Chapter 4). The obvious shortcoming of these two offenses was that they were misdemeanors carrying light penalties even though the receiving or concealing may have been a severe offense and the value of the property may have been great. In light of these and other reasons, many modern statutes have made the crime of receiving or concealing stolen property a felony. Some states classify it as either a felony or a misdemeanor depending on the value of the property involved.

Some common law courts apparently did hold that receiving or concealing stolen property could also make one liable as an accessory after the fact. This, however, conflicts with the generally accepted view at common law and with the universally held view today that an accessory after the fact must render aid and comfort to the felon personally (Chapter 4). The act of receiving or concealing stolen property does not constitute personal aid. Of course, an accused could render personal aid to a felon and also render nonpersonal aid by receiving or concealing stolen property. In such an instance, the person may be liable for several different types of criminal conduct.

10.13   ELEMENTS

Although not all the statutes on receiving and concealing stolen property are identical, all contain essentially the same elements. These elements include (1) the receiving or concealing, (2) of stolen property, (3) knowing the property to be stolen, and (4) with the intent either to gain personally from the act or to prevent the rightful possessor from enjoying the property.

Receiving

The word *receiving* does not necessarily mean the receiver must have manual and actual control over the property. The receiver need not hold the property in his or her hands for any length of time. All that is necessary is that the accused exercise dominion and discretionary control over the property. How does a person gain dominion over property without actually touching it? The person can direct the property to be delivered to a certain place over which the accused has control. If this is done, the individual has received. For example, Sam tells Joe to put the stolen radios in Sam's garage. Sam never sees or touches the radios before his arrest for receiving stolen property. He has received.

The accused may direct that the property be delivered to a third person who works for him or her. In such a case, the accused has received through the agent. Obviously, if the receiver takes manual custody of the property, this element is satisfied.

Although often the receiver of stolen property may give something in exchange for the goods, payment of any type of consideration is not a necessary part of receiving. One who does pay for stolen property is commonly known as a "fence." The fact that one accused of receiving stolen property intended to retain the goods only temporarily will not affect the character of the receiver. The fence received just as if there was intent to keep the goods permanently.

Under most modern statutes, husbands and wives may be criminally liable for receiving stolen property from each other. This differs to some extent from the common law rule that the wife could not receive stolen property from her husband because the common law considered husband and wife to be one person. Even at common law, however, a husband could receive stolen property from his wife. One further exception recognized at common law and still recognized in most states today is that one who steals property may not be charged as a receiver or concealer of that same property. This rule was interpreted by the common law courts to include all principals in either the first or second degree. However, the rule did not and does not extend to accessories before the fact. This exception creates a dilemma. Consider those states that have abolished the distinction between principals and accessories before the fact and now classify them all as principals (Chapter 4). Suppose that John procures and counsels Bill to steal certain goods from Sam. Bill steals the goods and gives them to John, who receives them. Although John would have been an accessory before the fact at common law, he is now a principal. Can he be charged with receiving stolen property, or can he be tried only for the actual theft as a principal? Actually, the prosecution may do either. The prosecution can elect to charge him with receiving, or it can charge him as a principal, but it cannot do both. The choice will be made on very practical grounds. Which will be the easiest charge on which to convict him? If the prosecution chooses to charge him with receiving, can John raise the defense that he is a principal and thus not subject to prosecution for receiving stolen property? The answer, of course, is yes. But by raising this defense, John is admitting his guilt as a principal, and he can then be tried on this charge. The converse is also true. If John were charged as a principal in the theft and based his defense on the rule that the thief may not be the receiver of property he stole, he would be confessing to a charge of receiving. A few states have avoided this problem by a finding that the assimilation of an accessory before the fact should be disregarded in cases such as this. In those states, the accused may be convicted of both receiving stolen property and the original theft.

The receiver of stolen property need not receive goods directly from the thief. Under early common law, this issue presented some problems. Today, it is fairly generally recognized that one may receive stolen property from another who was also a receiver rather than from the thief. Can a person be charged with the crime of receiving if he or she received the property from a person who did not know it was stolen? For example, John has heard that the Hope Diamond has been stolen. Billy, a 7-year-old child, finds the diamond but thinks it is a piece of glass, which he shows to his friend John.

John immediately recognizes it as the stolen diamond and gives Billy a nickel for it. John is guilty as a receiver of stolen property, because he knew the property was stolen. The fact that he received it from an innocent intermediary is immaterial. John does not even have to know the identity of the thief or who the rightful owner of the property is.

Concealing

Most state statutes, at least by implication, make receiving and concealing alternatives to satisfy the elements of this crime. Consequently, one may be guilty of concealing stolen property without ever having received it under any of the rules just discussed. At first, it might appear inconceivable that one could conceal property without having at least some limited dominion or control over it. Liberal interpretation of the statutes prohibiting concealment of stolen property has avoided this logical overlap by treating cases of aiding another in concealing stolen property as violations of the statute. This is true also because the courts have tended to construe strictly the word *receiving* within the context set out in the previous section. It must be remembered that one may be guilty of both receiving *and* concealing stolen property, but such conduct will be considered a single offense. There is some authority to the effect that, even though the thief cannot be guilty of receiving stolen property, the thief can be guilty of concealing property that he or she stole. The distinction can be justified on the basis that receiving is inherently a part of stealing, whereas concealing is not necessary when the theft takes place. Nor is concealment necessary later on, even though it is the natural thing to do. If, however, the thief does conceal the property, it is usually a separate and distinct act. In any case, if a particular jurisdiction allows a thief to be charged with concealing property he or she stole in a separate indictment or information, the authority will be based on the statutes of that jurisdiction.

A person does not have to hide stolen property under the bed to be convicted of concealing. Neither the statutes nor the courts have required a manual hiding of the property. All that is really required to satisfy this element of the crime is that the accused actively help the thief do something with the property that will either secure the goods to the benefit of the thief or in some other way prevent the true owner from using and enjoying the property. Perhaps the most notorious method of concealing stolen property is connected with auto theft. Persons who had no connection with the actual theft of the vehicle spend a great deal of time disguising the stolen property by changing the color of the car, removing identification numbers, switching engines, and more.

To illustrate the principles discussed in this section, let us consider the following situation. John steals a color television set. Bill, a friend of John's, knows that the set was stolen by John. Bill suggests a plan to him to help prevent the police from discovering the location of the set. Bill tells John to deliver the set to Bob's E-Z Pay Repair Emporium, pretending it needs service. Bill explains to John that this gimmick will put the set right under the noses of the cops and anyone else, for that matter. John delivers the set to Bob's and places it on a counter in the front of the store, where it can be seen by anyone passing the repair shop. This situation illustrates the two basic rules just discussed. The fact that Bill never exercised any physical control of the color television set does not prevent him from being charged with the crime of concealing stolen property. His active participation in helping John satisfied this element of the crime. Neither does the fact that the property was not hidden from the view of the public lessen his criminal liability. Bob may have a problem also. If a police investigation shows that Bob was aware of the fact that the color television was stolen, he could be guilty as a receiver of stolen property. This would be true even though he did not communicate this knowledge to John when the set was brought into his place of business. If Bob knew the property was stolen, he could also be guilty of concealing the stolen color television set. His act of allowing the set to remain in his shop for the purpose of depriving the true owner of the owner's rights in the set would constitute actively helping the thief conceal the stolen television.

Stolen Property

At common law, stolen property, for the purposes of this crime, could only be property taken through larceny. If the property in question was taken during commission of some other theft offense, such as false pretenses or embezzlement, or was taken during the commission of any number of closely related offenses such as burglary or robbery, receiving stolen property could not be charged. Recognizing the illogic of this situation, state legislatures and state courts expanded the scope of the definition of stolen property by statutes and judicial decision. In most states today, there are either statutes or court decisions holding that property taken during commission of any offense that would justify calling the property "stolen" is treated as stolen property for purposes of the receiving and concealing statutes.

Once property is stolen, does it always remain stolen? Can the property ever lose its character as stolen property? Can it lose its character as stolen property and still be possessed by the original thief? These and other questions are important in cases involving receiving or concealing stolen property.

Although it is possible for property to remain stolen forever, it can lose its stolen character. There are any number of ways for this to occur. If Al, the rightful owner of stolen property, decides, after the theft, that he wants to relinquish his rights to the thief voluntarily, he may do so. At that point, the thief obtains rightful possession and the property is no longer stolen. Of course, this does not cancel the original theft. The thief is still liable for larceny or some other theft offense, but now the thief may rightfully exercise discretionary control over the goods. Any person receiving goods from this thief or helping the thief conceal the property would not be liable unless the person received or concealed before the possessor relinquished possessory rights to the thief.

The same rule would apply if the true owner relinquished possessory rights to someone who had knowingly received stolen property. The receiver's liability for receiving would not be affected, but the receiver would become the rightful possessor of the property.

One further way in which property may lose its "stolen" character is by recapture. Once the victim regains control of property, anyone that the victim gives it to could not be convicted of receiving or concealing stolen property. Upon recapture by the victim, the property loses its character as stolen property. Recapture also includes voluntary return by the thief.

For someone to be liable for receiving or concealing stolen property, the property must be "stolen" as defined by the particular jurisdiction when it is received. If the property is not stolen when received, either because it has lost its character as stolen property or because it was never stolen in the first place, the accused cannot be convicted of the offense. This is true even if the accused thinks the property is stolen at the time and intends to receive stolen property. If the property is not, in fact, stolen, the crime has not been committed.

In light of the rules just stated, consider law enforcement antifencing or "sting" operations wherein police "sell" recovered stolen property to fences prior to returning the property to the rightful owner. Questions arise as to whether or not the property has lost its character as stolen because it has been recovered but has not yet been returned to the owner. In cases in which these operations have been successful, courts have supported the position that the property retains its stolen character.

Knowledge

One of the most controversial elements of the crime of receiving or concealing stolen property is knowledge. There is no doubt that the individual who actually knows property is stolen is guilty of receiving or concealing stolen property. However, would a person be guilty of this crime by believing or suspecting that the property is stolen but not actually knowing it? This is perhaps the most argued point. In answering this question, reference must be made to the wording of the statutes of the various jurisdictions. If the statute is worded to require the existence of actual knowledge, the accused may not be convicted unless it is proven the accused had actual knowledge. No mental status short of positive and affirmative knowledge will satisfy the requirements of such a statute. If the statute specifically requires "knowledge" on the part of the accused, there is little room for interpretation. There are not any states that statutorily require the defendant to have affirmative, actual knowledge because there would be few, if any, convictions. Not only would the actual knowledge be difficult or impossible to prove, but people would go out of their way to avoid learning if the property was stolen. However, some statutes are worded in such a way that it is unclear whether positive knowledge is required or whether mere suspicion or belief will suffice. Usually, statutes that require that the property be received, "knowing" it to be stolen, are the ones that create confusion. In states that use the word *knowing* as an element of the crime, some have interpreted this to mean actual knowledge, whereas others hold that a strong belief that the property is stolen will suffice to convict if, in fact, the property is stolen. Still others go so far as to say that, if a person receives property under circumstances that make the individual reasonably suspicious that the property might be stolen, that person may be found guilty of receiving stolen property. Again, the property must, in fact, be stolen. The suspicion may arise in many ways. For instance, the suspicion may arise from the nature of the property itself, the appearance of the seller, the time of day the proposition is made, the price, the location at which it is offered, and perhaps more. One might be suspicious of buying stolen goods if offered a television set at 3:00 A.M. in an alleyway, by a grubby-looking wino, at the unheard-of price of $25 with no state sales tax.

Depending on statutory wording, liability may attach to one who acquires stolen property without knowledge or reason to know its stolen character but subsequently learns or has reason to know the property is stolen. A Pennsylvania court upheld a state statute that punishes if a person receives, disposes of, or retains stolen property.

Intent

The intent required in the crime of receiving or concealing stolen property is the same as that required in larceny and the other theft offenses. The accused must intend to benefit personally by the act or must intend to permanently deprive the true owner of the property. Even though the person receiving stolen property knows it to be stolen, if that person takes it intending to return it to the true owner or rightful possessor, there is no crime. An investigator should not assume merely because an accused has received stolen property with knowledge of its stolen character, that the accused is automatically guilty. Intent must be proved.

V. OTHER THEFTS

10.14 GENERAL COMMENTS

One and a half million vehicles are stolen each year in this country. Despite reasonably high recovery rates, it cannot be said that most vehicles are stolen with only an intent to temporarily deprive the owner of his or her property. Although many vehicles are taken for joyrides, many are taken by complete strangers and many are taken specifically to be used during the commission of other crimes such as a drive-by shooting or a robbery. The clear intent in most of these cases is permanent deprivation.

Most of the unrecovered stolen vehicles are either exported to foreign countries, many of whom are not so concerned with previous ownership documents, or are processed through a chop shop where all usable parts are stripped off for later resale to dishonest parts stores or rebuilders. The remaining unusable parts are taken somewhere and abandoned or discarded.

As noted earlier, offenses that are relatively clear violations of existing laws are often given new titles or are treated distinctly for the purposes of enhancing punishment. So it is with auto theft. The theft of a motor vehicle satisfies all the elements of larceny. Because of the value of the property and the seriousness of the theft problem, it is treated as a separate crime. In some states the higher the provable fair market value of a vehicle, the more severe the punishment may be.

Thousands of boats and motors, together or separately, are stolen each year. Some are stolen for use by the thief after being disguised or are sold. Since boat hull identification number records and motor number records are not kept or are not so prevalent among state record-keeping agencies, it is often difficult to identify a stolen boat or motor. The theft of jet skis, wave runners, and other types of personal watercraft is increasing at alarming rates. Recovery and successful prosecution of watercraft thieves is limited by the sheer fact that these cases often require special investigative skills. When prosecuted, however, it is not that difficult to prove the elements of larceny in most instances.

Unauthorized access to computer files for the purpose of gaining something constitutes a theft in its clearest sense. What makes this relatively new activity interesting is the fact that the commodity being stolen is information. Whether this is the ultimate objective will require more analysis. Because computer crime is considered a commercial offense, it is covered in more depth in Chapter 22.

10.15 IDENTITY THEFT

Newspaper and magazine stories and television specials are continually alerting us to the dangers of **identity theft** and how this offense can be used to ruin reputations and credit histories. Theft can occur by someone stealing a wallet or purse containing identification but it can also be accomplished by the thief acquiring a Social Security number from some public record. Identity, like services such as electricity, is a nontangible item and thus does not as easily fit the standard elements of larceny. Consequently, legislation seeks to call attention to this offense, better define it, and create some specific penalties. As the problem becomes more rampant, and, in the opinion of many, reaching epidemic proportions, more and more states are joining the list of those who have enacted legislation making identity theft a crime. Consider the following Florida Statute: 817.568–Criminal use of personal identification information:

1. As used in this section:
   a. "Access device" means any card, plate, code, account number, electronic serial number, mobile identification number, personal identification number, or other telecommunications service, equipment, or instrument identifier, or other means of account access that can be used, alone or in conjunction with another access device, to obtain money, goods, services, or any other thing of value, or that can be used to initiate a transfer of funds, other than a transfer originated solely by paper instrument.
   b. "Authorization" means empowerment, permission, or competence to act.
   c. "Individual" means a single human being and does not mean a firm, association of individuals, corporation, partnership, joint venture, sole proprietorship, or any other entity.
   d. "Person" means a "person" as defined in s.1.01(3).
   e. "Personal identification information" means any name or number that may be used, alone or in conjunction with any other information, to identify a specific individual including any:
      1. Name, social security number, date of birth, official state issued or United States–issued driver's license or identification number, alien registration number, government passport number, employer or taxpayer identification number, or Medicaid or food stamp account number;
      2. Unique biometric data, such as fingerprint, voice print, retina or iris image, or other unique physical representation;
      3. Unique electronic identification number, address, or routing code; or
      4. Telecommunication identifying information or access device.
   f. "Scheme or artifice to defraud" means a systematic, ongoing course of conduct which is intended to defraud one or more persons, or to obtain property from one or more persons by false or fraudulent pretenses, representations, or promises or willful misrepresentations of a future act.
   g. "Specific use" means the particular transaction contemplated by the owner of the personal identification information when the

owner initially shared it, and may include continuous or repeated usage by the receiver of such personal identification information only if there is a knowing and voluntary agreement in writing or its electronic equivalent between the information's owner and the receiver of such personal identification information, allowing for continuous or repeated usage by the receiver of such personal identification information.

h. "Wantonly and maliciously" means willfully and purposely to the prejudice of an individual.

2. Any person who, in executing any scheme or artifice to defraud, willfully, knowingly, and without authorization uses or attempts to use personal identification information concerning an individual without first obtaining that individual's consent to the specific use commits the crime of fraudulent use of personal identification information, punishable as a felony of the third degree, as provided in § 775.082, § 775.083, or § 775.084.

3. Any person who, without legal justification or authorization, wantonly and maliciously uses or attempts to use personal identification information concerning an individual without first obtaining that individual's consent to the specific use commits the crime of malicious use of personal identification information, punishable as a misdemeanor of the first degree, as provided in § 775.082 or § 775.083.

4. This section does not prohibit any lawfully authorized investigative, protective, or intelligence activity of a law enforcement agency of this state or any of its political subdivisions, of any other state or its political subdivisions, or of the Federal Government or its political subdivisions.

5. In sentencing a defendant convicted of an offense under this section, the court may order that the defendant make restitution pursuant to § 775.089 to any victim of the offense. In addition to the victim's out-of-pocket costs, such restitution may include payment of any other costs, including attorney's fees incurred by the victim in clearing the victim's credit history or credit rating, or any costs incurred in connection with any civil or administrative proceeding to satisfy any debt, lien, or other obligation of the victim arising as the result of the actions of the defendant.

6. Prosecutions for violations of this section may be brought on behalf of the state by any state attorney or by the statewide prosecutor.

Other states including Arizona, California, Colorado, Georgia, Idaho, Kansas, Massachusetts, New Jersey, New York, Virginia, Washington, Wisconsin, and West Virginia have legislation punishing identity theft and, in many, the offense is a felony. Congress has also addressed the issue legislatively.

▓▓▓▓ ■ DISCUSSION QUESTIONS ■ ▓▓▓▓

1. A member of a private security patrol was assigned to guard a store that had been partially burned in a blaze of suspicious origin. He had keys to the store and was supposed to check both the inside and outside. During the night, the guard entered the store, took a couple of transistor radios, and put them in his car. The owner of the store discovered the loss shortly after the guard completed his tour of duty. The guard could be properly charged with what crime(s), if any, and why?

2. The defendant agreed to buy some apples from Smith for cash. Relying on the defendant's fraudulent promise that the cash would soon be paid, Smith let the defendant have the apples. Why would the defendant not properly be chargeable with larceny? What crime, if any, has the defendant committed?

3. Hotel X left a box of matches at the registration desk as a service to the patrons. A guest took the whole box of matches to his room. The hotel management had the house detective visit the guest in his room and inform him that the management did not regard the act in a kindly spirit and asked the guest to return the box of matches. The guest refused. An argument commenced, ending with many harsh words from both parties. The manager directed the house detective to obtain a warrant for the arrest of the guest. Would a charge of larceny be proper here?

4. If a diamond ring is lost and a person finds it and there is no clue as to the owner, it is not larceny if it is kept, although at the time it was found, the person intended to appropriate it for his own use. Why is this true?

5. Smith goes to Johnson's Bar and gives the bartender a $10 bill by mistake for a $1 bill. The bartender, believing it to be a $1 bill, rings it up. Smith soon discovers his error and the mistake in the change given to him. He asks the bartender to correct the mistake. The bartender refuses and an argument ensues, resulting in Smith's being forcibly ejected from the bar. Smith summons a deputy sheriff and demands that the bartender be arrested for larceny. Would the charge be correct?

6. Williams goes over to Sanduski's house next door to ask for his shovel to be returned. Williams needs it to do some gardening. Sanduski is not home but his garage is open. Williams goes and gets a shovel, believing it to be his. He has the intent to keep the shovel. In fact, Sanduski had returned Williams' shovel two months ago. Is Williams guilty of theft?

7. Joe Evergreen goes to Acme Used Cars to look for a "new" car. He sees a car he likes. The salesman tells Joe that the 42,000 miles reading on the odometer is all "highway" miles, not "stop and go" city miles and that it was a one-owner car. Joe buys the car believing the representations

made by the salesman. Joe later learns, after doing some checking when he ran into some unexpected repair bills, that the mileage is actual but the vehicle had three previous owners (He doesn't know how the vehicle was driven by those owners.), and Acme Used Cars had purchased Joe's vehicle at an auto auction. Has a crime been committed? If so, what crime and why? If not, why not?

▩▩▩ GLOSSARY ▩▩▩

Abandoned property – property which the owner no longer wants.

Animus furandi – the Latin term for the intent element in larceny.

Asportation – the Latin term for the carrying away element in larceny.

Bailee – the person who was entrusted with the possession of property.

Bailment – the result when a person entrusts another person with the possession of property.

Bailor – the person who entrusts another with possession of property.

Concealing stolen property – Putting property somewhere, knowing it was stolen.

Confidence game – obtaining property unlawfully by selling ones own personality.

Control – having physical custody of an item without having any right to do anything with the item.

Conversion – theft by converting someone else's property to one's own or another's benefit thereby depriving the owner of his/her rightful possession.

Custody – having physical control of an item without having the right to do anything with the item.

Dominion and control – trespassory possession unlawfully obtained by fraud.

Embezzlement – theft of property by someone to whom it has been entrusted.

False pretenses – theft of property by misrepresenting a past or present fact.

Identity theft – theft by using some else's Social Security number or drivers license to obtain goods.

Insurance fraud – filing a false claim with an insurance company.

Intangible – property without physical quality such as electricity.

Larceny – the common law offense of taking, carrying away, personal property of another, with the intent to permanently deprive the rightful owner or possessor of his/her property.

Larceny by trick – obtaining possession of property by artifice, trick, deception, or fraud.

Lost property – property, unintentionally parted with by the possessor under circumstances as to be unaware that it is missing.

Mislaid property – property that is placed in a certain location by the possessor who forgets to retrieve it.

Odometer fraud – reducing the mileage on the odometer of a motor vehicle to make it appear less used so as to increase the resale price to an unsuspecting customer.

Ownership – complete and unlimited discretion of the use of property by one who has legal title.

Personal property – anything capable of ownership except land or things permanently affixed to land.

Possession – the status of a person and property when the person has discretion in the use and handling of the property.

Receiving stolen property – one who exercises dominion and discretionary control over stolen property.

Shoplifting – theft of merchandise from a retail store.

Swindle – fraudulently obtaining property by gaining the confidence of the victim and presenting some idea that causes the victim to part with money or property; confidence game.

Tangible – property with a physical existence.

Theft – a general word for many forms of criminal stealing.

Title fraud – an offense against ownership of a motor vehicle or vessel by submitting falsified paperwork to a regulatory agency in order to obtain title to the vehicle or vessel.

Trespassory taking – when the taker has no right to possession of the property.

■ REFERENCE CASES, STATUTES, AND WEB SITES ■

CASES

People v. Wallace, 434 N.W.2d 422 9Mich. App. (1989).

People v. Weg, 450 N.Y.S.2d 957 (1982).

United States v. Werner, 160 F.2d 438 (NY 1947).

People v. Rife, 48 N.E.2d 367 (IL 1943).

Camp v. State, 89 P.2d 378 (OK 1939).

U.S. v. Prazak, 623 F.2d 152 (10th Cir. 1980).

Huggins v. People, 25 N.E. 1002 (IL 1890).

State v. Melina, 210 N.W.2d 855 (MN 1973).

Commonwealth v. Baker, 175 A. 438 (PA 1934).

Leonardo v. People, 728 P.2d 1252 (CO 1986).

Commonwealth v. Kelly, 446 A.2d 941 (PA 1982).

In re Crawford, 194 F.3d 954,958 (9th Cir. CA 1999).

Greidinger v. Davis, 988 F.2d 1344 (4th Cir. VA 1993).

In re Riccardo, 248 B.R. 717, 721 Bankr, S.D.N.Y. 2000.

Stevens v. First Interstate Bank, 999 P.2d 551, 552 (OR 2000).

STATUTES

United States: 18 USC 1028

Florida: 817.568 Florida Statutes

CHAPTER 11

Robbery

▓ KEY WORDS AND PHRASES ▓

Carjacking

Force

Larcenous taking

Robbery

Self-help doctrine

Threat of force

11.0 INTRODUCTORY COMMENTS

One of the first crimes we hear about as children is robbery. The "cops and robbers" game involves shootouts, wild chases, and "you'll never take me alive" scenes played with great gusto. Of course, a bank is always held up by gun-brandishing desperados. How surprised first-time students of the law are when they find out that robbery can be committed by persons other than gun-wielding thugs and in places other than a bank. We examine this common law crime and its statutory modifications element by element throughout this chapter.

11.1 DEFINITION OF ROBBERY

Robbery, a specific intent crime, is one of the hybrid or combination-type common law crimes. As a combination-type crime it involves all the aspects of assault as well as a larceny. Although personal property is taken, robbery is not classified as a crime against property but, rather, as a crime against the person. Any number of legislatures still put robbery in the midst of their crimes-against-property section. Whether this is done intentionally or not, the author does not know. When one illegally takes property with some value from another person or in the person's presence, by using **force** or by threatening to use force, the common law crime of robbery has been committed. Although the definition of common law robbery sounds simple and the combined elements of assault and larceny appear easy to apply, there are some unusual twists in the following pages.

11.2 THE TAKING IN ROBBERY

There must be a **larcenous taking** along with force or **threat of force** for there to be a robbery. Larceny, as was seen in Chapter 10, involves taking and carrying away the personal property of another intending to permanently deprive the owner of the property. If someone takes a gun and points it at the victim and says, "your money or your life," and nothing more happens, there is no robbery. This is assault and attempted robbery. Until the money has actually been taken and carried away, there can be no robbery. As in larceny, however, once the taker slightly removes property from the place it once occupied, the taker has done enough taking to commit robbery. If the gun-pointing accused removes a wallet from the victim's pocket but is caught before taking a step, the accused is guilty of robbery. It is not necessary that the robber actually take the property from the victim. There is sufficient taking if the victim delivers property to the robber under force or threat of force.

The property taken must belong to another and not to the accused. Although someone may have committed aggravated or merely simple assault, that person has not committed robbery when taking his or her own property from another by force or threat of force. This raises some interesting questions. Can a person who, in good faith, forcibly takes property thinking it is his or her own be convicted of robbery?

Suppose that Fred loaned John his lawn mower two years ago. John claims he does not have the mower. Fred sees his neighbor Sam using a mower identical to his own and he decides to get his mower back. Fred walks up to Sam with a hoe and tells Sam that he will hit him with the hoe if Sam does not give the mower to him right away. Sam complies and lets Fred take the mower home. Sam then reports the incident to the police. Fred is arrested, charged with robbery, and brought to trial. It is proved that the mower was, in fact, Sam's. Fred's defense is that he honestly thought the mower was his.

If the jury believes Fred's story, he should be acquitted of robbery. Mistaken identity is a commonplace occurrence that the law takes into account.

Because robbery is a specific intent crime, the defense of mistake of fact may apply (Chapter 6).

Another problem involves a repossessor of property or other zealous creditor-taker. Suppose that John owes Fred $700, which John has failed to pay. Can Fred forcibly take $100 from John as part payment of the debt and escape liability for robbery? Most states say yes as long as Fred had a clear-cut right to the money. When does a person have a clear-cut right to money? Basically, when there is a valid written contract that fixes the obligations and rights of the parties. What if Fred and John are involved in an automobile accident through Fred's fault? Can John go over to Fred and forcibly take money from Fred in settlement of the claim? No. The amount of Fred's liability has not yet been settled by a court judgment, nor have the parties entered into an enforceable settlement agreement. Therefore John has no clear-cut right to the money.

These situations raise some interesting questions. Can a person forcibly or by threat of force take property from someone else and escape liability for the crime of robbery? Yes. As illogical as this may sound, there are volumes of case law that support this view, even though some states hold to the contrary. Basically, these situations involve what is known as the **self-help doctrine**. One is entitled to self-help when one has a claim or rightful entitlement to property. This was the second situation. The $100 clearly belonged to Fred. The first situation also involved self-help, when Fred mistakenly believed he had a claim of right to the lawn mower. The third example, the auto accident, involved self-help with no claim of right. However, this does not mean that no crime has been committed in the first two situations. The law will not permit one to use violence or threat of violence to regain possession of property by self-help. In each of the first two examples, even though there was no larcenous taking to satisfy the elements of robbery, there was definitely an assault for which the taker would be criminally liable.

As in the case of all theft-related offenses, when consent is given or a claim of right exists, there is robbery if the accused takes more than he or she is entitled to under a claim of right or more than he or she has permission to take. If Fred were to take $800 from John instead of the $700 he was entitled to, Fred would be liable for robbery.

Property must be taken from the person or in the person's presence. What constitutes taking from the presence of the victim? The property does not have to be physically held by the victim or be on the victim's person. It merely has to be under the victim's control. Control in this sense means the right or privilege to use the property as the victim sees fit. Although the property has to be under the control of the victim, it is not essential that the property be visible to the victim when the crime is committed. Remember that robbery is primarily a crime against persons. John is guilty of robbery if he secures the keys to Fred's car by force or threat of force and makes off with the car even though Fred couldn't see the car at the time. The most important aspect is not the location of the property but the reality of the force or threat of force, which will be discussed in the next section.

With this in mind, let us consider the following problem. John wants Fred's valuable paintings that hang in Fred's home. John points a gun at Fred, forces him into a closet, and locks the closet door. John then removes the paintings and drives off. Is John guilty of robbery? Yes. The fact that Fred could no longer see John is immaterial, because Fred was threatened by John. It isn't essential that the victim and robber be side by side throughout the transaction.

Because the victim must only have control or the means of controlling the property, it is not essential that the victim have possession of the property. It is essential that the victim have custody of the property but no more. (The legal distinctions between possession and custody and control were fully discussed in Chapter 10.) It is possible for someone to be guilty of robbery even though the actor forcibly took stolen property from the original thief or from any other bare custodian of the property, such as a bus driver.

If we see robbery as an aggravated larceny, we might assume that the taking of one item by force or threat of force is one robbery no matter how many people are threatened while the one thing is taken. We also should have no difficulty in finding more than one robbery in the situation where the robber takes one or more items from each person present.

However, not all courts interpret their statutes this narrowly. The Pennsylvania Superior Court found multiple robberies where a robber stole by force from a business establishment. The threat to kill two employees while stealing from the business was found to be two robberies. Such an interpretation emphasizes the assault nature of the crime and not the theft aspect and confirms the "control or means of control" theory discussed earlier.

States continue to wrestle with the problem of whether larceny is a lesser included offense of armed robbery. North Carolina once said it was and in 1987 said it was not. In 1988, they changed their minds again and held that it is a lesser included offense. They recognized that there cannot be one without the other; that assault without larceny is only assault. Thus North Carolina rejoined the vast majority of states in their analysis.

He Wasn't Interested in All of the Elements

Michigan: The Ann Arbor News crime column reported that a man walked into a Burger King in Ypsilanti, Michigan at 7:50 A.M., flashed a gun, and demanded cash. The clerk turned him down because he said he couldn't open the cash register without a food order. When the man ordered onion rings, the clerk said they weren't available for breakfast. The man, frustrated, walked away.

Mega-Moron Award

Louisiana: A man walked into a Circle-K, put a $20 bill on the counter, and asked for change. When the clerk opened the cash drawer, the man pulled a gun and asked for all the cash in the register, which the clerk promptly provided. The man took the cash from the clerk and fled, leaving the $20 bill on the counter. The total amount of cash he got from the drawer? Fifteen dollars. (If someone points a gun at you and gives you money, was a crime committed?)

11.3 THE FORCE OR THREAT OF FORCE IN ROBBERY

We mentioned in the beginning of this chapter that robbery is a crime against persons. For an act to be a crime against persons, it must involve some physical or social force or threat of physical or social force directed at the victim's person, family, or property. Robbery must, however, involve physical force or the threat of physical force, not social force. Social force or intimidation is found in such crimes as extortion, blackmail, and criminal libel, for example, a threat to expose someone as a homosexual. This is a threat against one's social standing in the community rather than against one's personal physical safety.

Taking property without force is merely larceny. The force used to separate the victim from the property need not be great in robbery, however. Whether or not force was used is not a difficult question in most cases. Shooting someone to get property, hitting someone over the head to get it, or pushing a person down to get the property would all make the taker subject to prosecution for robbery. But is a purse-snatcher guilty of robbery? This depends on a factor known as resistance. Suppose that a woman puts her purse next to her on her bus seat but does not keep her hands on it. If a man quickly grabs the purse and runs, is he guilty of robbery? No. Neither the object nor the woman resisted. Would he be guilty of robbery if she had been tightly clutching her bag? Yes, because the victim resisted. Resistance does not have to be a fight to the death. Even slight resistance indicates that some force, greater than needed normally to take the article by simple larceny, was used.

Is the average, run-of-the-mill pickpocket guilty of robbery? Unless more force than is normally necessary to remove the article is used, the pickpocket would not be guilty of robbery. Taking property from a person by stealth is larceny, not robbery.

When must the force be used? If John takes Bill's wallet stealthily and then hits Bill over the head, is John guilty of robbery? No. Force must precede or accompany the taking. Force applied after the taking does not make the taker guilty of robbery. The accused is, however, guilty of the separate crimes of larceny and assault.

Suppose that property is taken by stealth, without force. The victim sees the thief making an escape and the victim chases the thief, trying to recover the property. Force used by the thief in the struggle that follows does not make this a case of robbery. Larceny and assault may be charged. This illustrates the principle of concurrence of act and intent (see Chapters 4 and 5).

To constitute a robbery it is clear that the theft must be accomplished while the victim is conscious. Rolling a drunken, unconscious victim over on his side to steal his wallet is only a simple theft despite the fact that the statute might punish "force however slight." The victim has to be aware of the threat or force used.

Thus robbery generally requires a preceding or contemporaneous use of force. However, not all courts apply the rule in the same manner. In a 1984 case the defendant had possession of the money box, was still on the victim's premises, and had just pried open the box when the owner–victim appeared. The defendant pushed the victim out of the way in order to escape while still clutching the open box. The Kansas Supreme Court held that the taking was not complete prior to the defendant's leaving the building. Not all courts would agree with this decision, but it does represent another way of looking at the issue of preceding or contemporaneous use of force.

One of the most frequently asked questions is whether a person can be convicted of robbery if the robber uses a toy gun. One federal court of appeals said that it can be a robbery. That court said its use places greater burdens on victims and police if it is not obviously a toy. If the toy gun's use created a risk that police and guards would respond to with deadly force, then that is enough to uphold the charge of robbery. But all courts do not agree, and some see this analysis as too subjective. Maryland says the armed robbery statute was intended to deter those actually capable of inflicting death or serious bodily harm. However, Maryland would concede a basic robbery as committed here using the toy gun but rejects only imposing the greater penalty for armed robbery.

Without using actual physical force, a person may be guilty of robbery if that person threatens force against the victim to secure his or her property. To repeat, force must be threatened against the victim or the victim's family's physical well-being, not against the victim's social position. If Taker tells Fred that he will tell everyone Fred is a bigamist unless Fred gives Taker money, there is no robbery even if Fred gives money to Taker. This is a threat to social position and not to physical well-being. If Taker points a loaded gun at Fred and says, "your money or your life," Taker has threatened force, so if Fred parts with his money there is a robbery.

Must the person threatened with physical harm be actually frightened to the point of panic? No. It is enough if the victim is reasonably apprehensive of harm—aware of the potential for injury. People can be apprehensive about harm to themselves, their families, or their property without actually being scared. However, not every fear entertained by the victim is sufficient to convert a larceny into a robbery. In one case a nervous and strangely acting defendant who made no threats engendered in the mind of the motel clerk that a robbery was about to happen in the lobby. On the basis of this fear, the motel clerk left the cash register area and locked herself in an adjacent office. The defendant went behind the counter and took all the money from the cash register. The Wyoming court said that one cannot focus on a victim's subjective overreaction when there is no purposeful threatening action on the part of the defendant.

On the other hand, if a real sense of fear is created, what role does "in the presence" play in a charge of robbery? Consider the following scenario. Fred enters Betty's house with her consent but after he is inside he takes out a gun and forces her to reveal where her valuable property is stored. She tells him that it is upstairs in a jewelry box. He leaves, goes upstairs, and takes the property. Can Fred properly be charged with robbery? One court that considered this problem said yes, it is still robbery even though the property was not taken from her person and Fred was no longer "in her presence." The court said that this was her home, her castle, and the threat continued as he took the property from the upstairs jewelry box.

According to some authorities, the threat may be directed against the physical well-being of the victim's property. If John threatens to blow up Bill's business building unless Bill turns over all his money, the force or threat of force element is satisfied according to these authorities.

I Give Up

New York: As a female shopper exited a convenience store, a man grabbed her purse and ran. The clerk called 911 immediately and the woman was able to give them a detailed description of the snatcher. Within minutes, the police had apprehended the snatcher. They put him in the cruiser and drove back to the store. The thief was then taken out of the car and told to stand there for a positive ID. To which he replied, "Yes, Officer, that's her. That's the lady I stole the purse from."

11.4 MODERN ROBBERY: STATUTORY MODIFICATIONS

As with most other common law crimes, robbery has been affected by most of our state legislatures. A few legislatures have put the common law crime in statutory form without any modification whatsoever. This is called codifying the common law.

Let us look at some of the changes made by a few states that illustrate the more common modifications found in all states. Illinois imposes a penalty of up to 20 years for common law robbery. A sentence of longer than 20 years can be imposed for violent or armed robbery. Illinois's neighbor, Indiana, also adopts the common law, but if a wound is inflicted a life sentence may be imposed instead of the 10 to 25 years imposed for ordinary robbery.

Michigan classifies robbery by name. It recognizes armed robbery, unarmed robbery, and bank safe and vault robbery. New York, although recognizing somewhat similar offenses, divides robbery by degree. North Carolina, a common law state, recognizes these separate offenses with greater penalties: armed robbery, train robbery, safecracking, and safe robbery.

Pennsylvania has ordinary general robbery, armed or violent robbery, robbery of bank vaults, and train robbery. A more severe penalty is imposed for robbery with an accomplice.

Although Wisconsin abolished the common law crimes, it adopted the common law definition of robbery and provided an additional penalty for armed robbery.

These have been some common changes in the definition of robbery made by state legislatures. Their primary purpose has been to increase the penalty for certain acts of robbery that seem more serious or pose more of a threat to society.

11.5 CARJACKING

Legislatures have reacted to reported incidents such as the one where men forced a woman out of her car to steal the car but the woman wouldn't let go after they removed her from the car because her infant child was in the back seat. As a result, she was dragged to her death. Although these acts contain all the elements of robbery, legislative reaction has been to increase penalties and call more attention to this robbery, now called **carjacking**. Two essential elements distinguish carjacking from conventional automobile theft: 1) the robbery occurs in the immediate presence of the automobile's driver or passenger; 2) the robbery is perpetrated by force or fear. These, of course, are the elements that distinguish all robberies from theft but is made a more severely punishable offense because it victimizes people in vulnerable settings and, because of the nature of the taking, raises a serious potential for harm to the victim, the perpetrator, and the public at large.

The California statute illustrates the severity with which this crime is treated:

§ 215. Carjacking

(a) "Carjacking" is the felonious taking of a motor vehicle in the possession of another, from his or her person or immediate presence, or from the person or immediate presence of a passenger of the motor vehicle, against his or her will and with the intent to either permanently or temporarily deprive the person in possession of the motor vehicle of his or her possession, accomplished by means of force or fear.

(b) Carjacking is punishable by imprisonment in the state prison for a term of three, five, or nine years.

Gun Totin' Granny Shoppin' for Justice

An elderly woman in California was returning to her car from shopping when she observed four males in it. Dropping her bags, she drew down on them. Screaming that she knew how to use her gun and would if she needed to, she kept her handgun pointed at them, ordering them out of the car. Apparently believing her, the four men got out and ran quickly away.

The woman picked up her shopping bags and loaded them into the back of the car. Sliding behind the wheel, she put her key into the ignition only to find that it wouldn't fit. Parked four or five spaces away sat her car, identical to the one she was in.

After transferring her shopping bags to the correct car, she headed for the nearest police station to tell them what happened. Upon hearing it, the desk sergeant was laughing so hard that he could barely point to the other end of the counter where four visibly shaken men were reporting a car jacking by a disturbed elderly woman. No charges were filed.

Bob Repik, Greenwich Police Department from, The APB, International Association of Auto Theft Investigators, November 2000, p. 20.

■ DISCUSSION QUESTIONS ■

1. Smith, Jones, and Walker approach Millie and falsely identify themselves as police officers. They claim that they have come to Millie's house under legal authority to repossess her furniture. They threaten to arrest her, handcuff her, and throw her into jail if she does not let them take the furniture. Millie submits and lets them take the furniture. Has a robbery been committed?

2. Lester was traveling through the country in his automobile. The engine became hot and he stopped his car to allow it time to cool. While waiting in his car he fell asleep behind the wheel. Al and Sam saw that Lester was asleep and went over to the car. Al grabbed Lester, shook him, and ordered him out of the car. Sam took the key from the switch. They told Lester that he was under arrest and that Al would take Lester before the judge unless Lester gave Al $100. Sam stood before Lester with clenched fists. Lester gave Al the $100. Can Al and Sam be charged with robbery?

3. Andy went to the Flea Bag Hotel and approached his good friend John, who was the desk clerk of the hotel, with a plan for a fake robbery of the hotel safe. Andy was to come in late one night and point a gun at John, whereupon John would hand over the keys to the safe. On the appointed night Andy showed up, walked up to John, pointed the gun, and

said, "This is a stickup—hand over the keys to the safe." At the same time, some 10 feet away there stood a hotel guest whose watch was in the safe. John handed over the keys and Andy cleaned out the safe. Was a robbery committed? Why or why not?

4. Al goes to an illegal casino. He gambles and loses heavily. He then pulls a gun and recaptures the money lost to illegal gambling. He is arrested and charged with robbery. Is he guilty? Why or why not?

▩ GLOSSARY ▩

Carjacking – taking a motor vehicle from another person by the use of force or fear.
Force – the use of physical violence against another person.
Larcenous taking – property must be taken by theft, with no right to possession.
Robbery – illegally taking property of some value from another person or in the person's presence, by using force or threatening to use force.
Self-help doctrine – one may recover one's own property from another.
Threat of force – the threat to engage in some physical violence against another person.

▩ REFERENCE CASES, STATUTES, AND WEB SITES ▩

CASES

Commonwealth v. Rozplochi, 561 A.2d 25 (Pa. Super. 1989).

State v. Hurst, 359 S.E.2d 776, 778 (N.C. 1987).

State v. White, 369 S.E.2d 813, 817 (N.C. 1988).

State v. Keeler, 710 P.2d 1287 (Kan. 1984).

State v. Bowman, 850 P.2d 236, 242-243 (Kan. 1993).

United States v. Martinez-Jimenez, 864 F.2d 664 (9th Cir. 1989).

Brooks v. State, 552 A.2d 872 (Md. 1989).

People v. Ortiz, 509 N.E.2d 633, 636-637 (Ill. App. 1987).

Lear v. State, 6 P.2d 426, 427 (Ariz. 1931).

People v. Antione, 56 Cal. Rptr.2d 530 (Ca. App. 1996).

STATUTES

California: Cal. Pen. Code §. 215 (West 2001).

Illinois: 720 Ill. Comp. Stat.5/18-2, 5/18-2 (West 2000).

Indiana: Ind. Code Ann. §. 35-50-2-4 (Michie 2001).

Michigan: Mich. Stat. Ann. §§. 750.529- 750.530 (Law Co-op 1999).

New York: N.Y. Penal Law §§. 160.00, 160.05 (2000).

North Carolina: N.C. Gen. Stat. §§. 14-87.1, 14-88, 14-89.1 (1999).

Pennsylvania: 18 Pa. Cons. Stat. §§. 3701, 3702 (2001).

Wisconsin: Wis. Stat. §. 943.32 (1999).

WEB SITE

www.findlaw.com

C H A P T E R

Burglary and Related Offenses

■ KEY WORDS AND PHRASES ■

Breaking

Curtilage

Dwelling

Entering

Nighttime

12.0 INTRODUCTION

People often say that they have been robbed when reporting a break in and theft of items from their home or car. In reality, they have probably been the victim of a common law burglary or a statutory modification of burglary.

12.1 BURGLARY DEFINED

The common law offense of burglary was defined as breaking and entering the dwelling house of another, in the nighttime, with intent to commit a felony therein. Each element of the offense is discussed in the following paragraphs.

Breaking

The element of **breaking** originally conveyed the idea that there had to be some type of forcible entry as the term, in its common usage, implies. Modern court decisions have somewhat extended this concept. Of course, if an accused breaks open a window or door of a dwelling house, this will satisfy the element, as it would if the accused merely had to turn a knob or twist a handle. There was some disagreement in the earlier cases as to whether pushing open a door that was already slightly ajar was sufficient to meet this requirement. Most later cases hold this is a sufficient breaking. The opening of a closed door or window will suffice regardless of whether it is locked. The modern test seems to be that, if force is used to remove or put aside something material that constitutes a part of the dwelling house, which is relied on to prevent intrusion, there has been a breaking. Pushing open a door that is ajar or tearing a screen over a window or opening a screen door—all would constitute breaking within the legal definition.

Breaking, as required in the crime of burglary, must be trespassory; that is, it must be unlawful. If someone breaks into a dwelling he or she has a lawful right to break into, that person cannot be charged with burglary regardless of the intent at the time. If Fred breaks into the house where his estranged wife lives, but which he owns, to take a color television set his wife purchased in her own name, this is not burglary.

The breaking must involve some part of the dwelling or of a building within the curtilage (building attached to or adjacent to the dwelling house used in support of the dwelling house). Burglary may not be charged when the intruder opens only a fence gate or other structure solely to enter the property on which the dwelling is situated.

The courts have not limited the element of breaking solely to breaking an outer portion of the dwelling. There are occasions when an accused may enter a dwelling house without breaking in, as when a door is left completely open, for example. If, however, an inner door or window must be opened to reach another room or other part of the house, and the accused proceeds to open the inner door, this act will constitute a breaking within the meaning of the offense. This should not be confused with a statutory crime called entering without breaking discussed later in this chapter. Breaking an inner door is considered breaking within the meaning of common law burglary.

When breaking out of the dwelling house will satisfy this element depends on the wording of the statutes in the various jurisdictions. Under the common law definition of burglary, breaking out of a dwelling was not a breaking, for it was not done with intent to commit a felony. Breaking out was done for the purpose of escaping. Under the interpretation of most modern statutes, breaking is also limited to breaking in. There are, however, a few cases on record for which the statute does not specify or imply the necessity of breaking in, for which it has been held that breaking out will suffice to charge a breaking.

The rules so far discussed have referred to an actual breaking, in which the accused acts with his or her own hand. Another means of breaking is recognized in most jurisdictions. This is called *constructive breaking*. When an accused manages to have the occupant of a dwelling open a door by using some trick or device and then forces his or her way into the dwelling, the law holds this to be a breaking within the meaning of this element. This does not mean that, if the accused enters with the owner's consent, the accused has broken in. If the occupant of the dwelling consents to the entry, there is no breaking. If the defendant forces another to break into the house or to open a door that the other has a right to open to gain access to the interior of the dwelling, there is a breaking. It is a constructive breaking if the accused has a confederate employed inside the house and the confederate allows the accused to enter without the consent of the occupant.

Almost Breaking!

Arkansas: Seems this guy wanted some beer pretty badly. He decided that he'd just throw a cinder block through a liquor store window, grab some booze, and run. So he lifted the cinder block and heaved it over his head at the window. The cinder block bounced back and hit the would-be thief on the head, knocking him unconscious. Seems the liquor store window was made of Plexiglass. The whole event was caught on videotape.

Entering

For an act to constitute the crime of burglary, there must be an entry as well as a breaking. The entry need not involve entry of the entire person through the break that was made. The slightest entry is sufficient to satisfy this element. The entry of a finger, hand, foot, or head is enough. This does not mean that only the entry of a part of the human body will constitute an entry. As in the commission of other crimes, the method used may also be a nonhuman agency, such as an animal or stick, or another innocent human agent. If an instrument is used to gain entry, it must be an instrument that is capable of helping the accused accomplish the felonious intent. It is not sufficient if the instrument is merely one that helps the accused break into the dwelling. If, to break into Steve's house to commit larceny, John throws a brick through a glass window, he has broken, but there is no entry into the house. The brick is not an entry within the definition of burglary. There is no way the brick can help John accomplish his intended purpose. It merely served as a tool to help him break in.

Can a person be convicted of a burglary with intent to kill by shooting through a screened window of the victim's house? The screen acted as a barrier, the barrier was penetrated by an object used by the accused to accomplish his or her purpose, and therefore the accused should be successfully prosecuted for burglary, assuming all other elements of the crime have been satisfied. The few courts that have considered this question agree with this conclusion.

Suppose, however, that in the previous example, the gun and the bullet are used merely to destroy the lock in the door to enable the defendant to commit a larceny in the house. Is there a burglary if the bullet enters the airspace? No, it was merely a tool used to gain entry.

Both the breaking and the entering must occur to warrant a burglary charge, and the entry must be a consequence of the breaking. The breaking must be done for the purpose of breaching the security of the house and allowing the perpetrator to enter. If, for example, Sam breaks a window in a dwelling house and then enters through an open door, there is no burglary. It is not necessary, however, that the breaking immediately precede the entry. It is burglary if the breaking occurs on Tuesday and the entry occurs on Wednesday, if it can be shown that this was all part of a single plan with one felonious intent.

Dwelling House

As with the crime of arson, described in Chapter 13, burglary is a crime against the habitation rather than a crime against the property itself. Therefore, the law is not concerned with the character of the property so much as it is concerned with the fact that it is occupied as a dwelling. A dwelling is defined in law as a place that is habitually used as a place in which to sleep. It can be seen that this is a rather broad concept and may include many structures that would not ordinarily be considered houses within the broad meaning of that term.

A structure is a dwelling house if it is occupied for the purpose of dwelling therein. An unfinished house that has never been and is not presently being occupied as a place in which someone habitually sleeps is not a dwelling even though it may be intended for that purpose at a later time. Until someone moves into the house and uses it as a dwelling, it does not take on that characteristic. It is not necessary, however, that construction be completed before the house will acquire the characteristic of a dwelling.

A building or other structure may serve a dual capacity and still be a dwelling. If Sam lives in the back of the store he owns, the premises are treated as a dwelling for purposes of the burglary laws. A structure that has once acquired the status of a dwelling may lose that status when it is permanently abandoned by the occupants. The occupants' temporary absence from the dwelling will not destroy the house's status as a dwelling. This is true even if the absence continues over a long period of time. If the occupant intends to maintain the house as a residence for dwelling purposes, it is a dwelling house.

At what point, then, can it be said that a dwelling has been abandoned as a dwelling? Consider the case of the defendant who broke and entered a house from which the residents were moving. Only a few bags containing some of the residents' possessions remained. The furniture was gone and the residents were not present. Was this house still a dwelling? Yes, said the Indiana Supreme Court. The house retains its character as a dwelling while the residents are in the midst of the move. Thus, only after the move is complete will a *dwelling* merely become a building we call a house.

Statutes have been added in some states to aid in the definition of the word dwelling in the context of burglary. Should a place where a nonowner sleeps be considered a dwelling, especially if the person sleeping there is a trespasser? Suppose that a building was formerly a dwelling but is now used for storage. Using the common law concept that a place can be abandoned as a dwelling, the storage building would not be a dwelling. Suppose, however, as in Texas, a statute defines a place of habitation as "a structure or vehicle that is adapted for the overnight accommodation of persons." Does the "lost character" analysis of the common law still apply? The Texas Court of Criminal Appeals felt that the lost character analysis should still be applied.

A dwelling house must be one that is capable of being occupied by human beings. The question arises as to whether or not breaking and entering a chicken coop is burglary under the common law definition. The answer has to be an emphatic no. But this leads to a discussion of another common law rule regarding the definition of a dwelling house. It was recognized at common law that more than one structure may serve as the dwelling house. As a result, the common law recognized that any structure within the curtilage may be subject to burglary. The **curtilage** includes any structures within the common enclosure that serve as part and parcel of the dwelling. This would include outhouses and, in more modern terms, garages. It is not necessary that the other structure be attached to the dwelling house itself as long as it is near enough and used as part of the dwelling situation. A barn may, under certain circumstances, be considered within the curtilage of the dwelling for purposes of common law burglary.

Perhaps the most famous case dealing with an outbuilding as part of the dwelling is the case known as the "corncrib" case. The defendant went upon the victim's land to a corncrib behind the victim's house. The defendant carried a burlap sack and a drill (auger and bit). He was apprehended just after the bit penetrated the air space in the bottom of the crib and before he could put his bag under the hole to capture the escaping corn kernels.

In convicting the defendant, the court held that the corncrib was a building within the curtilage that supported the house in its dwelling function (a place to eat) and also held that a breaking and entry had occurred even though no part of the defendant's body entered the crib because the auger and bit were essential to the planned larceny.

A **dwelling** is a place where people eat, drink, sleep, meet, and escape from the rest of the world. The curtilage doctrine recognizes that some outbuildings are necessary to support the home as a dwelling. Well-houses, outdoor rest room facilities, and for some, kitchens were located away from the house in separate buildings. Thus to break and enter these structures was to threaten and invade the dwelling.

However, there is a limit to the curtilage doctrine in spatial terms. Out-buildings separated from the house by a public road were not within the curtilage. Under modern law, courts consider use, proximity, and fencing in or out as factors in determining whether a building is within the curtilage of the home/dwelling.

The curtilage doctrine is still being explored. One court faced the issue of whether a person caught in the lobby of an apartment building could be convicted of burglary. It seems the defendant had to open a door to get into the lobby (breaking and entering); it was nighttime; he had burglar's tools, and admitted his intent to take property from an upstairs apartment. He was charged with burglary. The prosecution argued that a lobby is for guests, apartment dwellers, business invitees and licensees, and not for trespassers. The prosecution also argued that the lobby was under the shared control of both the apartment building owner and the apartment dwellers and should be considered part of the leased area for each dweller. Thus the lobby was in support of each dwelling and part of the curtilage. The court agreed and up-held the burglary conviction.

Of Another

Occupation rather than ownership is the essence of the subject of burglary. It is not necessary that the owner dwell within the dwelling house. If the owner also occupies the dwelling house, the owner may not commit bur-glary against it. On the other hand, if the property is leased to someone else who occupies it, the owner may be liable for burglary if the owner breaks and enters the dwelling intending to commit a felony. If the lawful tenant—occupant of a dwelling—breaks and enters it for an unlawful purpose, the owner may not complain that burglary has been committed, because the tenant is the occupant.

One area of much confusion to law enforcement officials seems to be the status of hotel and motel rooms and other such living quarters as a dwelling within the meaning of the burglary laws. The rules governing these situations go back to the definition of a dwelling house. It was stated that a dwelling house is a place habitually used as a place to sleep. A tran-sient who stays at a hotel for one night is probably not dwelling in that room, as it is not habitually used by the transient for that purpose. In such a case the landlord would be the victim of the burglary if the landlord resided on the premises, and only the landlord would have the right to complain. If it were the landlord who broke into and entered the room of a transient, burglary would probably be an improper charge. On the other hand, if the guest is not transient, the guest's hotel room is a dwelling if the guest intends it to be one, and a breaking and entering of the premises with the intent to commit a felony therein by the landlord would be a burglary.

In the Nighttime

At common law, the offense of burglary had to be committed at night. This meant that both the initial elements of the offense—the breaking and the entry—had to occur at night, but both elements did not have to be committed on the same night. If Fred broke into John's house on Monday night but did not enter, and returned during daylight on Tuesday to enter, the crime of burglary was not committed regardless of Fred's intent.

The definition of **nighttime** created problems for the courts. At early common law, nighttime was established as anytime between sunset and sunrise of the following day. Alterations of this rule produced the standard accepted by the later common law, to the effect that nighttime was when there was not sufficient natural daylight to make out a man's features. The fact that there was moonlight or artificial light available, so that features and identity could be established, was not material to this determination. If there was not sufficient natural daylight for this purpose, it was nighttime.

Some modern jurisdictions have kept this common law rule, but many states have changed the definition of nighttime. In some states, nighttime is defined as the period between one-half hour after sunset until one-half hour before sunrise. In other states, it is one hour after sunset to one hour before sunrise, and so forth. Still other states have resorted to the early common law rule and established the hours between sunset and sunrise as nighttime.

Several states no longer require that the offense be committed in the nighttime to charge burglary. Breaking and entering the dwelling house of another with the required intent carries the same penalty regardless of when it occurs. Another group of states has kept the element but has created a separate offense of breaking and entering the dwelling house of another in the daytime with intent to commit a felony therein. Usually, in these states, a daytime burglary carries a less severe penalty than does a nighttime burglary. In some states that have eliminated nighttime as a requirement of burglary, a presumption arises if the act did, in fact, occur at night. It is presumed that the intruder entered with a criminal intent but that intent would be to commit only a misdemeanor. This statutory presumption could be used by the prosecution if it was unable to prove the accused intended to commit a felony.

In states that have kept nighttime as an element of the offense, a problem often arises. Burglaries often occur when there is no one present in the dwelling. In such a case, it is often difficult to pinpoint the time when the burglary occurred. Because the element of nighttime must be affirmatively proved by the prosecution in those states, factual situations arise that may very well bar prosecution for this crime. Suppose that the Smith family leaves their house for a weekend visit. They depart on Friday afternoon and return early Sunday evening to find their house has been broken into and their furniture is gone. There are no witnesses, but Wilson is arrested for the crime after his fingerprints are found at the scene. The state is unable to establish when, during that weekend, the offense occurred. There is little chance the state could convict Wilson of burglary in the nighttime. There may be some other statutory offense for which he is liable, but it will not be nighttime burglary.

With the Intent to Commit a Felony Therein

To be chargeable with common law burglary, the intruder must break and enter intending to commit a felony. The intent is a specific intent (Chapter 5) that must be affirmatively proven by the prosecution. To prove this element, the specific felony the accused intended to commit must be shown, and it must be alleged in the indictment or information. If the prosecution cannot prove the intended felony, burglary is an improper charge. Any felony at common law would have satisfied this requirement. Today, any felony under the statutes of the state will satisfy this requirement. It is not necessary that the felony be larceny. This is a popular misconception due, in part, to the fact that larceny is the most common object of the burglary. If the felony is, in fact, committed after the breaking and entering, there is no merger. No other felony has elements in common with breaking and entering. Consequently, both the committed felony and the burglary are separately chargeable offenses.

The intent to commit a felony must be simultaneous with the elements of breaking and entering. Even if the breaking and entering occur on different dates, the intent to commit a felony must continue to exist along with both acts. This is so because the instant when the breaking and entering occurs determines whether or not the crime of burglary is complete. If the requisite intent exists at that time, the offense is complete despite any further conduct on the part of the accused. If this seems to imply that it is not necessary for the target felony to be completed, that is exactly correct. The fact that the intruder does not commit a felony once gaining entry to the dwelling is immaterial to a charge of burglary. The burglar's intent when breaking and entering completes the crime. If Fred breaks and enters a dwelling to commit rape or larceny, but once inside is prevented from accomplishing his objective or changes his mind, he has still committed the crime of burglary. Of course, the fact that he did commit a felony would lend additional weight to the state's contention as to his intent, but it is not an essential ingredient of the prosecutor's case. In fact, it is not even necessary that the accused could have committed a felony upon entry. Suppose that Jack enters a dwelling house to commit grand larceny, but upon entering finds there is nothing to steal. If he intended to commit the felony when he entered, this unknown fact will not serve as a defense.

Burglary is a specific intent crime. It is often difficult affirmatively to prove intent to commit a felony. As a result, legislatures in some states, to prevent this problem from being used as an escape mechanism, have created a presumption that, if intent to commit a felony cannot be shown, it will be presumed that the accused intended to commit a misdemeanor when breaking and entering. This, of course, is not common law burglary, but it is made a separate offense by these statutes. As with burglary, it is generally a felony. A similar problem concerns intent to commit larceny. Larceny was not divided into degrees at common law. All larcenies were felonies. The statutory distinctions between grand larceny (a felony) and petit larceny (a misdemeanor) would cause havoc with the burglary laws if no provision were made to eliminate this problem: How to determine whether a burglar who broke in intended to commit grand or petit larceny. Whether burglary is a proper charge may depend on the answer to this question. Several alternative solutions to this dilemma are followed by the various states. In some states, the problem has been solved by adding a few words to the burglary statute so that it reads, "... with the intent to commit a felony or petit larceny." Vermont and Michigan adhere to this method. Other states have eliminated larceny as one of the crimes that can be the object of a burglary. Still others word their statutes so that they read, "... with the intent to commit a felony or larceny." North Carolina and Illinois have solved the issue in this manner. Many jurisdictions have reached the same result through court interpretation of the burglary laws. These states have ruled that, even if the accused committed only petit larceny, if there had been more property to take or if the accused could have carried more out, more would have been taken, so the accused intended to commit a felony.

Where's The Remote?

Men have to have their remote controls! Just ask two burglars who wound up in jail. Police say a woman called to report a pair of men skulking around a neighbor's house. The caller told officers she had seen the men loading a television into a sport utility vehicle. When police arrived they discovered one suspect behind the wheel of the SUV and the other hiding behind the vehicle.

In the SUV, officers found two TVs, screwdrivers, a toy gun, a knife/brass knuckles combination, and a Halloween mask.

Under questioning, the suspects admitted stealing the pair of TVs. As they were leaving, they realized they forgot the remotes. When they went back, the police were there.

Tallahassee Democrat

12.2 STATUTORY MODIFICATIONS

A number of statutes have been enacted in each jurisdiction to expand the common law burglary offense. A number of these statutes deal with conduct that would amount to burglary except for the fact that one or more of the elements of burglary are missing. As a result, there are statutes providing criminal penalties for (1) entering without breaking (District of Columbia), (2) breaking and entering the dwelling house of another in the daytime instead of nighttime (Maryland), (3) breaking and entering a building, structure, or conveyance other than a dwelling house (Massachusetts), and (4) breaking and entering with intent to commit a misdemeanor (Massachusetts). In each of these statutes, all the elements but one are identical to the common law definition of burglary. For example, in (1), all elements of burglary are satisfied except that there is no breaking; in (2) it is not done in the nighttime; in (3), it is not required that the building be a dwelling house; and in (4), the intent to commit a felony need not be shown. The holes left by the common law definition of burglary are fairly well closed by the enactment of these and other statutes.

To avoid confusing any of these statutory crimes with common law burglary, they are usually classified as "breaking and entering" crimes or home invasion crimes (when the object is a home), rather than as a burglary. Actually, the difference is one of semantics. Most people use the terms *burglary* and *breaking and entering* synonymously to designate the entire group of offenses, including common law burglary. Technically, the distinction between these statutory offenses and the common law crime of burglary should be maintained. For example, the distinction might be important in a felony–murder case. Suppose a state statute provides that a homicide occurring during commission of, or attempt to commit, burglary is murder. Does this mean burglary as the crime was defined at common law, or does it include all the breaking and entering offenses? The answer to this question will depend on whether the courts of each state use the terms *burglary* and *breaking and entering* interchangeably. Some states distinguish the offenses by recognizing different degrees of burglary.

No matter how the courts interpret the statutes, one thing is certain. The interpretations given the elements of each of these crimes are identical to the meaning of the elements of burglary. In the crime of entering without breaking, entering means the same thing as it does in the crime of burglary, and so forth.

Some additional statutory crimes include breaking and entering an automobile and breaking and entering (or burglary) with the help of explosives or similar devices. The use of explosives generally increases the penalty.

12.3 HOME INVASION

There is a developing offense called *home invasion* that is almost a hybrid between burglary and robbery. Home invasion may be thought of as a specific type of burglary that involves the robbery of a home. Florida has such a law that punishes as a very serious (1st degree) felony, "any robbery that occurs when the offender enters a dwelling with the intent to commit robbery, and does commit a robbery of the occupants therein."

In many jurisdictions, home invasion differs from common law burglary in that it lacks the requirements of breaking and nighttime entry. However, it does require both a specific type of action (robbery) and a specific intent (the intent to rob).

▬▬ DISCUSSION QUESTIONS ▬▬

1. Mr. and Mrs. Green have occupied their old homestead for the last 50 years. Last week, their property was condemned by the State Road Department and taken to be used as part of the right-of-way for the new interstate highway. The Greens were to have been out yesterday morning, but, unknown to the state officials, they are still sleeping in the house. Last night, Paul Jones broke into the house after seeing the condemnation sign and thinking the house was vacated. Paul's purpose was to steal the plumbing and light fixtures before the house was torn down. Has Jones committed burglary?

2. Smith intends to rob Brown. Smith knows Brown goes for a nightly walk. One night, while Brown is out of his house, Smith hides in the bushes near the front door of the house. When Brown returns and puts his key in the door to unlock it, Smith springs from the bushes with a gun in his hand. Without a word, Smith pushes Brown through the partially open door and enters behind him. Smith proceeds to rob Brown of his valuables, which are worth an enormous sum of money. Has Smith committed burglary?

▬▬ GLOSSARY ▬▬

Breaking – unlawful means of accomplishing an entry.
Curtilage – buildings attached to or adjacent to a dwelling house and used in support of the dwelling house; sometimes, within a common enclosure.
Dwelling – a place that is habitually used as a place to sleep.
Entering – physical intrusion into the dwelling.
Nighttime – at common law, the time between sunset and sunrise on the following day; courts have varied the definition.

▬▬ REFERENCE CASES, STATUTES, AND WEB SITES ▬▬

CASES

Silverstein v. State, 963 P.2d 1069 (Alas. App. 1998).

Young v. State, 286 So.2d 76 (Ala. App. 1973).

State v. Chappell, 193 S.E. 924 (S.C. 1937).

Byers v. State, 521 N.E.2d 318, 319 (Ind. 1988).

Blankenship v. State, 780 S.W.2d 198 (Tex. Crim. App. 1988).

State v. Emry, 18 S.W.2d 10 (Mo. 1929).

State v. Moore, 95 S.W.2d 1167 (Mo. 1936).

United States v. Dunn, 480 U.S. 294, 334–335 (1987).

People v. Wilson, 209 Cal. Rptr. 808 (Cal. App. 1989).

State v. Lefebvre, 609 A.2d 957 (R.I. 1992).

Commonwealth v. Kingsbury, 393 N.E.2d 391, 392 (Mass.1979).

People v. Davenport, 332 N.W.2d 443, 446 (Mich. App. 1982).

People v. Figgers, 179 N.E.2d 626, 627 (Ill. 1962).

Gazaille v. State, 235 A.2d 306 (Md. App. 1967).

STATUTES

District of Columbia: D.C. Code Ann. § 22–1801(a) (2001).

Florida: Fla. Stat. § 812.135 (2000).

Massachusetts: Mass. Ann. Laws ch.266, § 16A (2001).

North Carolina: N.C. Gen. Stat. § 14–54(a) (2000).

Vermont: 13 Vt. Stat. Ann. § 1201(a) (2001).

WEB SITE

www.findlaw.com

CHAPTER 13

Arson

▓ KEY WORDS AND PHRASES ▓

Arson

Burning

Dwelling house

Willful and malicious intent

13.0 INTRODUCTION

Arson was considered a very serious common law offense. Today, arson and its modern counterparts are still considered heinous offenses. Arson consisted of willfully and maliciously burning the dwelling house of another. The common law definition of arson contained four essential elements to be proved.

13.1 WILLFUL AND MALICIOUS INTENT

As in other common law crimes, *mens rea*, or intent, was an essential element of arson. **Willful and malicious intent** had to be proven to support a charge of arson. If someone acted voluntarily and without a justifiable excuse, that person acted with the necessary mental state. Although it was said to be a general intent crime, arson did not require that the actor have any ill will toward the owner of the property. If the fire was set negligently or accidentally, however, the necessary intent could not be proven. Deliberate conduct was a necessary ingredient.

A rule similar to the felony-murder rule is applied to arson. In fact, it could probably be called the felony-arson rule. If one set fire to or burned a house while committing another felony, that person was liable to prosecution for arson. This was true even if the felon-arsonist did not intend to burn the house. The commission of the felony supplied the intent necessary to support an arson charge.

13.2 BURNING

The requirement of a **burning** would seem to present no problem at all. However, it presented, as it still does, one of the more complex problems of arson. A roaring blaze that consumes the entire structure is not needed to complete the crime, although, of course, it is sufficient to justify a charge of arson. Smoke damage is not enough. The distinctions between a scorch and a char are used to determine whether or not there is any burning. Scorching involves only discoloration of the surface. There is not enough of a chemical change in scorching to constitute a burning. On the other hand, charring involves ignition of a flammable substance to the point at which there is a noticeable chemical change.

For the law of arson, it is essential to show evidence of charring to prove burning. The question then is, how much charring is essential? Not much. If any part of the house, no matter how little, is consumed by the fire, it is enough. But it must be a part of the house or its fixtures. Burning personal property, such as furniture or drapes inside the house, is not arson unless these items, in turn, ignite and char the house itself. If a law enforcement officer catches someone in the act of setting a fire, the charge cannot be arson unless there is some charring. If there is none, the charge should be attempted arson.

13.3 DWELLING HOUSE

Almost as troublesome as burning is the term *dwelling house,* the second element to be proved. The definition of a **dwelling house** has two important aspects that must be understood. First, what is a dwelling house? Second, how far does it extend? Does the definition include other buildings and property beside the dwelling house itself?

The first question may come up when an unfinished house has never been and is not presently occupied as someone's home. Is this a dwelling house? No. Even if the builder-owner intends to reside permanently on the property, the house has not yet become a dwelling house. On the other hand, an unfinished house that is occupied is a dwelling house even though it is incomplete. Would it be arson to burn a store building that also serves as the storekeeper's home? Yes. The fact that a building has a dual purpose does not mean it is not a dwelling house.

Once a building acquires the status of a dwelling, can it ever lose that status? Yes. Permanent abandonment of a house will cause it to lose its status, but the occupants' temporary absence from their house does not make it lose its dwelling house status. Suppose John and his family decide to go on a world tour for a year but intend to return to their house once the tour is complete. Will intentional burning of the house make the defendant guilty of arson? The length of the occupants' absence is not the key. Their intent to return or not is the determining factor. This intent is determined as of the time of the burning. It follows, then, that the dwellers need not be present in the house at the time of the burning. The common law crime of arson was a crime against habitation or possession and not against property or, strictly speaking, the person.

The question arises as to whether or not certain modern structures and some not-so-modern structures in which people dwell can be classified under the common law as dwelling houses so that it would be arson to burn them. For example, would it be arson for someone to burn the mobile home or trailer of another? Because arson is a crime against habitation and not against property, the burning of such a structure would be arson as long as all other required elements are present. Wouldn't the same be true of a tent? The answer would no doubt be yes if the tent was someone else's dwelling.

Would this be true if the structure burnt is a camper being temporarily used as a dwelling? Permanency of residence is not a key factor. The test is whether the structure is being used as a dwelling at the time it is burned. It is important here to note that although the U.S. Supreme Court has determined that a motorized camper used as a dwelling is an automobile for search and seizure purposes, the Court's decision does not settle the dwelling issue for arson or burglary purposes. Criminal justice practitioners have to be very careful not to confuse search doctrines with substantive crime doctrines.

The second set of problems involves the territorial aspect. The definition of a dwelling house for purposes of the law of arson is identical to the definition of a dwelling house under burglary law. As a general rule, any structure within the curtilage of the home falls within the definition. The other buildings do not have to be attached to the house, but they have to be close enough so that it is possible for a fire to spread to the dwelling house if it is set in the other building or buildings.

13.4 OF ANOTHER

Because arson is a crime against possession or habitation, it is not essential that the owner dwell in the house as long as someone dwells there. If the owner was the possessor, the owner could not, at common law, be guilty of arson against his or her own dwelling. If the owner rents the house as a dwelling to another person, can the owner be guilty of arson against the property? Obviously yes. Arson is a crime against habitation. The tenant would be the rightful inhabitant. Can a tenant be guilty of arson against the house that he or she is renting? No, because the tenant is the possessor and inhabitant. The inhabitant need not be rightfully on the premises. Even a building inhabited by a trespasser is protected by the arson laws.

13.5 STATUTORY MODIFICATIONS OF ARSON

Of course, there were obvious shortcomings to such a limited burning crime. Other burnings, because they were only attacks against possession, were criminal trespasses. Unless the fire spread from a nondwelling to a dwelling that then burned, there could be no arson charged. Second, with the advent of fire insurance, there were those who burned for profit. Finally, why should a person escape liability for burning merely because that person burned his or her own dwelling? Responding to these and other problems, many state legislatures have modified their arson laws by adding to the type of buildings and property that can be burned, by making it a crime to burn with intent to defraud an insurer, or by doing both.

Even though a number of legislatures have in some way extended or modified arson laws, certain common problems and concepts remain constant. Arson statutes that retain the dwelling house element still involve the problems of what constitutes a dwelling house. The problem of what constitutes *burning* under modern statutes is still settled on the basis of the common law tests.

A number of states now label the burning of other buildings besides dwelling houses as arson. There can be some interpretational problems with some of these statutes. For example, one state punished the burning of "other buildings." The defendant burned a structure that had four walls but no roof. Was this a building? The appellate court felt that without a roof, this structure was not a building. The legislature of that state later cured the problem by adding the words *or structures* to the statute.

Many states label the burning of someone else's personal property as arson. In most of these statutes, the value of the other's personal property is specifically set out.

The common law did not specify the time when the burning had to take place. However, modern statutes in some states impose a more severe penalty if the burning takes place at night.

A few states including Maine and New Jersey have made it arson to burn one's own property. This represents an extension of the common law. The only time the common law allowed prosecution of the resident for burning one's own dwelling was when the dweller hired another to burn the dwelling or aided and abetted the burning. (See the discussion of parties to crimes in Chapter 4.)

Maine requires a specific intent to commit arson. New Jersey requires a specific intent in its aggravated arson statute.

Some states have created a specific crime labeled *attempted arson*. At first, one would not think this was necessary because of the existence of general attempt statutes in all jurisdictions. However, the attempted arson statutes generally impose a more severe and definite penalty including a fine. In most of these states, attempted arson is a felony. Oklahoma is among these states.

Many jurisdictions have enacted separate statutory punishments for burning schools, churches, or other public buildings. In these statutes, the punishment provided is more severe than it is for general arson. The punishment in Louisiana for simple arson of a religious building includes a fine of not more than $15,000 and no less than two years of hard labor. The burning of a religious institution in Tennessee constitutes the crime of felony aggravated arson.

13.6 BURNING WITH THE INTENT TO DEFRAUD AN INSURER

Perhaps the most innovative change in this field is the enactment of statutes making it a crime to burn or attempt to burn any building or personal property of a certain value to defraud an insurer. Unlike arson, this is not a crime against habitation.

Although the statutes are not exactly the same in each state, certain common principles have been developed.

Primarily, the crime is a specific intent crime because the fundamental basis of the offense is intent to defraud. The mere fact that there is an insurance policy on the property does not, of itself, bring the burning under these statutes. If a person burns his or her own property, knowing it to be insured, for the purpose of collecting the insurance, there is no doubt that the individual has breached this statute. Suppose, however, that someone believes his or her property is insured but, in fact, the policy is not in force. Does this burning of property to gain the proceeds make that person guilty of this crime? Yes. Under most, if not all, of these statutes, the validity of the insurance is not an element as long as the defendant thought the insurance was valid. Michigan requires, however, a valid and in-force policy for such a conviction.

Some states still recognize the common law classification of parties to crimes as principals in the first and second degree and as accessories before and after the fact. In those states, there can be only one principal in the first degree for this crime. That is the person or persons who will directly benefit financially from the proceeds of the insurance policy.

Some of the states have made this crime a degree of arson, making it a lesser included offense. The others have maintained the arson statutes separately, so it is possible in some states to be charged with and convicted of both. Wyoming makes the use of arson to defraud an insurance company a second degree felony.

A common burning to defraud scheme involves what appears to be underinsured buildings, but a close look at such cases reveals a series of sales that escalate the paper value of the building. The last owner then insures the building for considerably less than the apparent (on paper) value. Thus A sells to B, who sells to C, and so forth, in a series of "sales" taking place over a few months. Each sale is for several thousand dollars more than the previous sales price. After the final owner insures it for less than it is "worth," the burning occurs and the owners think that the "underinsurance" will turn the focus of suspicion away from them. However, insurance companies are wise to this game and it is becoming a less common method of attempting to defraud insurers.

13.7 PROBLEMS OF PROVING AN ARSON CASE

As economic times get harder, it follows that the number of arson cases will rise. It is the belief of fire chiefs that one-third to one-half of all fires are deliberately set. Nearly 1000 people annually lose their lives in fires purposefully set.

Agencies responsible for the investigation of arson cases have insufficient personnel to handle the cases. There appears to be a direct inverse ratio of the number of investigators to the number of arson cases. Communities with the highest numbers of investigators have the fewest intentionally set fires. But there are also problems in defining the appropriate agency to handle arson investigations and pursue arsonists. In many places the fire department is charged with the responsibility, whereas at the same time the police believe it should be part of their function. Another problem is the need for more experts in forensic science for the gathering and analyzing of evidence in arson cases.

To prove an arson case, it must be established that the fire was not accidentally caused, thus showing that the fire was intentionally set or was of an incendiary nature. Once the *corpus delicti* of the crime has been shown, the guilt of the defendant must be proven. Too often this has to be done by circumstantial evidence. Circumstantial evidence must show more than mere presence of the defendant at the scene. The evidence must exclude every reasonable hypothesis of guilt, and all the evidence must be consistent with the hypothesis of guilt. This is where motive can play an important part in the proof of a crime. For instance, credible evidence of a defendant's financial difficulties may be used to support a finding of guilt when coupled with other sufficient evidence.

■■■■ DISCUSSION QUESTIONS ■■■■

1. John owns a home that he decides is no longer of any use. He sets up an elaborate electrical system hooked to an alarm clock. When the alarm clock rings, the electrical current will set off a spark that will ignite a five-gallon can of gasoline. He sets the alarm and leaves the house. A severe electrical storm moves across the city. Lightning hits the house, ignites the gasoline, and the house burns down. Under the common law could John be charged with arson? Under modern law can he be charged with arson?

2. Fred goes to Ben, the insurance agent, and fills out an application blank for home insurance that covers losses arising from a fire. Fred leaves a check with Ben. Ben forgot to tell Fred that he has no authority to insure the house until the home office approves the policy. That evening Fred returns to his house after going to a movie. Fred sees a wastebasket on fire in the house and decides to let the house burn because he thinks it is insured. The application blank is in the mail and has not yet been received by the company. Can Fred be charged with any crime under common law? Under modern law?

GLOSSARY

Arson – a crime against habitation; at common law willfully and maliciously burning the dwelling house of another.
Burning – ignition of a flammable substance to the point where there is a noticeable chemical change.
Dwelling house – a place where a person habitually sleeps.
Willful and malicious intent – deliberate and unlawful.

REFERENCE CASES, STATUTES, AND WEB SITES

CASES

California v. Carney, 105 U.S. 2066 (1985).

STATUTES

Louisiana: La Rev. Stat. Ann. § 52.1 (West 2000).

Maine: Maine Rev. Stat. § 802 (2000).

New Jersey: N.J. Stat. Ann. § 2C: 17-1 (2001).

Oklahoma: Okla. Stats. Ann. § 1404 (2000).

Tennessee: Tenn. Code Ann. § 39-14-302

Wyoming: Wy. Stats. § 6-3-102 (2001).

WEB SITE

www.findlaw.com

CHAPTER

14

Forgery and Related Offenses

■ KEY WORDS AND PHRASES ■

Counterfeiting

Forgery

Legal force

Uttering

Worthless checks

14.0 INTRODUCTION

At early common law, there was a group of misdemeanors called *cheats*. Among these offenses was the crime of forgery. Counterfeiting and uttering forged instruments were separately recognized common law crimes. The worthless check violations were not known at early common law, primarily because checks are a comparatively recent innovation. In this chapter we will discuss each of these crimes, their common law backgrounds, if any, and the effect of modern decisional and statutory law.

14.1 FORGERY

Forgery is defined as falsely making or materially altering a writing that, if genuine, would be of legal efficacy or the foundation of legal liability intending to defraud or prejudice the rights of another. To illustrate the difference between intent to defraud and intent to injure, let us consider this example. John is the registrar of Big University. Fred, a graduate of that institution, wants to take postgraduate work at Small College and writes John to send a transcript of his work to Small College.

John does not like Fred and forges a transcript of what appears to be Fred's grades. The forged transcript shows that Fred (who was an honor student while at Big University) barely graduated. In this situation, John intended to injure Fred.

Using the same basic facts, suppose that John was a friend of Fred. John forges a transcript that shows Fred to be a perfect "A" student even though he was not. Here, John's intent is to defraud Small College. Admittedly there is but a shade of difference between the two, but there is a difference.

The Intent to Defraud or Injure Another

Unless intent to defraud or to injure another is proven, there is no crime. People in everyday life write on and change written instruments, many of which are the basis of legal liability. There are times, for instance, when people inadvertently leave blanks on checks. A person cannot be prosecuted for filling in these blanks unless the intention is to get something he or she is not entitled to or unless that person intends to injure another. Intent, as in other crimes, is proved by the surrounding circumstances. If the date blank was filled in, could a fraudulent purpose be found? It would not be fraud to fill in the correct date or in most instances, to fill in any date the parties agreed upon. However, what if the true date of issue would make the check or document "stale" so that it would not have any legal efficacy or force anymore? Would it be forgery to fill in a date that would make it "fresh" and negotiable? Unless the maker (who wrote the check) agreed, it would be forgery. The same rule would apply to changing a date to make it a good instrument.

Bill writes a check for $25 naming Sam as the payee. The check is written and dated February 1, 2001. Under the laws of the State of Blackacre, "No check shall be honored or paid if it is more than one year old. Such a check is null and void." On January 1, 2002, Sam changes the last number "one" in the year numerals to make it look like a "two" so that the date line reads February 1, 2002. Sam did this to cash the check. This is a forged instrument because Sam changed the date to make a stale check fresh and negotiable. Filling in the amount on a document could also be criminal if it is done with intent to defraud or injure. Whether filling in the payee's name would be forgery depends on the name placed there. Every day checks are given to people with the payee line blank so that the payee can affix a store's rubber stamp. When this is done with the consent of the maker, there is no forgery. If an employee of the store were to put his own name in the blank instead of the rubber stamp of the store, forgery is likely. The next step would be to inquire whether the intended payee, the store, gave the employee the right to insert his own name.

Although the accused must intend to defraud or injure another, the law does not require that the fraud or injury intended be successfully concluded. Only the completed act and the necessary intent are required for there to be forgery. As a matter of fact, it is not essential to prove that the forger would have personally benefited if the scheme had been successful. It is sufficient if someone else is injured or defrauded.

Instruments Subject to Forgery

Only a written instrument or document can be forged. But the definition of an instrument or document is as broad or as narrow as the courts and the legislature want it to be. The category of instruments commonly subject to forgery includes bills rendered upon completion of services; checks, canceled and uncanceled; money orders; cashier checks; bonds; deeds; bank notes; bank account books; bank deposit slips; promissory notes; mortgages; and all other commercial paper. Besides these commonly known instruments, there are some that would not normally come to mind. For example, a fraudulent letter of recommendation or character reference would render the maker liable for forgery. A high school or college diploma can be subject to forgery. Tickets to a circus, theater, or other events can be forged.

It is not forgery to imitate a manufacturer's label, because labels are not treated as legal documents. For example, Art, a local producer of jams and jellies, is having a hard time selling his products because they are not well-known. He decides he would sell more if they bore the same label as that of the leading brand of jams and jellies, called Spanky's. Art buys a printing machine and engraving equipment and prepares labels that are the exact duplicate of the Spanky label. He puts the labels on his jars and offers them for sale. This is not forgery, but it would probably be counterfeiting, as we shall see later in this chapter. Would it be forgery for someone to paint a picture and then sign some great painter's name to it? Generally no. This would be a counterfeit, but paintings, sculpture, and so on are not subject to forgery, because they are not documents.

Legal Force of the Instrument

The document or writing must have some apparent **legal force** or efficacy. It must have the potential to be legally binding, to create or destroy legal liability, or be liable to defraud someone. Unless a document has this potential, it can be forged to the hilt without making anyone liable for the crime of forgery.

How is it determined whether the forged writing or document has this potential? If the instrument is invalid on its face, it has no legal significance and there can be no charge of forgery. For example, most states require at least two witnesses to a valid will. Would a charge of forgery be proper if an accused offers a forged will for probate that has no witnesses' signatures? No.

No one can be defrauded by this instrument, because it is void, without legal effect. The omission of some item without which the instrument would be absolutely void prevents the charge of forgery.

Do mistakes in spelling or mere form prevent the fraudulent maker from being charged with forgery? Formal errors such as these do not relieve the maker of criminal liability. Sloppiness counts. In the case of signature forgeries, all that is required is that the name bear some resemblance to the actual name sought to be forged. This is not to say that there must be a resemblance in handwriting. This is not necessary. There is one limitation, however. A jury is permitted to consider whether the attempted forgery is so imperfect or inaccurate that it could not possibly deceive a person of ordinary prudence.

If some required stamp or seal is missing from the instrument, can the fraudulent maker be prosecuted for forgery? If the missing item does not make the instrument void, a charge of forgery is proper. However, if the instrument has no legal effect without this missing item, no charge of forgery can be upheld.

The Act Required in Forgery

There are any number of ways in which an instrument may be forged, referring to either the actual device used to make the forgery or the method employed. As to devices, forgery can be accomplished by ink or ball-point pen or by pencil or other writing devices. It can also be accomplished by any printing method such as lithography or mimeography. One can also forge by engraving.

Alteration of the written instrument is one of the methods that may be employed. An alteration occurs when a material portion of the instrument has been crossed out or erased. One can alter an instrument by adding material terms to the instrument, by writing or typing additional terms on the instrument, or by improperly filling in blanks, intending to defraud.

Perhaps the most popular method of forgery known involves the use of signatures. It is forgery to sign another person's name to an apparently legally effective instrument, intending to defraud. If Sam somehow secures one of Fred Smith's checks and then signs as Fred Smith with the necessary intent, there has been a forgery.

There are times when a person has the power to sign on behalf of someone else. Corporate officers are agents authorized to sign papers on behalf of their corporations. The corporation is considered a "person" or legal entity under the law. If an officer exceeds his or her authority in signing something, with intent to defraud, there has been a forgery. If the officer exceeds such authority without intent to defraud, he or she may be civilly liable but has committed no forgery. Other examples of agents with power to sign on someone else's behalf are partners, those holding powers of attorney, and public officials.

A forgery may be committed when a person signs with a fictitious name. Use of an assumed name or alias with intent to defraud on a document having potential legal efficacy is forgery. The authorities are split over whether it is forgery to get someone to sign something through fraud. Some states, such as New York, which has a statute prohibiting fraudulently obtaining a signature, say this is forgery, whereas others say that it is not. If it is not forgery, what crime would it be? (See Chapter 10.) Arizona refers to fraudulently obtained signatures as obtaining a signature by deception and charges violators with a misdemeanor.

Would it be forgery if John signs his own name to an instrument that he pretends is a receipt for payment when in fact it is not a receipt at all? Yes. This is a false instrument even though the signature is good. On the other hand, if someone fills in a financial statement with false facts and signs his real signature, would this be forgery? Yes, and the same would be true if a person took someone else's financial statement and added his own name or took a financial statement prepared by an accountant and made changes in it to make it look better.

If two men have the same or similar names, under what circumstances could a forgery arise? Suppose Joe is a bank officer with authority to grant loans. Joe knows that a certain businessman in town is a good credit risk, but he has never met this man. A man by that name comes into the bank and asks for a loan, saying that he is the businessman but, in fact, he is not. Joe grants the loan and the man signs his real name to a promissory note. Has there been a forgery? Yes, because he has intentionally created a false instrument by deceptive practices. Could he be charged also with false pretenses? (See Chapter 10.)

14.2 UTTERING A FORGED INSTRUMENT

At common law anyone giving or attempting to give another person an instrument knowing it to be false, with intent to defraud, was guilty of the common law misdemeanor of uttering a forged instrument. The intent required for this crime was the same as required for forgery.

An Instrument Known to Be False

The person passing or attempting to pass the instrument must know of its ungenuine character to be charged with the crime. Someone who knows the instrument is false can give it to another innocent of that knowledge to pass it on. In this situation, the innocent person cannot be charged, but the one who knows the character of the instrument, using the innocent, may be charged with the crime of uttering.

The Utterer

Uttering was a common law offense separate from forgery. It was not and is still not essential that the utterer be the forger. But a forger was almost always an utterer except when the one doing the forgery passed the forged instrument to another guilty party for use in a fraud scheme. But even in that situation, the forger could be charged as a principal and was truly an utterer also. So it is a problem of semantics. A number of states have the two offenses, one called *forging* and a separate offense called *uttering a forged instrument.*

The Act of Uttering

As with other crimes, there must be an act. The act of **uttering** occurs when a person, who knows an instrument is false, attempts to pass it as the real thing. No one need take the forged instrument. Simply offering the instrument as genuine, by word or conduct, is all that is needed. The key word here is *offer.* The law considers mere offering of a forged instrument, in an attempt to get property or other rights, enough of a social harm and, therefore, does not require that the final object be obtained or accomplished.

Although property does not have to be exchanged for the forged instrument, the person to be defrauded has to receive the forged instrument for there to be a sufficient act. Until there is actual receipt by the intended victim, no uttering has been committed. The word *receipt,* as used here, may appear to contradict the explanation of the term *offer* in the previous paragraph. What is meant is that the person to whom the instrument is offered must have the option of accepting it or not. If the person does have the option, he or she has received it within the meaning of the uttering laws. For example, suppose that John negotiates a loan with a bank on the strength of a promissory note that he says he holds, and he gets the loan without ever showing the note. John does, in fact, have a promissory note on his person and it is, in fact, forged. Has John committed the crime of uttering? No. No offer of the instrument has been made, and the bank is not in receipt of the instrument. But would the result be the same if John had placed the note on the banker's desk in full view of the banker who read the note and, on the strength of what he saw, made the loan? Would there be an offer within the limits of this crime? Yes.

Bill gets a forged deed from Sam, and knows it is forged. He gives the deed to his son Fred to take to the county courthouse to be recorded. Fred does not know the deed is forged. Before he can get to the courthouse, he is picked up by the police. Can Bill be charged with uttering a forged instrument? No. The forged document has not been offered to the person intended to be defrauded. Could Bill be charged with uttering if Fred gave the deed to the clerk for recording? Yes, at that point the necessary offer has been made.

14.3 MAKING AND UTTERING WORTHLESS CHECKS

Unknown at common law, the crime of making and uttering **worthless checks** was created by statute. A person who made out a check on a nonexistent account and attempted to pass it as valid could always be charged with forgery at common law. Similarly, a person who made out a check on the existing account of another person could be charged with forgery. But could a person who made out a check on an existing account of his or her own, knowing there was not enough in the account to cover the check, be charged with forgery? Most authorities said no. Yet there were thousands of merchants throughout the country who each day took such bad checks in good faith. What could be done to help deter this flood of bad checks? The answer was to make it a crime knowingly to draw a check on an account that had insufficient funds to cover the check. This is the gist of most worthless check statutes. However, some states have included within their worthless check statutes the problems raised by the first two examples in this section: writing a check on a nonexistent account or writing a check on the existing account of another person.

All the statutes require insufficient funds, but not all agree as to when the insufficiency must exist for the offense to be chargeable. Some say that if one issues a check when there are insufficient funds, the issuer is liable to prosecution if there is the required intent. Other states say that until the check is presented to the bank for payment and there are no funds to cover it, liability does not arise. Suppose John knowingly issues Bill a bad check. Bill, not knowing the check is bad, calls the bank and asks if there is enough money to cover it. The bank says no. Can John be charged? In those states that say mere issuance is enough, the answer is yes. He is chargeable at that point. But in those states that require actual dishonor of the check by the bank, the charge would be premature.

Whether property has to be obtained or not varies from state to state. Some states require that something of value be given in exchange for the check, whereas others merely require the uttering of a bad check. Checks that are uttered for the purpose of obtaining money or any article of value are classified as worthless checks in Tennessee. Suppose a state requires that the victim actually part with something of value in reliance upon the validity of the check. Would there be a violation of the statute in this instance? John borrows money from Fred and promises to pay it back in twelve months. John pays by check with good checks for nine months, but, when the tenth payment is due, he knowingly draws a bad check that is dishonored by the bank. The courts of the states that require that the victim part with something of value conflict as to whether the statute has been breached. Some say that, because the check was offered or given as payment for a debt already in existence, no crime can be charged. On the other hand, some states would pay no attention to the fact that the debt was already in existence and would hold John liable for the crime.

Does the issuance of a postdated check render one liable under the worthless check statutes? A postdated check is one that is issued on one day but is dated for some later time. For example, John gives his grocer a check for $10 on March 1 but it is dated March 10. This is a postdated check. Here again, the states' decisions vary. Most postdated checks are worthless in the sense that most people give them in the hope of having money in the bank when the check is presented to the bank. Some states flatly prohibit the issuing of postdated checks and make it a crime to issue one when, at the time of issuance, there are not sufficient funds in the bank to cover the check. Other states say that even if given in good faith, if there is no money on deposit, the crime has been committed. A few states would not follow this line of reasoning unless the issuer never intended to have the proper amount of funds in the bank on the date the check was to be presented.

The foregoing discussion of bad checks illustrates the confusion that exists in this area of the criminal law. Until such time as a uniform worthless check statute is passed by all legislatures, this confusion will continue. The student is urged to consult local statutes to determine the scope of his or her state's worthless check statutes.

In some states the amount stated on the worthless check will determine whether the person issuing the worthless check is guilty of a misdemeanor or felony. For example, in North Carolina, if the check is greater than $2,000, the offense is a felony. In a number of these jurisdictions, the prosecution may not bring a single felony charge for the issuance of a number of small checks in the misdemeanor category even though, when added together, they total an amount equal to or greater than the amount required for a felony charge. This is true even though a single intent can be shown. On the other hand, in Wisconsin, although the punishment for a worthless check under $1000 is classified as a misdemeanor, if more than one worthless checks are written within a fifteen day period, the amounts can be aggregated to increase the penalty to a felony if the total of the worthless checks exceeds $1000. But the reality is that most states and the Model Penal Code still only consider bad check writing a misdemeanor. As it stands now, in some states, including West Virginia, a person can "paper" the town all day with worthless checks and escape the stigma of being convicted of a felony. Colorado, Kansas, Louisiana, Ohio and Oregon make it a felony for a prior violator to give a bad check. Kansas also makes it a felony to utter a worthless check over $500.

14.4 COUNTERFEITING

Because forgery covered only the making of false documents, or the falsification of existing documents, the common law crime of **counterfeiting** arose to cover instances in which something other than a document was falsely made with intent to defraud. One who duplicated a famous statue and sold it as the original could not be guilty of forgery, but would be guilty of counterfeiting under the common law. Under forgery, there does not have to be a genuine original in existence. The document forged is usually the only one of its kind, or, in the case of alteration, it is the original. Although, in counterfeiting, there does not have to be a valid or genuine original in existence, there usually is. The existence of a genuine Picasso or Rembrandt or $10 bill is what gives the counterfeit an aura of respectability. In forgery, the victim usually relies on the representations of both the document and the issuer. This is usually true in counterfeiting also. However, in counterfeiting the victim often relies more on the genuineness of the object than on the reliability of the person offering it.

Practically every item of tangible personal property that cannot be subject to forgery can be subject to counterfeiting. There are areas in which counterfeiting overlaps forgery even though forgery does not overlap counterfeiting. Bank stock that can be forged can also be counterfeited. At this point, the two crimes appear to be alike. With which crime should the men in the following example be charged—counterfeiting or forgery? John Jones is the president of the XYZ Bank, a corporation organized under the laws of state B. Bill Smith is the secretary of the bank. These men have the authority to sign stock certificates that the board of directors wants issued. All the stock of the bank was properly issued, and all the printed certificates were used. John Jones and Bill Smith decide they need more money. They have a printer print up new stock certificates that are identical to the ones that have been issued. They sign the certificates and sell them to people who do not know that they are bogus duplicates. In this situation, the men are chargeable with either counterfeiting or forgery and uttering, because the elements of each of these crimes have been fulfilled.

One federal circuit court sees counterfeiting as an attempt to pass as genuine a reproduction of money that so resembles the real thing that an honest, sensible, and unsuspecting person would be fooled. Therefore, "feeding" black and white photocopies of one-dollar bills into a change machine does not constitute the felony of passing counterfeit bills. The most the defendant could be convicted of is the misdemeanor of using something similar to money to procure anything of value from any machine designed to use or deliver coins or currency.

Counterfeiting is most commonly thought of in the context of bogus money. Only Congress has the power to print money and mint coins under our form of government. Similarly, only Congress has the power to punish those who would print or mint bogus money. But both Congress and the state legislatures have the power to create laws punishing the person who passes or utters counterfeit money.

The states have statutes that punish those who would counterfeit items other than money. Such statutes usually cover the counterfeiting of state and municipal bonds, corporate securities, and so forth. Many states make it a crime to counterfeit labels that are attached to consumer goods. The reader should consult local statutes to see what items may be subject to counterfeiting.

▓▓▓ DISCUSSION QUESTIONS ▓▓▓

1. Brown enters the First National Bank of Perryville in the State of Roxy. He opens a checking account with a deposit of $500 in cash and receives a checkbook. Brown proceeds from the bank to Ace Television Sales and Service. He purchases a color television set and writes a check for the full amount of $457. He suggests that the salesman call the bank to cover the check. The salesman calls and is advised that there are sufficient funds to cover the check. Brown leaves with the television. Shortly thereafter he enters White's Television Sales Company and repeats the same procedure, purchasing another television for an identical amount. Again the salesman calls the bank and verifies that there are sufficient funds to cover the check. Brown repeats this four more times and each time the check is verified. Brown loads all the televisions into a trailer, stops by the bank and closes his account before any of the checks are presented, and leaves town. Of what crime or crimes is Brown guilty?

2. Smith is arrested for burglary. At the time of his arrest, he is carrying false identification and gives the arresting officers a fictitious name to match the identification. Smith decides that he wants to be bonded out. He calls I. Letugo, a local bail bondsman, who agrees to bond Smith out. Smith gives Letugo the same false name and signs the papers the same way. He later forfeits the bond. Of what crime or crimes is Smith guilty?

3. Bill Kylar sends a $500 money order to his 18-year-old son, Ryan, who is a freshman in a university in another state. Ryan lived in an off campus apartment complex. Before the money order arrived, Ryan had moved to a different apartment complex with new roommates. When the money order arrived at his former complex, the new manager did not recognize the name but knew that there was a single mother, Angela Jones, whose two-year-old son was named Andrew Kylar. Assuming the envelope was meant for Jones' son, the manager convinces the mailman to put the envelope in Jones' mailbox. Jones opens the envelop, changes the payee's first name to Andrew from Ryan, and cashes the money order. What crimes, if any, have been committed?

▓▓▓ GLOSSARY ▓▓▓

Counterfeiting – falsely making something other than a document, with the intent to defraud.

Forgery – making or altering a writing, that if genuine, would have legal efficacy or be the foundation for legal liability, intending to defraud another person.

Legal force – legal efficacy; be of legal significance.

Uttering – passing a document to another person, knowing it to be false, with the intent to defraud.

Worthless checks – a statutory crime; generally, writing a check on an account with insufficient funds or on a nonexistent account.

■ REFERENCE STATUTES AND WEB SITES ■

STATUTES

Arizona: Ariz. Rev. Stat. Ann. § 13-2005 (West 2001).

Kansas: Kan. Stat. Ann. § 21-3707 (2000).

New York: McKinney's Cons. Laws of N.Y. Ann. § 170.15 (2001).

North Carolina: N.C. Gen. Stat. Ann. § 14-107 (2000).

Tennessee: Tenn. Code Ann. § 39-14-121 (2000).

West Virginia: W.Va. Code § 61-3-39a (2000).

Wisconsin: Wisc. Stat. Ann. § 943.24 (2001).

WEB SITE

www.findlaw.com

CHAPTER

False Imprisonment, Abduction, and Kidnapping

▨ KEY WORDS AND PHRASES ▨

Abduction

False imprisonment

Inveigled

Kidnapping

Unlawful restraint

15.0 INTRODUCTORY COMMENTS

The English recognized early the necessity of protecting each person's freedom of mobility. To prevent interference with this fundamental right, three offenses were created. These were misdemeanors known as false imprisonment, abduction, and kidnapping. Each of these crimes still exists in the United States in one form or another. In this chapter we examine each of these crimes and their modern counterparts.

15.1 FALSE IMPRISONMENT

One of the fundamental rights of a person is the freedom to move about freely from one location to another. The framers of the Constitution, recognizing the necessity of this right, included it in that document. Of course, as with all other rights, society must impose certain limits to prevent chaos when these rights are exercised. When a person breaks a law, society demands that his or her freedom of locomotion be restricted through a process called *arrest.* The law imposes limits on how fast one may travel and, to some extent, where he or she may travel. But, because freedom of movement is a fundamental right, no person may detain, imprison, or arrest another without legal justification or they are guilty of **false imprisonment.** The common law recognized this right early by establishing the crime of false imprisonment, defining it as the unlawful restraint of one person's physical liberty by another. The key words that spell out the two essential elements of the crime are *unlawful* and *restraint.*

Unlawful Restraint

The victim must be detained against his or her will to constitute unlawful restraint, but that is not all. Every person serving time in prison is probably there against his or her will. This detention is lawful because it is done within the framework of society's processes. Even an innocent person who is mistakenly convicted cannot successfully prosecute a charge of false imprisonment. The law recognizes that this situation is unjust but not unlawful. The victim must be restrained against his or her will and unlawfully.

Unlawful restraint is an act that deprives another of freedom of locomotion without legal authority. An invalid arrest by a peace officer or private citizen without proper authority or in excess of authority is unlawful restraint, in other words, false imprisonment. In fact, any person who decides that someone else should not be allowed to move about and who restrains the other without lawful authority commits the crime of false imprisonment.

There is a corresponding civil action for false imprisonment that requires unlawful restraint. In a criminal action, however, the person accused of false imprisonment may base their defense on good motive, good faith, or the honest belief that he or she was doing the right thing in view of the facts. Although these defenses are permitted, they do not affect the unlawful character of the restraint. These defenses are allowed because a person may intentionally restrain someone, thinking it is being done lawfully, when in fact, it is being done unlawfully. Society will not accept these excuses when only civil damages are involved; however, mistake of fact is an acceptable defense when a person is faced with the possibility of imprisonment.

Law enforcement officers should be aware that they are subject to prosecution for false imprisonment if they unlawfully restrain someone. However, the law treats law enforcement officers more liberally because they are compelled to make quick decisions in many instances and, as a result, are more susceptible to error. This is not meant as a commentary on the quality of law enforcement training and law enforcement officers. It is simply a recognition of the complexities of the job.

A law enforcement officer may be guilty of unlawful restraint when he or she makes an arrest using too much force; arrests someone outside the jurisdiction; or makes an arrest without a warrant, probable cause, or witnessing the violation. In general, procedural defects in the manner in which an arrest is made may make the restraint involved unlawful. The guilt or innocence of the person arrested has no bearing on this question.

There are other requirements involved in making a lawful arrest under various state laws. Failure to comply with these regulations will invalidate an arrest and may subject the arresting officer to prosecution for false imprisonment. For example, if a plainclothes officer fails to produce identification when making an arrest, he could be prosecuted for false imprisonment. The likelihood of prosecution is slim, but it should be realized that most states, by statute, require law enforcement officers to announce their identity and purpose to the persons they seek to arrest.

Restraint that begins as lawful can become unlawful. Suppose that Officer Smith arrests William Johnson for driving while intoxicated and decides that William needs to sober up before he can go before the judge. Officer Smith jails William Friday night and goes home, expecting to return Saturday morning. Smith forgets about William for two months. In this situation, the once lawful restraint has become unlawful. Likewise, failure to release a prisoner once he or she has served the required sentence constitutes unlawful restraint and false imprisonment.

The crime of false imprisonment is most often prosecuted against nonpolice officers. Mr. Johnson may decide that his neighbor's son, Art, should not take the usual shortcut through the Johnson's yard. Johnson takes aim with his rifle and tells Art to stop where he is and not to move. This is unlawful restraint, because Johnson exceeded the amount of force he could lawfully use to prevent trespass. An overzealous store clerk who locks a suspected shoplifter in the rest room may be guilty of false imprisonment if the suspect turns out to be innocent. Another type of unlawful detention might involve a group of escaped convicts who break into John Doe's house and tie him and his family up. This last example is more likely to lead to a conviction than would the case of the overzealous store clerk. The type of restraint that does not involve a good faith motive is what the law seeks to prohibit. In essence, restraint is unlawful if it is done without legal justification or beyond the point of legal justification. However, understanding only the nature of unlawful restraint solves only half of the problem. The next part of this section deals with the restraint itself and the elements that constitute detention.

The Restraint Itself

As the old adage says, "Stone walls do not a prison make, nor iron bars a cage." It is not necessary that the victim actually be locked behind bars for false imprisonment to be committed, although this type of restraint would qualify if done unlawfully. It is possible for a person to be confined in an open field. Suppose a very strong person threatens to mutilate a very weak person if the weak person moves. Even if this takes place in the middle of an open field, the weak person is imprisoned. Realistically, the weak person would be foolish to move; the limits on freedom of motion are as real as any four walls could be.

This example raises two problems. The first concerns the area in which confinement takes place. Two examples have already been given indicating that confinement with or without a structure, such as bars or stone walls, may satisfy the element of restraint. In open-area restraint cases, the courts have been presented with situations in which force or threat of force reasonably caused the victim to feel compelled not to move. The area to which mobility was limited was immaterial because of the presence of the threat. As in other crimes that involve threats, the threat must be designed to cause reasonable apprehension of immediate bodily harm.

The second problem area involves the element of escape. In the last example, if the weak person had a reasonable way to escape, that person could not say he or she was imprisoned. However, if the only means of escape would expose him or her to danger or injury, confinement is complete. For example, it would be unreasonable for the weak person to try to escape by jumping from a cliff. Similarly, if a person is confined to a room located on the tenth floor of a building and the only means of escape is through an open window, an escape attempt would be unreasonable. In both cases, imprisonment is complete because there has been more than a mere blockage of the victim's passageway and there is no feasible avenue of escape.

Would false imprisonment occur in the following example? John takes Mary to his apartment for dinner. At the end of the meal, Mary gets up to leave but John steps in front of her and says, "The door is locked and I won't unlock it until you make love with me." Mary refuses, screams, and is rescued by the police. Because John has no right to demand what he demanded, the fact that Mary could have secured her release by complying does not cancel out the imprisonment. This illustrates the rule that if the price of freedom of locomotion is compliance with an unlawful demand, imprisonment is complete. If, however, a person has a legal right to demand what has been demanded, the imprisonment, as long as it is not excessive, will not render the imprisoner liable.

For example, a person gets on an Amtrak train in San Francisco and just before arriving in Dallas, the conductor learns the passenger has no ticket. The conductor refuses to let the person disembark without first paying the fare.

In addition to the completeness and reality of the confinement, the law requires that the restraint be against the will of the victim. As in the case of forcible rape (Chapter 9), people who are intoxicated, incompetent, or unconscious may not legally consent to false imprisonment. Lack of legally recognizable consent implicitly satisfies the requirement that restraint be against the will of the victim.

It should be obvious from the foregoing discussion that every case of false imprisonment involves assault. When actual force is used to restrain the victim physically, battery may also be involved. If there is insufficient evidence of false imprisonment, assault or battery may still be chargeable. Conversely, if the imprisonment can be proved, the assault or battery will merge into that crime.

The Intent

At common law, the misdemeanor of false imprisonment required only a general intent. In some states today, the crime is merely a restatement of the common law definition, and it requires only general intent. However, a few states now require specific intent to confine a person secretly or to imprison a person against his or her will.

15.2 ABDUCTION

Before we discuss kidnapping, let us take a look at the crime of abduction that was created around A.D. 1500 by statute in England. Abduction was defined by Parliament as the carrying away of a female without the consent of the parents, guardians, or the victim herself for the purpose of causing her to marry someone against her will or for the purpose of prostitution against her will.

Although our states did not adopt the English statute word for word, most states did enact abduction statutes patterned on the English law. There are two primary differences between modern abduction statutes and kidnapping statutes. Abduction can only be committed against females. In kidnapping, the victim's sex makes no difference. In addition, in line with the English law, the purpose of abduction differs from kidnapping. A kidnapper may take a person for any reason, from revenge to monetary gain. **Abduction** occurs when someone takes a female either for illicit sex or to compel her to marry someone. The person she is compelled to marry may either be the abductor or someone else. Because the intent in each crime is different, it is possible for one to be guilty of both kidnapping and abduction. If Fred forcibly takes Mary from her home for the dual purpose of exacting sexual intercourse and ransom, he could be charged with both crimes. If, however, Fred merely wants sexual intercourse, he is chargeable under modern law with either crime but not with both. In the latter instance there is only one intent that would satisfy either crime, but that makes the abduction a lesser offense included in the kidnap.

The element of carrying the female away by force also distinguishes abduction from the crime of seduction (Chapter 9). In seduction, the victim is usually lured away from her house by a promise to marry or the like. Force is not an element of seduction. The legal term for luring away is **inveigled.**

Aside from these differences, the crimes of abduction and kidnapping are very similar. The detention or taking, the unlawfulness of the taking, consent or lack of consent, whether or not force is essential, and other factors under the crime of kidnapping will be discussed in the next section. What is covered there will apply equally to the crime of abduction.

15.3   KIDNAPPING

At Common Law

At common law, the misdemeanor of kidnapping was committed when a person was forcibly taken and sent from his or her country into another country. Because this was considered a more serious form of false imprisonment, the elements of the crime of false imprisonment had to be met. Although a number of states have combined the crimes of false imprisonment and kidnapping, the requirement that the victim be taken into another country has been dropped. Some form of carrying the victim away still remains, however, and this element distinguishes kidnapping from false imprisonment.

Kidnapping Today—Defined

By modern statutes, **kidnapping** is the unlawful taking of a person against his or her will. Many states have separate statutes to prohibit the unlawful taking of a person against his or her will for ransom, providing a more severe penalty. The statutes are not in complete harmony as to whether or not kidnapping is a felony. Some say that simple kidnapping is a felony and that more serious forms of kidnapping, such as kidnapping for ransom, carry greater penalties; in Tennessee aggravated kidnapping is a felony. For the more serious kidnappings, a number of states provide the death penalty or life imprisonment. If the violator commits a crime against the victim during the act of kidnapping in New Jersey, the violator could face life in prison. A few states consider simple kidnapping merely a misdemeanor and make the more serious forms of kidnapping felonies.

As already noted, some states combine kidnapping and false imprisonment in the same statute and make abduction a separate crime. Tennessee's aggravated kidnapping constitutes false imprisonment if the violator uses a deadly weapon, if the victims is under 13 years of age, if the victim is held for ransom or in an effort to extort money, or if the victim suffers bodily injury. On the other hand, there are a few states that combine abduction and kidnapping in the same statutes and keep their false imprisonment laws separate. As can be seen, no two statutes may be exactly alike. For this reason, the reader is urged to consult local statutes for the variations in effect in any particular state. Even though the statutes are not exact duplicates of one another, they do have certain key elements in common that are discussed in the following paragraphs.

Unlawful Taking and Carrying Away or Unlawful Taking and Detention

First, there must be an unlawful taking of a person. This means that the kidnapper must do so without legal right or authority. Suppose that an officer has a legal arrest warrant for John Doe. The officer sees John Doe and arrests him, not intending to take him to jail but intending to run him out of the state instead. The officer drives Doe to the state line. At this point, the officer exceeded his legal authority and could be charged with kidnapping.

Mary and Richard Roe are married and have a child, Bill. Mary gets a divorce and is awarded custody of the child by the court. Richard is given visitation rights. On one visit Richard takes Bill and leaves the state to keep Bill away from his mother. Would this be an unlawful taking? In most states it would be. However, a few state courts have said it is not, because Richard, as a parent, has a legal right to enjoy his child. In those states where this is not an illegal taking, the legislatures have responded by making it a separate crime to take a child away from the custody of a parent or guardian when that custody has been granted by court order.

Does the act meet the need or problems involved? Does it give enough leeway to the noncustodial parent who has a genuine but mistaken belief in the threat to the child if the child remains where he or she is? Under what circumstances can a parent be convicted of kidnapping his or her own child? A biological or adoptive parent who has been stripped of all parental rights by court action is no longer the parent of that child and, thus, is treated like any other kidnapper. A parent who voluntarily terminates parental rights by surrendering a child for adoption could also be convicted of kidnapping.

The more common case, however, centers on the parent who, though not stripped of parental rights, is not given custody as a result of a divorce. In such a case the noncustodial parent would "snatch" the child, take the child to another state, and get the second state to award the "snatching" parent custody. The federal kidnapping law immunized such a parent from prosecution. State courts did not have to recognize the validity of another state's custody orders; therefore, the "fugitive" state often found no violation of its kidnapping statute. Some states recognized parental immunity; some did not, but found juries unwilling to convict such a parent. Even in states that punish such conduct, criminal proceedings do not address the basic problem that states create by not recognizing each others' custody orders. Two statutes have been passed to remove the hope that parental kidnapping might result in a new custody order. The Uniform Child Custody Jurisdiction Act is a state law that requires state courts to decline such custody suits. The federal act is the Parental Kidnapping Prevention Act (PKPA). The federal law accomplishes several goals. It requires states to give full faith and credit recognition to another state's custody orders, it allows for the F.B.I. to investigate, and it provides for the establishment of the Federal Parent Locater Service. These laws have encouraged most states to now punish kidnapping by parents.

The Model Penal Code has made an attempt to deal with the noncustodial parent taking. In Section 212.4 it states:

1. Custody of Children. A person commits an offense if he knowingly or recklessly takes or entices any child under the age of 18 from the custody of its parent, guardian, or other lawful custodian, when he has no privilege to do so. It is an affirmative defense that:
   a. the actor believed that his action was necessary to preserve the child from danger to its welfare; or
   b. the child, being at the time not less than 14 years old, was taken away at its own instigation without enticement and without purpose to commit a criminal offense with or against the child.

Proof that the child was below critical age gives rise to a presumption that the actor knew the child's age or acted in reckless disregard thereof. The offense is a misdemeanor unless the actor, not being a parent or person in equivalent relation to the child, acted with knowledge that his conduct would cause serious alarm for the child's safety, or in reckless disregard of a likelihood of causing such alarm, in which case the offense is a felony of the third degree.

Most kidnapping charges are brought against people who take, carry away, and hide their victim for some monetary gain or for some other selfish motive without any real or pretended right to the victim. A number of years ago, we saw a case in which the daughter of a wealthy man was taken and buried alive. The kidnappers were not relatives and had no legal right to take the girl, who survived. They merely wanted money and, perhaps, publicity. Probably the most famous kidnapping involved ace-flyer Charles Lindbergh, whose son was kidnapped and later found dead.

Let's Talk First... Or Last!

Kidnappers who abducted Gildo dos Santos near his factory in a suburb of Sao Paulo demanded $690,000 but Santos escaped. The next day, Santos got a phone call asking for $11,500 to defray the cost of the abduction. After negotiating a discount of 50 percent, Santos called police who were waiting when the defendant showed up to collect payment.

www.dumbcriminalacts.com

Carrying Away the Victim

The main difference between false imprisonment in its pure sense and kidnapping is the fact that in kidnapping the victim is removed from one place and brought to another. This is not a factor in false imprisonment. Most states require for a kidnapping charge that the victim be taken from the place of seizure and transported elsewhere. The distance traveled is not significant in most states. Taking a person from a sidewalk and putting him or her in a car would be sufficient to justify a charge of kidnapping. One or two states do require that the victim be taken with intent to transport out of the county or state in which the victim is found. But, even in these states, there is no requirement that the victim actually be taken out of the county or state as long as the intent could have existed at the time of the taking.

New Jersey, for example, has a "substantial distance" requirement in its statute, and the court was asked whether dragging a victim from her car and down an embankment out of sight of passersby was in fact a "substantial distance." The court held that it was criminally significant and not merely incidental to the underlying crime of sexual assault.

Taking without Consent

If the victim consents to the taking, there can be no charge of kidnapping. Of course, the victim must be able to consent. Children below certain ages—set either by statutes or court decisions—cannot give consent. Insane persons are deemed incapable of giving consent. Consent obtained by fraud is no good either (Chapter 6). In such a case, the fraud vitiates the consent, and the person taking the victim is chargeable for the crime of kidnapping.

Assuming that minor children and insane persons are incapable of giving consent, can one who has legal custody of such a person consent to the taking of the incompetent, thus relieving the accused of criminal liability? Yes, it is possible that no kidnapping charge could be brought. For this reason, some state legislatures have eliminated lack of consent as an essential element of kidnapping. The elements of the crime in Kentucky are unlawful restraint and the intent of the violator.

Where two parents are still married and living under the same roof, would the consent of only one parent relieve the kidnapper of criminal liability? No cases could be found dealing with such a situation. However, the law enforcement officer, we feel, would be on safe grounds in assuming that the consent is not binding on the other parent. Consent in such a case would be a matter of defense (Chapter 6) and would not affect the officer's determination of probable cause to arrest.

Consent is not normally given by the victim. The victim is usually taken by force. When the victim is taken by force, the question of consent usually does not arise, because lack of consent is implied from the circumstances of the taking. How much force is necessary to justify this implication? Can the threat of force satisfy the element of taking without consent?

As to the first question, any force used will be sufficient to justify arresting the kidnapper. This is so because most statutes do not require that force be used. The answer to the question of whether or not the threat of force will suffice is "yes." Threats of force, or even threats to a person's reputation in a given case, may be enough to overcome the will of the victim so that it may be said that the victim went unwillingly. This is a question for the jury.

Can a person knowingly consent to be "kidnapped?" Yes. But the consent will eliminate any possibility of kidnapping being charged. If one consents to being kidnapped and later changes his or her mind, however, a kidnapping charge may be proper if, when the victim changes his or her mind and tells the "kidnappers" he or she has done so, they decide to continue the transportation and detention against the victim's will. On the other hand, if the victim does not change his or her mind until after being carried off, there is no kidnapping. For example, if Sam consents to being kidnapped, is taken to a house in another town and changes his mind after arriving at the house, but does not communicate this change of mind to his kidnappers until they are in the house, there is no kidnapping. If Sam does communicate his change of mind and the kidnappers refuse to let him leave the house, however, a charge of false imprisonment is proper. This would be so because the transportation element was not contemporaneous with the other elements of kidnapping.

If, however, Sam consents to go on a trip and, during the trip, his companions decide to detain him without his consent, the consent given by Sam applies only to the original purpose of the trip. His companions' later decision will be grounds for a charge of kidnapping. If the decision to detain Sam is made after the trip is over, a charge of false imprisonment would be proper for the reason discussed in the preceding paragraph. If the decision was actually made before the trip but not communicated to Sam until after the trip was over, then kidnapping may be the proper charge. The investigating officer should look for proof of fraud used to obtain Sam's consent. In one case, the victim accepted a ride from the defendants to take her to her home. They passed the residence, and she asked why. They stopped at another house, dragged her out of the car and raped her. The appellate court held the evidence was sufficient to sustain the kidnapping conviction. The fact that the victim got into the car voluntarily is not consent.

Finally, and obviously, a person cannot give consent unless he or she is aware of what is going on. Taking and carrying away an unconscious person or someone who is asleep would be sufficient to satisfy this element of the crime.

The Mental Element

Although a few states require specific intent for the crime of kidnapping, most states require only general intent for simple kidnapping. The kidnapping-for-ransom statutes, for the most part, do require a specific intent: that the victim be taken with the intent to gain ransom money.

The Federal Kidnapping Statute

The federal statute is found at 18 USC _ 1201, and it reads as follows:

_1201. Transportation

(a) whoever knowingly transports in interstate or foreign commerce, any person who has been unlawfully seized, confined, inveigled, decoyed, kidnapped, abducted, or carried away and held for ransom or reward or otherwise, except in the case of a minor, by a parent thereof, shall be punished (1) by death if the kidnapped person has not been liberated unharmed, and if the verdict of the jury shall so recommend, or (2) by imprisonment for any term of years or for life, if the death penalty is not imposed . . . [Sections (b) and (c) omitted.]

The federal statute is designed to make kidnapping across state lines illegal. The statute does not require that the kidnapped person be harmed in any way. However, the issue of harm and the amount of the injury may affect the punishment that can be meted out to the offender.

It is not necessary that the government prove that the defendant meant or intended to go across state lines. All the government has to show is that the defendant did, in fact, go across state lines with the victim. In fact, part (b) of the statute raises such a presumption if the kidnapped victim is not returned within 24 hours after the taking. The statute also makes conspiracy to kidnap for ransom a separate crime.

Statute 18 USC _ 1202 makes it a separate crime to knowingly receive, possess, or dispose of any money or property that has been delivered as ransom for a kidnap victim.

▤▤ DISCUSSION QUESTIONS ▤▤

1. Angus entered his boss's house one night, intending to kidnap his 3-year-old daughter for ransom. He crept up the stairs and stole into the child's room, where he picked up the sleeping girl. As he turned to leave the room, he dropped the child and woke everyone in the house. The police were summoned and Angus was arrested. Of what crimes, if any, is Angus guilty?

2. Bill owns a house on Ninth Street. As Bill arrives at his home, he learns that three escaped convicts are in the house and will not let Bill in. Bill goes to the police station and tells the officer in charge that he wants to swear out a complaint charging the three with false imprisonment because he is being restrained from going into his home. Is Bill correct in his assessment of the case? Why or why not?

■■■■ GLOSSARY ■■■■

Abduction – carrying away a female without her consent or parental consent, for the purpose of causing her to marry someone else or for prostitution.
False imprisonment – unlawful restraint of one person's physical liberty by another.
Inveigled – luring away.
Kidnapping – unlawful taking of a person against his or her will.
Unlawful restraint – any act that deprives another of freedom of locomotion without legal authority.

■■■ REFERENCE CASES, STATUTES, AND WEB SITES ■■■

CASES

State v. Niska, 514 N.W.2d 260 (Minn. 1994).

STATUTES

Kentucky: K.Y. Rev. Stat. Ann. § 509.040 (2000).

New Jersey: N.J. Stat. Ann. § 2C:13–1 (2001).

Tennessee: Tenn. Code Ann. § 39-13–305 (2000).

WEB SITE

www.findlaw.com

16

Crimes Involving Narcotic Drugs and Alcoholic Beverages

▓ KEY WORDS AND PHRASES ▓

Controlled Substances Act
External possession

Local option
Schedules

16.0 NARCOTICS LEGISLATION: INTRODUCTORY COMMENTS

Perhaps one of the greatest problems our complex and growing society has faced and continues to face is drug use and abuse. Although the problem receives disproportionate attention through mass media today, it also caused great concern in the early years of the twentieth century. Our society had two choices open to it: either permit the traffic in drugs to go on uncontrolled and hope that the incidence of addiction and its corollary problems would be stable and few, or place legislative controls on drug traffic and, by direct government action, induce the stability that was felt to be needed.

The United States chose the latter course of action. Both federal and state governments enacted laws to control the traffic in drugs. Prior to 1971, federal and state laws were not in complete harmony. This created some problems when federal and state agents wished to mount an effective campaign.

The primary state law enacted in the states was the Uniform Narcotic Drug Act of 1932, amended in 1942 and 1958. Additionally, states have other "homegrown" laws. The primary federal statutes were not part of an overall control scheme but were found piecemeal in taxing laws and the Pure Food and Drug Act.

Congress enacted the comprehensive Drug Abuse Prevention and Control Act of 1970, popularly called the **Controlled Substances Act.** This created a need for a new approach by the states. The Commission on Uniform Laws drafted the Uniform Controlled Substances Act in 1972. In the prefatory note to this act, the commissioners summed up the needs as follows:

> This Uniform Act was drafted to achieve uniformity between the laws of the several states and those of the federal government. It has been designed to complement the new federal narcotic and dangerous drug legislation and provide an interlocking trellis of federal and state law to enable government at all levels to control more effectively the drug abuse problem.
>
> The exploding drug abuse problem in the past ten years has reached epidemic proportions. No longer is the problem confined to a few major cities or to a particular economic group. Today it encompasses almost every nationality, race, and economic level. It has moved from the major urban areas into the suburban and even rural communities, and has manifested itself in every state in the Union.
>
> Much of this increase in drug use and abuse is attributable to the increased mobility of our citizens and their affluence. As modern American society becomes increasingly mobile, drugs clandestinely manufactured or illegally diverted from legitimate channels in one part of a state are easily transported for sale to another part of that state or even to another state. Nowhere is this mobility manifested with greater impact than in the legitimate pharmaceutical industry. The lines of distribution of the products of this major national industry cross in and out of a state innumerable times during the manufacturing or distribution processes. To assure the continued free movement of controlled substances between states, while at the same time securing such states against drug diversion from legitimate sources, it becomes critical to approach not only the control of illicit and legitimate traffic in these substances at the national and international levels, but also to approach this problem at the state and local levels on a uniform basis.
>
> A main objective of the Uniform Act is to create a coordinated and codified system of drug control, similar to that utilized at the

federal level, which classifies all narcotics, marijuana, and dangerous drugs subject to control into five schedules, with each schedule having its own criteria for drug placement. This classification system will enable the agency charged with implementing it to add, delete, or reschedule substances based upon new scientific findings and the abuse potential of the substance.

Another objective of this Act is to establish a closed regulatory system for the legitimate handlers of controlled drugs in order better to prevent illicit drug diversion. This system will require that these individuals register with a designated state agency, maintain records, and make biennial inventories of all controlled drug stocks.

The Act sets out the prohibited activities in detail, but does not prescribe specific fines or sentences, this being left to the discretion of the individual states. It further provides innovative law enforcement tools to improve investigative efforts and provides for interim education and training programs relating to the drug abuse problem.

The Uniform Act updates and improves existing state laws and ensures legislative and administrative flexibility to enable the states to cope with both present and future drug problems. It is recognized that law enforcement may not be the ultimate solution to the drug abuse problem. It is hoped that present research efforts will be continued and vigorously expanded, particularly as they relate to the development of rehabilitation, treatment, and educational programs for addicts, drug-dependent persons, and potential drug abusers.

Although the Act has been replaced with a later version in 1994, the philosophy and objectives have not changed and the majority of states adopted the Controlled Substances Act and its 1994 version with only minor modifications.

16.1 AN ANALYSIS OF THE UNIFORM ACT

The act provides five lists, known as **schedules,** that determine the scientific names of the controlled substances. The first schedule includes all opiates that have a high potential for abuse and no acceptable medical use or are unsafe even under medical supervision. The second list contains those substances that have a high potential for abuse but are medically acceptable under severe restrictions. This schedule, however, includes those items that may lead to severe psychic or physical dependence. The third list contains those items that have a lesser abuse potential but may lead to some physical or psychological dependence. The fourth list includes those substances that may lead to a limited physical or psychological dependence. The fifth list presents those items that create an even lower possibility of dependence.

Each of these schedules and each item in each schedule is established by meeting an eight-point test. These eight points are:

1. The actual or relative potential for abuse

2. The scientific evidence of its pharmacological effect, if known

3. The state of current scientific knowledge regarding the substance

4. The history and current pattern of abuse

5. The scope, duration, and significance of abuse

6. The risk to the public health

7. The potential of the substance to produce psychic or physiological dependence

8. The status of the substance as an immediate precursor of a substance already controlled under this article

The act clarifies some definitional areas that caused problems under the prior narcotics laws. For example, manufacturing includes all production, preparation, propagation (growing), compounding, converting, or processing of a controlled substance. The act also defines production as the manufacturing, planting, cultivating, growing, or harvesting of a controlled substance.

As with previous acts, it determines who may lawfully possess or otherwise deal in controlled substances. And, as with previous acts, those who wish to deal in controlled substances must meet certain rigid criteria regarding training and past conduct. A person convicted of certain crimes can be prohibited from dealing in such drugs. This would include those owners of corporations as well as those who actually produce the substances.

Any person who violates the provisions of the act can then be prosecuted. The act makes it unlawful for any person to knowingly or intentionally possess a controlled substance unless it was received through proper channels. The act also penalizes those registered drug manufacturers who refuse to furnish records or refuse to allow an inspection of their premises. It is also a crime to knowingly keep or maintain any type of building, vehicle, boat, or aircraft that is resorted to by persons using controlled substances or for the purpose of selling controlled substances. This prevents opium dens and would appear to keep those who open their houses to drug users from claiming innocent activity.

One section of the act permits a judge to place a first offender on probation instead of sending the offender to prison. However, this applies only to cases of simple possession. Even though the act gives this bit of forgiveness, it provides that a person convicted of second and subsequent offenses may be imprisoned for terms and fines twice those provided in the act.

The act also provides for the exchange of information among state, local, and federal law enforcement agencies. The intent is to ensure that all agencies work harmoniously and to prevent duplication of effort. This law appears to declare war on illegal drug traffic.

16.2 ACQUISITION, POSSESSION, AND USE OF NARCOTIC DRUGS

There are several methods by which a person may obtain drugs. A person may genuinely need drugs and honestly acquire them upon a written order from a physician to a pharmacist or receive them honestly and in good faith directly from the doctor. Second, a person may get them from a legitimate source by lying to a doctor or pharmacist. Third, a person may secure them through so-called underground or illegitimate sources. Finally, a person can steal them. The first method involves no crime on the buyer's part. The second and third methods will in some way cause the person to be subject to penalty for violation of the act. The last method will subject the person to penalties under the general criminal laws of the state.

The act makes it a crime to obtain or attempt to obtain narcotics by fraud, deceit, misrepresentation, or subterfuge. Drug addicts are no less resourceful when it comes to getting a "fix" than alcoholics are at getting something to drink. Drug addicts have been known to see several doctors in the same building in a two- or three-hour period, getting a prescription from each. Each doctor was unaware of this fact. The addict would then take the prescriptions to several pharmacies that were also unaware of the plot. Addicts have been known to steal prescription pads, writing out their own prescriptions and signing the doctor's name. They have even carefully removed directions from prescriptions directing that they not be refilled. The methods that can be employed are limited only by one's imagination.

When a drug user purchases drugs on the black market, the user is not criminally liable for the purchase in most states, but upon receipt of the drugs, the user is liable to prosecution for the crime of possessing or having narcotic drugs under his or her control. The most difficult problems raised by the act involve what constitutes possession. Most states have determined that possession means that the narcotic drug is under the actual control, care, or management of the person charged and that the defendant must have more than a passing control. The perplexing problem is this: Suppose an officer discovers a group of marijuana smokers sitting in a circle passing the "joint" or "blunt" to one another. Who has control? Does the person who has the cannabis at the time of the discovery have the possession or control necessary for the charge? Or should the officer charge all who are there for possession or control? Or is the one who brought the "joint" to the group the only one who can be charged? Or did his act of sharing destroy the possession?

Some states have solved the problem of possession by finding that narcotic drugs may be jointly possessed by two or more persons under a theory called *constructive possession*. Therefore, all the participants in the "pot" party would be subject to arrest. The jury would be permitted to imply from these circumstances, if there is no proof to the contrary, that each had discretionary control over the marijuana. It is important to note that this rule is applied in only a few states.

Some states have rejected the joint possession concept and have put the entire burden of proving which person or persons actually had possession or control on the state. In several cases dealing with the problem, the participants have been said by the courts to have "fleeting" control, which is not prohibited by the act. It does not appear, however, that the state in these cases made any attempt to prove actual possession of any individual or individuals involved, but merely rested its case on the facts discovered at the time of arrest.

Most states agree that the narcotic drugs do not have to be in the hands of, or on the person of, the accused as long as the narcotics are found in a place or thing the defendant has control over, such as the defendant's home or automobile. Similarly, as long as the defendant has the right to control the use, distribution, and disposal of the drug, the defendant can be charged with possession. For instance, if the accused hides drugs in the middle of a city park, the fact that the drugs are not found on the person, or on something he or she owns, will not prevent the person's being charged with possession or control of narcotic drugs. The law recognizes constructive possession. A jury is permitted to infer that a defendant had control of the narcotic drug when the narcotics are found on his or her premises.

The defendant must know that the drugs in his or her possession are illegally possessed. The state must prove the defendant knew or should have known the nature of the substance possessed. This may be proved by showing all the surrounding circumstances. Unless the jury is convinced that there was no way the defendant could have known what was in his or her house, car, or person, the defendant can be convicted on such circumstantial evidence.

Only **external possession** is chargeable. There is no penalty for internal possession. Let us illustrate the concept of internal possession. Suppose John has possession of a narcotic. He is sitting at home one evening and hears a police officer outside say, "This is the police. Open up. We have a warrant for your arrest." John takes all his narcotics and swallows them so the police won't find them. By ingesting the narcotics, John has changed his possession from external to internal. In a majority of the states, John could not be charged with possession or control of narcotics. The only possible crime chargeable may be habitual use unless the state has an internal possession statute.

Can a state constitutionally punish a user of narcotic drugs? The answer to this question is a qualified yes. As we saw early in this book (Chapter 4), no state can make a person's state of being or status criminal without serious constitutional problems. In the case of *Robinson v. California*, the Supreme Court held invalid the California statute that made drug addiction a crime. But the same Court made it clear that a state could constitutionally punish a habitual user of drugs for being under the influence of drugs. Why the difference? In the latter cases, the court was able to find a sufficient act with clearly defined terms. In the former case, the Court was presented with a statute that was so vague that it could not determine whether or not the addict got to be an addict of his or her own free will. The court recognized that not all addicts voluntarily reach that condition. The more specific the statute, the more likely it is to be held constitutional.

16.3 CRIMES INVOLVING THE USE, SALE, AND MANUFACTURE OF ALCOHOLIC BEVERAGES: INTRODUCTORY COMMENTS

Perhaps some of the best stories comedians tell are about alcohol and its effects. There is an old story that goes like this: "Our town is so small that we don't even have a town drunk and so poor we couldn't afford one." Many children, whether they are from the city or not, learn to sing the old folk song, "Mountain Dew." If the effects of alcohol were not so serious, all this would be very romantic and funny. Statistics show that in a large percentage of fatal traffic accidents, alcohol plays a major part. Who knows how many broken homes result from overuse of alcoholic beverages or how many crimes have been committed because of the influence of alcohol? But drinking is a social problem, and experience shows that morals cannot be legislated. Prohibition, the "Noble Experiment," rather than having the good effect it was supposed to have, did just the opposite. More people drank than ever before. People openly and notoriously broke the law. Centuries from now, historians will probably point to Prohibition as the event that started an attitude of disrespect for the law and that caused the problems of the 1940s through the 1990s. Of course, hindsight has always been better than foresight.

Some states still try to control the problem of alcohol consumption by allowing each county to decide whether to permit the sale of alcoholic beverages. This is called the **local option.** Dry areas exist in name only. People leave their dry county and go to a wet county to buy their favorite whiskey, wine, or beer.

Most states have a more realistic approach to the problem and seek to regulate it from a different point of view. The states regulate who makes the beverage to make sure that those who wish to drink can do so without fear of going blind or dying from drinking poisoned alcohol. States try to keep drinkers off the roads and provide severe penalties for those who cause accidents on the road. States also provide penalties for habitual drunkards.

16.4 CRIMES INVOLVING THE MANUFACTURE AND SALE OF ALCOHOLIC BEVERAGES

All states and the federal government make it a crime to produce alcoholic beverages in excess of a limited amount. These laws are aimed at preventing the sale of adulterated alcohol that could cause death or serious injury. Usually, under some sort of revenue law, the manufacturers must be registered with the state and federal governments. The success of these laws is in direct proportion to the integrity, diligence, and size of the enforcement body. Illicit stills continue to operate. Most of these are put together with such haste and lack of safety devices that the liquor produced, sometimes called "white lightning," can cause death, permanent blindness, or other severe physical problems.

All states require that anyone who sells alcoholic beverages be licensed, and all make it a crime for anyone to sell liquor without a license. In addition, the states make it a crime for anyone to sell or give alcoholic beverages to minors. A few states, California being among them, have enacted laws prohibiting the sale or gift of alcoholic beverages to persons known to be alcoholic. Some states, such as Montana, provide criminal penalties for the sale or delivery of alcoholic beverages to someone who is already intoxicated. All states delegate to municipal and county officials the right to regulate the hours during which licensed sellers may operate. Penal ordinances are provided for those who do not observe these hours. The only possible exception to this delegation of power may involve a general statewide Sunday closing statute, or blue law, like the one that exists in Mississippi. In addition, some states prohibit the sale of any alcoholic beverage by the bottle or drink, and some prevent bars only from selling on Sunday. Others do the opposite and prevent only the sale of "packaged goods" and allow the bars to be open. Connecticut law allows establishments to dispense alcohol in "glasses or other receptacles suitable to permit the consumption of alcoholic liquor by an individual" on Sundays between the hours of 11:00 a.m. and 1:00 a.m., but prohibits the sale of liquor in "places operating under package store permits any time on Sundays." Any violation of these provisions will cause the violator to be punished. These offenses are all *mala prohibita* offenses and require no showing of criminal intent.

16.5 PUBLIC DRUNKENNESS

Drunkenness is a status and as such it is not criminal. A number of states provide, however, that being drunk in public areas is a criminal offense. A few states make it a criminal offense to be a habitual drunkard. Some states make it a crime to drink in public places. Such a statute is constitutional because the drinking is enough of an act for the completion of a crime. In these states, the places in which drinking is not allowed are listed in the statutes. Because the variations are so great from state to state, the reader should consult both the state statutes and municipal ordinances for the approach taken in that area.

Normally, the act of voluntarily drinking and then appearing in public is enough of an act to keep the sanction from constituting the punishing of a person's status. As a Washington court once said:

> We see no reason under the constitutions of the state and the United States why the communities of this state may not adopt and enforce legislation to protect themselves from the nuisance, offensive misconduct and serious dangers at times associated with public drunkenness.

When the issue of constitutionality of such statutes came before the U.S. Supreme Court, the Court held:

> We are unable to conclude, on the state of this record or on the current state of medical knowledge that chronic alcoholics... suffer from such an irresistible compulsion to drink and to get drunk in public that they are utterly unable to control their performance... and cannot be deterred at all from public intoxication.

Thus the issue could become whether or not the defendant acted voluntarily in pursuing alcoholism. This point came up in a Minnesota case in which, after considering the defendant's conduct and health status, the court held:

> that the words "voluntarily drinking," as used in the statute, making intoxication by voluntarily drinking intoxicating liquors a criminal offense, meant drinking by choice, and that where the defendant was a chronic alcoholic whose drinking was due to his illness and was involuntary, he could not be convicted under the statute.

The issue of public drunkenness statutes came up as a collateral issue in a 1972 U.S. Supreme Court case. Citing a special American Bar Association report, the Court included the following in the opinion:

> Regulation of various types of conduct which harm no one other than those involved (e.g., public drunkenness, narcotics addiction, vagrancy, and deviant sexual behavior) should be taken out of the courts. The handling of these matters should be transferred to nonjudicial entities, such as detoxification centers, narcotics treatment centers and social service agencies. The handling of other nonserious offenses, such as housing code and traffic violations, should be transferred to specialized administrative bodies. Such a solution, of course, is peculiarly within the province of state and local legislatures.

As a result, a few states, for example, New Jersey and Illinois, have repealed their public intoxication statutes. These states now provide detoxification centers and continuing treatment for those suffering from alcoholism. The use and effects of alcohol in traffic offenses are covered in Chapter 19.

■■■■■ **DISCUSSION QUESTIONS** ■■■■■

1. John has a diagnosable illness that requires a certain narcotic drug for its control. Bill, a friend of John's, is a narcotics addict and relies on the same drug that John needs for his illness. Bill tells John that he will pay him twice the cost of the drug if John will get some of it. John agrees, visits his doctor, and after diagnosis, the doctor gives John a prescription for the narcotic. John takes the prescription to the drugstore and has it filled. He then sells the drug to Bill at the agreed-upon price. For what crime or crimes are John and Bill liable?

2. Police lawfully enter a motel room. In that room are three people, A, B, and C. A is seated on the bed. B is seated on the only chair. C is standing beside B. B was in the process of giving himself a shot. As the police enter all three run for the bathroom. The police catch the three in the bathroom where B was flushing the commode. Further investigation of the room produced in a suitcase an eyedropper and bulb with a needle attached. There was also morphine in the medicine cabinet. C was the person to whom the room was registered. Could C be convicted of possession?

3. Police enter a bar. Two patrons get up and run out. The police discover a package of heroin taped under the table where the two were seated. Is this sufficient to justify their arrest for possession? Does it make any difference that this is a public place?

■■■■■ **GLOSSARY** ■■■■■

Controlled Substances Act – common name for the Drug Abuse Prevention and Control Act of 1970.

External possession – referring to the only way possession can be charged; internal possession is not chargeable.

Local option – a decision left to each county of a state whether to permit the sale of alcoholic beverages.

Schedules – five lists of drugs ordered on the severity of impact based on an eight-point test.

■■■■■ **REFERENCE CASES, STATUTES, AND WEB SITES** ■■■■■

CASES

Tamez v. State, 251 S.E.2d 159 (Ga. App. 1978).

Robinson v. California, 370 U.S. 660 (1962).

Seattle v. Hill, 435 P.2d 692,699 (Wash. 1967).

Powell v. Texas. 392 U.S. 514, 535 (1968).

State v. Fearon, 166 N.W.2d 720 (Minn. 1969).

Argersinger v. Hamlin, 407 U.S. 25, 38 (1972).

STATUTES

California: Ca. Bus. & Prof. Code § 25602(a) (Deering 2001).

Connecticut: Conn. Gen. Stat. §§ 30-91(a), (d) (2001).

Mississippi: Miss. Code Ann. § 67-1-83 (2001).

Montana: Mont. Code Ann. § 16-3-301(3)(b) (2000).

WEB SITE

www.findlaw.com

CHAPTER

17

Extortion, Blackmail, and Bribery

■ KEY WORDS AND PHRASES ■

Blackmail   Misfeasance

Bribery   Nonfeasance

Extortion

17.0 INTRODUCTORY COMMENTS

Most of the crimes we have examined have shown humans capable of physical brutality that society seeks to eliminate or at least hold to a minimum. We have seen the more commonly known crimes such as larceny and false pretenses in which one secures money or property from another. In this chapter, we examine nonviolent crimes in which money or property is secured not by trick, physical force, or stealth, but is extracted by misuse of public office. In addition, we examine a crime in which one mentally forces another to part with money, property, or influence. Finally, we see a crime in which money or property is not necessarily involved but in which one uses inflamed words to damage another's reputation.

17.1 EXTORTION

At common law, **extortion** was a crime for any public officer corruptly to demand money or property to which neither the officer nor the office was entitled. Today the crime exists in most states as a separate and distinct crime, but extortion is found in the bribery statutes of some states.

To be guilty of the offense of extortion, the accused must be a public officer. A public officer, simply put, is one who works for a government or holds elective office. This would include all municipal, county, state, and federal officials. No branch of government is exempt. Administrative, executive, judicial, and legislative officials are all subject to prosecution for the crime of extortion if all the elements of the crime are met.

The crime may be committed by one who is properly appointed or elected to office—a *de jure* officer. It can also be committed by one whose appointment or election was not technically perfect in all respects—a *de facto* officer. There are even cases where a "phony" officer has been convicted. The courts, in these cases, have prevented the self-appointed public officer from denying the office, thus making that individual pay for the pretense. There is, however, one qualifying point that must be met in this situation. If the office the person pretends to fill is nonexistent, there is no extortion. The office must be one that actually exists.

The money or property demanded by the public officer must be demanded in connection with his or her official capacity. It would not be extortion for an officer to demand money in an unofficial capacity not connected with the job held. For example, it would not be extortion for a police officer to demand money from another person for getting a building permit okayed. Such a demand is well beyond the scope of the person's duties as a police officer. It would be extortion for the building inspector to demand excessive fees for granting a building permit, because this is within the building inspector's scope of employment.

The common law recognized three circumstances in which extortion was said to exist. First, an official was guilty of extortion by demanding fees in excess of the fees set by law for the service to be rendered. Next, it was extortion to demand a fee for a service for which the law set no fee because it was to be rendered at the general expense of the government. Finally, it was extortion to extract a fee corruptly before the service was rendered. This was true even if the service was later performed by the official. In studying the old common law cases, it appears that, at the time of the demand and payment by the victim, the victim was apparently unaware that the fee demanded was illegal. When the victims later learned of the falsehoods, they reported them.

The crime today is basically one of official corruption. The law labels it a crime and attempts to stifle a system of private favors for public services. Nowhere does it appear that the modern victim has to be unaware of the illegality of the fees demanded. To complete the crime, all that is needed is the demand for and receipt by the official of money or property to which that official is not entitled for the performance of a duty.

Although the money has to be received by the extortionist, it does not have to be placed directly in his or her hands. Delivery to an agent appointed by the extortionist will be enough. Similarly, delivery to a post office box or bank account belonging to the extortionist will satisfy the delivery element.

If the extortionist receives nothing but a promise of payment, the crime is not complete. Receipt of any item, with even the slightest value, will be enough. Thus, a written promise to pay placed on a properly executed negotiable promissory note or a personal check could be enough. Some states have enacted laws that say that a promise given to pay when the demand is made will suffice. This represents an extension of the common law.

The common law required that the fee be demanded by the official with knowledge of its illegality. Thus the common law would excuse an official who negligently miscalculated or who by some other method made a good-faith mistake. Although this is the rule applied in most states today, there are some states that, by statute, permit no excuses. This is done to eliminate the possibility of sham defenses thought up after the official is caught. Oklahoma and Virginia do not provide defenses to extortion by public officials.

17.2 BLACKMAIL

Demands by public officers in their official capacities for illegal fees are extortion. Illegal fees demanded by private citizens are likewise punishable and are popularly known as **blackmail.** Many states call blackmail extortion by a private person. Blackmail occurs when a person, using written or oral threats of force or fright, demands money or other property to which he or she is not entitled.

The most common method of blackmailing a person is to threaten to expose the victim to the public by telling the public about certain past misdeeds, whether real or pretended. The crime thrives on the basic tenet that each person would like to protect his or her reputation or the reputation of a member of the family and will pay to do so. The threat can also be to the physical well-being of the victim or a member of the family. Some states even say that the threat can be to the property of the victim.

The statutes from state to state are not harmonious, but they all agree that there must be intent to extract or extort money from another person. Here the words used constitute the act needed for the crime. Unless a local statute requires receipt of the money or property demanded, the crime is complete upon communication to the victim of the threat to person, reputation, property, or family.

17.3 BRIBERY

Perhaps the best-known crime of official corruption is the crime of bribery. Stories have been written and continue to be written concerning the subject. It has been said that if bribery were stopped, organized crime would cease to exist. Whether that statement is true is not for us to debate. Suffice it to say that, if bribery were completely eliminated, it would certainly hurt the operation of crime syndicates.

There are two types of bribery. First, it is **bribery** to give a public official money or property of any value in exchange for an agreement by the public official to do or refrain from doing something that is against or in contradiction to the official duty. Second, it is **bribery** for a public official to agree to do something that is in contradiction of his or her duty in exchange for money or property of any value.

The agreement, not the act called for in the agreement, is essential to completion of the crime. That is why the crime would be complete even if the agreement were to affect a possible, not even certain, transaction in the future. In a few states, there has to be a tender or offer of the bribe money or other property and receipt by the official. If there is a mere offer of a bribe but no agreement on the part of the official, only attempted bribery can be charged. In states that require an offer and receipt of the money, if an agreement is reached but the official refuses the money or property, or if the money is not offered, there would also be only an attempted bribery. This is not the case at common law, because the crime was the offer of a bribe with an agreement or the acceptance of bribe money with an agreement. An official who refused money after an agreement was reached could be charged with bribery and so could the offeror. Because the statutes have changed the law somewhat, many legislatures make it a punishable offense to offer a bribe, but continue to recognize bribery when an officer approves the agreement without accepting the money or other property. These states find a completed bribery when there is an agreement and either acceptance or tender of the bribe money along with the completed agreement.

Contrary to the common law, statutes have made it bribery to give or receive money or other property to influence sports officials, athletes, municipal officers, county officials, members of juries, whether grand or petit, and witnesses. The states' statutes differ on this point and should be consulted to determine the scope of the statute in the reader's state.

Unlike extortion, the thing agreed to by the officer need not be within the scope of his or her authority or employment. Extortion involves illegal payment for doing the right thing, but bribery calls for doing or not doing the wrong thing—something contrary to the powers of the office. This is the other main distinction between the two crimes. At common law, bribery was a misdemeanor, and it continues as such in a few states. However, a large percentage of the states have made bribery a felony.

17.4 OTHER CRIMES CHARGEABLE AGAINST PUBLIC OFFICIALS

In addition to extortion and bribery, there are other crimes a public official can commit in an official capacity. Some of these come from the common law. They are set out briefly in this section to help the reader become aware of the possibilities. The crimes cover a multitude of official "sins" and serve as a catchall in this area.

Nonfeasance

Whenever an official purposely fails to perform a duty that he or she is commanded by law to perform, that official is guilty of **nonfeasance** in public office. The only defense available is that the dangers were greater than any normal person could be expected to face.

Misfeasance

If the official is not acting under a genuine mistake of fact, **misfeasance** occurs when any public officer abuses a discretionary power, commits any fraud or act of oppression, or does any other act in an illegal manner.

DISCUSSION QUESTIONS

1. The Warrensville County School Board hires the accounting firm of Smith, Smith, and Smith to audit the books of all the schools in the county. Smith, Jr., a member of the firm, discovers a shortage of funds in the Warrensville High School account. He corners the principal of the high school, and after ascertaining that the principal is aware of the shortage, tells him that for the sum of $500, he 'Smith' will overlook the shortage and straighten the books. The principal agrees. It is agreed that, to avoid any possibility of their plan's being discovered, the principal will leave the money in a booth at a particular restaurant that night where Smith will pick it up. The money is left, but before Smith can get to the restaurant, Joe, a local wino, discovers the package and takes it for his own. What is the liability of each party involved in this situation?

2. A bank officer is charged with embezzlement. He would like to avoid prosecution and trial and accept probation. The prosecuting attorney is running for re-election and needs campaign funds. Hearing that the bank officer would prefer probation, the prosecutor goes to the banker's home. Without anything being said about the embezzlement, the prosecutor suggests that the banker might want to make a contribution. The bank officer makes the contribution. Several delays in the criminal process occur due to the prosecutor's efforts. The banker is ultimately tried and sentenced. The banker's wife attempts to get the contribution back but is refused. She relates these facts to you. Has any crime been committed?

3. State A has a statute that provides that the tax collector would get, as a fee, a payment equal to 25 percent of "the fine imposed." The tax collector goes to a taxpayer and tells him that if the taxpayer would pay 25 percent of what the fine would normally be, there will be no prosecution for the offense of failing to register his motor vehicles. This amount is paid and there is no prosecution. Has a crime been committed?

GLOSSARY

Blackmail – illegal fees demanded by a private citizen.

Bribery – to give a public official money or property of any value in an exchange for an agreement by the public official to do or refrain from doing something that is against or contrary to the official duty; if the official agrees, the official has also committed the offense.

Extortion – illegal fees demanded by a public officer in his or her official capacity.

Misfeasance – in an official capacity, when a public officer abuses a discretionary power, commits a fraud or an act of oppression, or does any other act in an illegal manner.

Nonfeasance – when an official purposely fails to perform a duty he or she is required by law to perform.

REFERENCE CASES, STATUTES, AND WEB SITES

STATUTES

Oklahoma: Okla. Stat. Ann. § 1484 (2000).

Virginia: Code of Va. § 18.2-470 (2000).

WEB SITE

www.findlaw.com

Offenses By and Against Juveniles

■ KEY WORDS AND PHRASES ■

Battered child syndrome
Child abuse
Child molesting
Delinquency
Dependency
Direct file
Discretionary waiver
Juvenile
Mandatory waiver

Neglect
Once an adult/always an adult
Parens patriae
Persons in need of supervision (or service) (PINS)
Reverse waiver
Status offenses
Statutory exclusion
Vicarious liability

18.0 INTRODUCTION

The key distinction between a crime and any other form of prohibited conduct is the state of mind with which a person acts, commonly referred to as *intent.* The common law established a set of arbitrary rules governing when and how well a person was capable of forming this intent, based partially on chronological age. Under 7 years of age, a child was conclusively presumed incapable of forming the intent to commit a crime. This presumption could not be rebutted. Children between 7 and 14 years of age were rebuttably presumed incapable of forming the intent. For persons over 14 years of age, the presumption was in favor of capability. As a result of these arbitrary rules, children of tender years often escaped punishment for conduct that, if committed by an adult, would have constituted heinous crimes. On the other hand, for a mere 14-year-old child, the punishment meted out for the commission of an offense was as harsh as it would have been for an adult.

Many people felt that both alternatives provided by the common law system were unjust. They argued that it was wrong to allow a child under 14 years of age to possibly escape all liability for his/her conduct. On the other hand, they felt it equally bad to subject a child, possibly as young as seven years of age, to the harshness of the legal punishment of the times, especially in light of the poor penal systems. Over the long range, failing to segregate youthful offenders from adults took its toll.

It was not until 1899, in Cook County, Illinois, that any major steps were taken to rectify this situation. In that year, the first juvenile court was created for the handling and disposition of cases involving juveniles. From that court grew the widespread juvenile court system. The system was developed with a philosophy of *parens patriae*, lets do what's best for the child. Treatment and rehabilitation were the reforms intended by the new system of justice. To foster this philosophy, most juvenile court proceedings are held in closed session (not open to public viewing).

To avoid the stigma of the adult judicial system, a whole new vocabulary developed within the framework of juvenile justice. A new language was developed using new concepts to describe the juvenile version of comparable adult processes. Among the terms that gained popularity in this context were petition instead of complaint, summons instead of warrant, initial hearing instead of arraignment, findings of involvement instead of conviction, and disposition instead of sentence.

18.1 WHO IS A JUVENILE?

Of course, the most important term to define is *juvenile.* Who is a juvenile? To repeat, the key to successful prosecution of someone charged with criminal conduct is the ability to prove intent. Children are sometimes not old enough or mature enough to be capable of forming a criminal intent, so they may not be held liable for acts that would otherwise be criminal. With this in mind, the common law set some arbitrary rules regarding the capability of children to form a criminal state of mind. The rules have no foundation in the actual abilities of children, nor do they take into consideration the actual abilities of the individual child except within the framework of the rules.

A child under the age of 7 years was conclusively presumed incapable of entertaining a criminal intent and therefore could not be held accountable for otherwise criminal conduct. It could not be shown in these instances that a particular child was sufficiently intelligent to know the act was wrong.

A child between the ages of 7 and 14 years was rebuttably presumed incapable of entertaining a criminal intent. This meant that, although the child was presumed incapable, the prosecution could offer evidence to show that the particular child did possess sufficient intelligence to know the difference between right and wrong and did have the capacity to act with an evil state of mind. Unless the prosecution could show the child had this capability, the presumption remained and the child could not be convicted.

Children over 14 years of age were presumed capable of forming a criminal intent. This presumption was also rebuttable, as in the case of adults, by showing that the accused suffered from some deficiency that prevented the person from acting with an evil state of mind. This breakdown by age bracket has been modified by statute in many states, and the student should become familiar with local law.

As in the case of most legal matters under modern law, there is little uniformity among the states. Each state was and still is free to define and interpret its own juvenile court law. A juvenile is, of course, determined by chronological age, but not all states are in agreement as to what that age should be. In Connecticut, New York, and North Carolina, a juvenile is a child under the age of 15 years. Georgia, Illinois, Louisiana, Massachusetts, Michigan, Missouri, New Hampshire, South Carolina, Texas, and Wisconsin set 16 as the upper age of juvenile court jurisdiction at the time the act is committed. In all the other states, a **juvenile** is a person under 17 years of age. The exact age is in all cases established by legislation. It is an arbitrary decision derived from the best judgment of the lawmakers, who must determine when a child has been sufficiently socialized by the community to know right from wrong. Once that goal is achieved, the person is considered an adult liable for the legal consequences of his or her actions.

What about a minimum age for juvenile court jurisdiction? In light of the common law conclusive presumption that a child under 7 years was conclusively presumed incapable of forming the intent to commit a crime, and a child between 7 and 14 years was rebuttably presumed incapable, has this impacted any state's determination whether there should be a minimum age under which a child cannot be declared delinquent and should not be subject to juvenile court jurisdiction? The answer is "yes". Sixteen states have set a minimum age for juvenile court jurisdiction. North Carolina has set the lowest at 6 years of age, followed by Maryland, Massachusetts, and New York at age 7. Arizona is alone at age 8 years. Eleven states have set 10 years of age as a minimum including: Arkansas, Colorado, Kansas, Louisiana,

Mississippi, Minnesota, Pennsylvania, South Dakota, Texas, Vermont, and Wisconsin. The remaining states have no minimum age set for juvenile court jurisdiction.

18.2 MAKEUP OF THE JUVENILE COURT SYSTEM

It must be pointed out that even though all states have a juvenile court system, the makeup of the court may vary from area to area within the same state. This is dictated by practical population and economic factors. In the densely populated urban and suburban areas, there may be separate staff, buildings, and other facilities serving the separate juvenile court. The court may be composed of one or more judges whose sole function is to serve in juvenile matters. In the less densely populated rural areas of a state, the court may work out of the same facilities and share the staff or other county offices. In fact, the judge may wear two or more hats. The judge may be the county judge, or the like, and serve as the juvenile judge also. As a result, the quality of juvenile justice may vary greatly from one area of the state to another. The quality of justice administered to juveniles is also dependent upon the qualifications of the juvenile judges, their length of service, the diversity of the judicial role, and the size of the court's jurisdiction. Most importantly, however, it will be affected by the resources available to the judge in making appropriate dispositions of cases. For example, a part-time juvenile judge in a rural area may have only a few options available, such as release on probation or referral to a state training school. Judges in urban or suburban areas, on the other hand, may have a variety of local treatment and probation programs from which to choose. The majority of juveniles sent to state training schools from these areas are, in fact, the "hard-core" delinquents.

18.3 THE CHANGING PHILOSOPHY

Under the philosophy of *parens patriae*, the primary purpose of juvenile courts was accountability, so juveniles were not really treated as criminals. All that began to change in 1966 when the United States Supreme Court began deciding a number of cases which guaranteed the same procedural safeguards to juveniles as were provided to adult criminal defendants. The Court was critical of the philosophy and operations of the juvenile court system because it said that the outcome of juvenile court proceedings carried the same stigma as convictions in adult court, and juveniles were often subject to a loss of liberty.

18.4 JUVENILE COURT JURISDICTION

As with any other court referred to in Chapter 3, the juvenile courts of each state must meet the tests of jurisdiction before they are permitted to dispose of any case.

Territorial Jurisdiction

The juvenile court has territorial jurisdiction only within a specific geographic area, usually a county or circuit, and it has no authority to dispose of any cases outside its jurisdiction. However, there may be exceptions to this rule in light of the philosophy of the juvenile court system. Courts may, by legislative action, transfer a case to the juvenile court of another jurisdiction closer to the place where the juvenile lives so that, once disposition of the case is made, the court will be able to keep track of the youth better. This procedure is perfectly permissible because of the nature of a juvenile court proceeding. However, this is permitted only between courts of the same state.

Jurisdiction over the Person

Juvenile court statutes generally specify that they have jurisdiction over youths who were juveniles at the time the prohibited act was committed. Thus, in most states, even if a youth has passed the age limit for juvenile court jurisdiction by the time the hearing is held, he or she may still be subject to juvenile court jurisdiction if, at the time the act was committed, he or she was within the age limit.

Transfers to Adult Court

All states and the District of Columbia allow adult criminal prosecution of juveniles under some circumstances. State transferring mechanisms differ from one another primarily in where they locate the responsibility for deciding whether a given juvenile should be handled in front of a juvenile court or sent to an adult court with criminal jurisdiction. There are other factors considered, such as the age of the offender, the seriousness of the offense, the juvenile's previous delinquency, the views or desires of the victim, or combinations of any of these factors.

Transfer provisions fall into one of four major categories or combinations of those categories. **Discretionary waiver** gives juvenile court judges the discretion to transfer a juvenile to be tried as an adult if the criteria established by the state for transfer are met. **Mandatory waiver** requires a juvenile court judge to transfer the case to adult court. **Statutory exclusion** involves cases where the state law requires the case to be initially filed in adult court even though the offense was committed by a juvenile. The fourth category—**direct file**—gives the prosecutor the option of proceeding in juvenile or adult court. These categories are not always clear-cut. There are many individual differences among the jurisdictions. Local laws need to be consulted.

In broad terms, forty-six states have waiver provisions which leave the transfer decision to the juvenile court judge, but even those with waiver provisions differ from one another in the degree of decision-making flexibility allowed for the juvenile court. Some states make the waiver decision entirely discretionary. Fifteen states set up a presumption in favor of a waiver and another fourteen states specify circumstances under which a waiver is mandatory.

Fifteen states al'ow a prosecutor to make the decision to file directly, thus giving the prosecutor the authority to determine whether to initiate a case against a minor in juvenile court or in criminal adult court.

Twenty-eight jurisdictions have a statutory exclusion provision which grants the criminal courts original jurisdiction over a whole class of cases involving juveniles. Under statutory exclusion, state legislatures take the decision out of the hands of both the prosecutors and the courts.

As may be noted, these numbers do not add up to fifty-one jurisdictions (including the District of Columbia). This is due to the fact that many states have a combination of these mechanisms. Nebraska has only direct file; thus, in all cases, only the prosecutor can determine whether to initiate the case against a youth in juvenile court or in adult court. New Mexico and New York have only statutory exclusion provisions, meaning the legislature has dictated under what circumstances a juvenile may be prosecuted as an adult.

States that provide for either discretionary, mandatory, or presumptive waiver only include: California, Connecticut, Hawaii, Kansas, Kentucky, Maine, Missouri, New Hampshire, New Jersey, North Carolina, North Dakota, Ohio, Rhode Island, Tennessee, Texas, and West Virginia.

In Massachusetts, there are statutory exclusions and some cases in which the prosecutor has the discretion to initiate a direct file.

A combination of waiver provisions and direct file exists in Arkansas, Colorado, District of Columbia, Michigan, Virginia, and Wyoming.

States that combine waiver provisions with statutory exclusions include Alabama, Alaska, Delaware, Idaho, Illinois, Indiana, Iowa, Maryland, Minnesota, Mississippi, Nevada, Oregon, Pennsylvania, South Carolina, South Dakota, Utah, Washington, and Wisconsin.

There are also a few states that use all of these mechanisms: Arizona, Florida, Georgia, Louisiana, Montana, Oklahoma, and Vermont.

The laws of twenty-three states provide for **reverse waiver,** a mechanism whereby a juvenile who is being prosecuted as an adult in criminal court may petition to have the case transferred to juvenile court for disposition. Among the jurisdictions with reverse waivers are Arizona, Arkansas, Colorado, Delaware, Georgia, Iowa, Maryland, Mississippi, Nebraska, Nevada, New York, Oklahoma, Oregon, Pennsylvania, South Carolina, South Dakota, Vermont, Virginia, Wisconsin, and Wyoming.

A special transfer category has been created in thirty-one jurisdictions for juveniles who have once been convicted in criminal court of an offense. This **once an adult/always an adult** classification mandates that once a juvenile has been tried (and, in most cases, convicted) in adult court, any new offense will automatically be handled by the criminal court rather than the juvenile court. States that have this provision are Alabama, Arizona, California, Delaware, District of Columbia, Florida, Hawaii, Idaho, Indiana, Iowa, Kansas, Maine, Michigan, Minnesota, Mississippi, Missouri, Nevada, New Hampshire, North Dakota, Ohio, Oklahoma, Oregon, Pennsylvania, Rhode Island, South Dakota, Tennessee, Texas, Utah, Virginia, Washington, and Wisconsin.

Statistics reflect that the average age of those who commit felonies has dropped. This increase in youthful felons has driven the reexamination of the juvenile justice system and the issue of whether juvenile felons should be prosecuted in the adult system. As can be ascertained from the preceding information, legislative emphasis is on the transfer of juveniles to adult courts in serious criminal cases.

Most states still provide age distinctions. Some states will not transfer a 13-year-old. Other states require the child to be at least 15 years old at the time of the commission of the crime before any consideration will be given to trying the juvenile as an adult and, if convicted, punishing the juvenile as an adult.

Jurisdiction over the Subject Matter

Juvenile courts have jurisdiction over two basic types of situations: offenses committed by juveniles and offenses committed against juveniles. In the former case, offenses committed by juveniles can be subdivided into two categories: status offenses and delinquency. Offenses committed against juveniles are covered in Section 18.10.

Status Offenses

Juvenile **status offenses** are violations of rules of conduct established by society to protect the interests and the process of socialization of youths. They are not considered criminal violations of the laws of the state, county, or city in which they are applied. Unlike criminal violations, status offenses lose their relevancy to a person once that person becomes an adult. Among the status offenses are such behaviors as truancy from school, curfew violations, running away from home, and a category of behavior commonly known by such titles as stubborn or unruly child, ungovernable behavior, conduct injurious to a juvenile's person, or conduct unbecoming a juvenile.

Despite the noncriminal nature of status offenses, the jurisdiction over them has been awarded to the juvenile courts in most jurisdictions. There are many who feel that this jurisdiction should be transferred to other public agencies not associated with the judicial aspects of the criminal justice system. In the alternative, in which juvenile courts maintain this jurisdiction, efforts are being made to dissociate this kind of conduct from more serious acts of delinquency. One method of accomplishing this has been to attach the label **"persons in need of supervision"** or **"persons in need of service"** (PINS) to the juvenile status offender. In fact, some jurisdictions have urged that PINS be extended to all juveniles coming within the jurisdiction of the juvenile court for whatever reason.

Delinquency

Delinquency is conduct engaged in by a juvenile that, if committed by an adult, would violate the penal laws of the state. Boiled down to simple language, delinquency is a crime committed by a juvenile. Some jurisdictions also attach arbitrary labels to distinguish between serious and nonserious delinquency acts. The distinction is based upon the nature and seriousness of the offense and the severity of the disposition that can or should be imposed by the juvenile court judge.

18.5 CHILD ABUSE

Statutes prohibiting and punishing child abuse have existed for many years in most jurisdictions and often provide for a more severe penalty than battery because of the inability of children to defend themselves. Nevertheless, the offense is a form of assault and battery (see Chapter 7) unless death results. The offense is closely tied to neglect as described in section 18.10 and often consists of physical, sexual, or emotional mistreatment, or neglect of the child. **Child abuse** has been defined as an act or failure to act (omission—see section 4. 5) on the part of a parent or caretaker that results in the death, serious physical or emotional harm, sexual abuse or exploitation of a child, or which places the child in an imminent risk of harm.

Child abuse laws raise some difficult legal and political issues, pitting the right of children to be free from harm against the right of privacy for families and the rights of parents to raise their children as they want to or believe to be right without government interference. Despite parental rights, the conduct described here is not legally tolerated and action is taken. Investigations are usually done by child welfare agents of the local or state government and, where appropriate and necessary, legal action will be taken against those who commit criminal acts. But, caution is, or should be, exercised before jumping to conclusions. Prosecutors must look very closely at the motives behind the reporting of child abuse. Claims are sometimes made because of intra-family squabbles or as a method of seeking revenge for some other dispute between adult parties.

Despite the growing concern about child abuse and the legislation which seeks to protect children by severely punishing those who abuse, there are still some horrific examples. Parents of two young girls, one 9, the other 4 years old, went to Mexico for a nine day Christmas vacation, leaving the girls behind to care for themselves. The girls had frozen dinners and cereal to eat. The parents left a note telling the girls when to go to bed. A neighbor found the girls barefoot in the snow after a smoke alarm had gone off in the girls' house. The parents were placed on probation for two years after thirty days of house arrest, even after a grand jury found evidence that the parents had beaten, kicked and choked the children to discipline them. Meanwhile, the girls were in a foster home. Four months later the parents gave up all parental rights and placed their daughters for permanent adoption. In another example, after his 6-year-old son began fighting with another child in the car, the defendant pulled over, stopped the car, removed his 6-year-old son from the car and left the child on the side of the road. The court said these actions were sufficient to constitute child neglect.

18.6   CHILD MOLESTING

Although very similar in most respects to the child abuse laws, **child molesting** is the offense more often thought of as committed by a stranger or someone who is not a parent or caretaker of the child victim. Child molesting is prohibited by statute in most jurisdictions and is broadly defined to include most sex offenses committed against children. All other sex offenses are drawn from and come within the purview of this statute, including handling and fondling. In a number of states, special procedural rules are established for the treatment of persons convicted of child molesting.

18.7   BATTERED CHILD SYNDROME

Akin to the battered spouse syndrome (see § 6.19, Battered Person Syndrome as a Defense) is the **battered child syndrome.** Statutes prohibiting and punishing child abuse have existed for many years in most jurisdictions as a form of assault and battery. Statutes often provide a more severe penalty as an assault or battery because of the inability of children to defend themselves.

As for using the battered child syndrome as a defense to a homicide charge, not as many states have had an opportunity to consider the issue as they have had with the battered spouse problem. Some states, while accepting the battered spouse syndrome, have not fully accepted such a perfect defense for the child. Rather, the discretionary use of such information in the hands of the prosecutor has been readily used to not charge at all; or when charging, to treat such a death as manslaughter rather than murder. Judges have used such information relative to sentencing and thus have been more lenient. In one Massachusetts case, a 15-year-old, tired of the beatings received by his siblings and himself, grabbed his father's gun and shot him when the father began another screaming tirade. The prosecution charged manslaughter, the jury convicted, and the judge sentenced the youth to one year in a juvenile facility, even though he had been tried as an adult and could have received a life sentence.

One final thought on these syndromes. The more premeditated the killing, the less likely the defendant will be able to use this branch of the self-defense doctrine as a defense. For example, in a Wyoming case, while a child's parents were at dinner, he gathered several guns, hid his sister in the basement, and lay in wait in the garage for his father's return. When the father opened the garage door, the son shot his father. The court felt this was not an appropriate case for self-defense.

18.8   JUVENILES AND THE INTERNET

Pornography, sexual solicitation, fraud, drug use, bomb making, tobacco use, crime commission, and more explicit material can be found on the Internet, and much of it is targeted toward young people who are inexperienced, naïve, and susceptible. There may be in excess of 28,000 hard- and soft-core pornography sites on the Internet.

The number of young people using the Internet is growing rapidly. There are about 45 million kids online, and that number is expected to reach well over 75 million by 2005.

Efforts are underway to try to protect young people from being victimized by predators and others seeking to profit from the youth. Solutions to the legal issues involved are not easily supportive of what many view to be ideal goals. The sidebar entitled "The Internet and Kids" is illustrative of the problem.

Public libraries have tried to help by filtering sexual materials using blocking software so young users can't gain access at those locations. A federal district court ruled that such an attempt by a library in Virginia constitutes a violation of the First Amendment guarantee of free speech. The filters, as the blocking software is called, prohibited anyone, adults included, from having free access to the Internet. The court did suggest that there might be some less restrictive means of accomplishing the library's goal to protect children, but didn't go into specifics. Some ideas may be filtering only certain computers reserved only for youths, turning filters off for adult users, educating users, or establishing a use policy in the library.

As far as home computers are concerned, parents must take a great deal of responsibility for what their children access. There are over a dozen "blocking" software programs on the market for installation on home personal computers. Their effectiveness varies and one thing is certain, they won't work on a newly created web site or one not part of the software package.

The Internet And Kids

The Internet is an unregulated conglomeration of information from hundreds of thousands, if not millions, of independent sources worldwide. Many people are concerned about the easy access to pornographic materials on the Internet and have been seeking methods to protect children from such exposure. In response to this problem, Congress passed the Communication Decency Act of 1996 which provided federal regulation of the Internet. There was an outpouring of opposition, and the U.S. Supreme Court ruled the law unconstitutional as being in violation of the First Amendment guarantee of freedom of speech. Most people did not want regulation by the government, believing that there are other ways to protect the vulnerable besides depriving people of freedom of speech, and they certainly did not want to give any jurisdiction over a worldwide nonsystem to the United States federal government. Congress is trying again with the passage of the Child Online Protection Act. The act requires commercial Web sites to collect a credit card number or some other access code as proof of age before allowing Internet users to view online material deemed "harmful to minors." Violators could be sentenced to up to six months in jail and a $50,000 fine. A federal district court granted a preliminary injunction against enforcement of this act and in 2000, a circuit court of appeals affirmed the district court. The court said it thought the ACLU, the group challenging the constitutionality of the statute, had a good chance of winning on the merits. The court said that since the standard against which material "harmful to minors" is evaluated is "contemporary community standards," the Web publishers can't be held to different standards for every different community. To hold otherwise would impose an unconstitutional burden upon them. The court went on to say that technological limitations may make it impossible to place restrictions on such material. Technological advancements may make it possible in the future. Meanwhile, the U.S. Supreme Court has granted certiori to hear an appeal in the case. The name of the case has changed from Reno to the new Attorney General, *Ashcroft v. ACLU.*

18.9 ASSAULTS ON SCHOOLTEACHERS

A growing problem that affects not only large city schools but rural schools as well is that more and more teachers are being assaulted on school grounds by students. Recent exposés by the media portray the daily fear that is experienced by many teachers. Studies show that teachers are actually suffering from "battle fatigue" and need psychiatric help to continue in their profession.

The causes of the growth of classroom violence are many and beyond the scope of this book. The fact remains that assaults on teachers in all areas of the nation are up and continue to rise. As with the area of spousal assaults, legislatures are beginning to react with both civil and criminal remedies. Illustrative of legislative reaction to this growing problem is the statute passed by the West Virginia legislature. It reads as follows:

_61-2-15 Assault, battery on school employees; penalties

(a) If any person commits an assault by unlawfully attempting to commit a violent injury to the person of a school employee or by unlawfully committing an act which places a school employee in reasonable apprehension of immediately receiving a violent injury, he shall be guilty of a misdemeanor, and, upon conviction, shall be confined in jail not less than five days nor more than six months and fined not less than fifty dollars nor more than one hundred dollars.

(b) If any person commits a battery by unlawfully and intentionally making physical contact of an insulting or provoking nature with the person of a school employee or by unlawfully and intentionally causing physical harm to a school employee, he shall be guilty of a misdemeanor, and upon conviction, shall be confined in jail not less than ten days nor more than twelve months and fined not less than one hundred dollars nor more than five hundred dollars.

(c) For the purposes of this section, "school employee" means a person employed by a county board of education whether employed on a regular full-time basis, an hourly basis or otherwise if, at the time of the commission of any offense provided for in this section, such person is engaged in the performance of his duties or is commuting to or from his place of employment. For the purposes of this section, a school employee shall be deemed to include a student teacher.

18.10   VICARIOUS LIABILITY AND PARENTAL RESPONSIBILITY

States have been enacting laws holding parents criminally liable for the delinquent acts of their children for 100 years. Many of these **vicarious liability** statutes are amendments to state juvenile court laws that subject parents, guardians and other adults to the jurisdiction of the juvenile court system. The philosophy is geared toward treating or providing services to the whole family. In some instances, it results in punitive sanctions against multiple family members.

Juvenile courts have jurisdiction over cases involving offenses committed against juveniles. Dependency and neglect cases fall within this classification. The juvenile court often extends its authority to youngsters who have not committed any overt act that demonstrates that the juvenile has offended society, but, if circumstances exist that indicate the failure on the part of a parent or guardian to provide properly for the welfare of the juvenile, the court is considered to have sufficient reason to assume jurisdiction of the case.

Dependency and neglect are often used interchangeably, and in many jurisdictions the authority of the court to dispose of such cases is identical. However, a technical distinction does exist. In cases of **neglect** there is generally some parental fault. Deliberate abandonment, physical and emotional abuse, general failure to provide for the minimum needs of the child's welfare, and the outright physical abuses associated with the battered child syndrome all fall within the court's jurisdiction over neglect cases. Cases classified as dependency generally involve no deliberate parental irresponsibility. Instances of **dependency** include the total absence of a parent or guardian to look after the welfare of a child, lack of proper care by virtue of physical or mental incapacity on the part of the parent or guardian, cases where a parent or guardian, with good cause, wishes to give up parental responsibility, or instances where parental poverty prohibits the child from receiving minimum care.

To a great extent, enforcement of many of these laws slackened during the middle part of the twentieth century but, in this day of increased serious criminal activity by children, many states are reconsidering whether parents should be made vicariously liable for the acts of their children. In the past, many states have held such attempts to be unconstitutional even for traffic violations and misdemeanors. On the other hand, courts did not have such great difficulty in holding parents directly liable for their own criminally negligent conduct (if applicable statutes existed), conduct which had a bearing on the offense committed by the child. For example, some state laws punish the parent for failing to keep a firearm out of reach when a child uses the weapon and causes injury or death.

Whether under juvenile court jurisdiction or subject to the jurisdiction of criminal courts, offenses have been created to punish parents, guardians and others for transgressions against juveniles. Contributing to the delinquency of a minor is one of the most prevalent statutes of this type.

The Model Penal Code attempts to identify the very real problems that face modern society. Two provisions cover other offenses against juveniles. One is a mild child abuse statute entitled "Endangering the Welfare of Children," and the other is "Persistent Nonsupport."

The first makes it a misdemeanor to knowingly do anything that endangers a child's welfare by violating a duty of care, protection, or support. The child has to be under the age of 18, and the duty is one owed by parents, guardian, or any other person supervising the welfare of the child. The second offense makes it a misdemeanor to fail persistently to provide support for a wife, child, or other dependent. Of course, the defendant must have the ability to provide support.

18.11   IMPOSITION OF THE DEATH PENALTY ON JUVENILES

The death penalty was not much of a concern in juvenile cases when *parens patriae* was still the predominate philosophy in juvenile courts. In 1972, however, the United States Supreme Court decided a case that consolidated two cases from Georgia and one from Texas. In two of the cases, the juvenile defendants were charged with rape and murder was charged in the other case. All three defendants were convicted in their respective trials and the convictions were upheld all through the state court systems. In these cases, the Supreme Court held that carrying out the death penalty in these three cases constituted cruel and unusual punishment, in violation of the Eighth and Fourteenth Amendments to the United States Constitution.

Since the court decision, seventeen juvenile offenders have been executed in the United States; nine in Texas, three in Virginia, and one each in Oklahoma, Louisiana, Missouri, Georgia, and South Carolina. These executions are only a fraction of the 200 total juvenile death sentences imposed. One hundred and ten have been reversed on appeal, and seventy-three juvenile death row inmates await execution or further litigation. Of the 200 sentences, seventy percent were given to seventeen-year-old offenders, the other thirty percent to sixteen-year-old offenders. Half of the 200 death sentences have occurred in just three states: Texas has had fifty, Florida has imposed thirty, and Alabama has sentenced twenty-one juveniles to death.

Currently, both the civilian court system and the military judicial system at the federal level, along with thirty-eight states, allow for the death penalty in capital cases. Sixteen of those jurisdictions chose 18 years of age as the minimum allowed for execution, five jurisdictions will not allow the death penalty for anyone under 17 years of age, and 16 is the minimum execution age in nineteen jurisdictions.

The arguments for and against the death penalty in juvenile cases are the same as the arguments put forth by the proponents and opponents of capital punishment in adult criminal proceedings. Those in favor of the death penalty cite the increasing violence perpetuated by youthful offenders, deterrence, and retribution. The opponents argue that the retributive urge cannot be satisfied by imposing the death penalty on people who are not fully responsible for their actions and who have less capacity to control their conduct than adults. Further, they blame failures of the family, school and social system for the criminality of juveniles who do earn the death penalty.

18.12 THE FUTURE OF THE JUVENILE COURT SYSTEM

There are proponents and opponents to maintaining the juvenile court system. The opponents argue that it should be abolished. From its inception the juvenile justice system has sought to rehabilitate children, rather than punish them, for their criminal behavior. Critics of the juvenile court system contend that rehabilitation has not worked and serious criminal behavior by young people is not being punished. The defenders of the system argue that, for the vast majority of juveniles who come in contact with the system, it is working and that the public should not be pressured into believing the system to be a failure because of the relatively small number of violent juveniles who have committed some horrendous crimes. The proponents further contend that the problem with the juvenile court system is that it, and social programs, are under funded.

Opponents, calling for abolition of the juvenile court system, urge that the difference in treatment of an adult and someone who is almost an adult for the commission of the same unlawful conduct is not fair. Opponents urge abolishing the juvenile court system and providing full constitutional due process rights, including the right to a trial by a jury.

Time will tell the ultimate outcome but, for now, the juvenile system remains.

DISCUSSION QUESTIONS

1. Under what circumstances can parents or guardians be subjected to the jurisdiction of a juvenile court?

2. What transfer provisions exist in your state? Under what circumstances do they operate?

3. Eleven-year-old Willie has been physically beaten for as long as he can remember by his father. One night, he couldn't take it anymore and, while his father was sleeping, Willie got a sledgehammer from the garage and hit his father seven times in the head. Father died. How, and in what court system, if any, should the case be processed? Does Willie have any defenses?

GLOSSARY

Battered child syndrome – an affirmative defense used in homicide cases when an abused child kills the abuser.

Child abuse – physical, sexual, or emotional mistreatment of a child or neglect.

Child molesting – sexual mistreatment of a child normally by someone other than a parent, guardian, or caregiver.

Delinquency – an act committed by a juvenile that would be classified as a crime if committed by an adult.

Dependency – a child whose parents or guardians do not have the ability or capability to provide the basics—food, clothing, shelter—for a child.

Direct file – the authority of a prosecutor to try a juvenile as an adult.

Discretionary waiver – the authority of a juvenile court judge to order a juvenile to be tried as an adult.

Juvenile – a person under the age of majority (age varies by state) and subject to the jurisdiction of a juvenile court.

Mandatory waiver – requirement that a juvenile court judge direct a juvenile to be tried as an adult in certain circumstances.

Neglect – failure of parents or guardian to provide the basic necessities of food, shelter, and clothing to a child; also includes child abuse and other acts; also includes child abuse and endangering welfare violations.

Once an adult/always an adult – policy in some jurisdictions that once a juvenile is tried as an adult for any offense, the child will be treated as an adult for all future violations.

Parens patriae – Latin term describing the paternalistic philosophy of the juvenile court system to do what is best for the juveniles subject to the court's jurisdiction.

Persons in need of supervision (or service) (PINS) – A manner of handling status offenders in the juvenile court system.

Reverse waiver – the process of transferring a juvenile from the adult criminal courts to the juvenile court for disposition of the case.

Status offenses – violations that are only violations because the offender is a juvenile.

Statutory exclusion – situations where the law dictates that a juvenile shall be tried as an adult.

Vicarious liability – a person criminally liable for the acts of another, as such a parent found criminally liable for the acts of the child.

▓ REFERENCE CASES, STATUTES, AND WEB SITES ▓

CASES

In re Gault, 387 U.S. 1, 87 S.Ct. 1428, 18 L.Ed.2d 527 (1966).

Kent v. United States, 383 U.S. 541 (1966).

In re Winship, 397 U.S. 358, 90 S.Ct. 1068, 25 L.Ed.2d 368 (1977).

State v. Q.D., 685 P.2d 557 (WA 1984).

Furman v. Georgia, 408 U.S. 238 (1972).

In re Kemmler, 136 U.S. 436 (1890).

Eddings v. Oklahoma, 455 U.S. 104 (1982).

Thompson v. Oklahoma, 487 U.S. 815 (1988).

Stanford v. Kentucky, 492 U.S. 361 (1989).

Jahnke v. State, 682 P.2d 991 (WY 1984).

Reno v. ACLU, 217 F.3d 162 (2000).

Ashcroft v. ACLU, 121 S.Ct. 1997 (2001).

State v. Wynne, 26 FLW D1376a, 2dD.C.A.

STATUTES

United States: 42 U.S.C.A. §§ 5105-5106

Alabama: Ala. Code § 12-15-13 (1995).

California: Cal. Penal Code §§ 270-272

Florida: Fla. Stat. Ann. 784.05 (1999).

Indiana: Ind. Code § 34-4-31-1 (1995).

Kentucky: KY Rev. Stat. § 530.050; KY. Rev. Stat. Ann. 530.060 (Baldwin 1995).

Louisiana: La. Rev. Stat. Ann. § 14:92.2 (West Supp. 2000).

Massachusetts: Mass. Gen. L. ch. 119, S. 67 (1994).

N.C. Gen. Stat. § 14-316.1 (1995).

New York: N.Y. City Dom. Rel. Ct. Act § 61(2) (McKinney 1936); N.Y. Penal Law § 26.10 (McKinney 1989 & Supp. 1996).

Oregon: Or. Rev. Stat. Ann. § 419B.160 (Michie Butterworth 1995).

Pennsylvania: 42 Pa. Cons. Stat. § 6326(a)(1) (1982).

Wisconsin: Wis. Stat. Ann. § 948.40 (West 2000).

WEB SITES

www. law.onu.edu/faculty/streib/juvdeath.htm (or pdf)
www.hrw.org/children/
www. Florida lawWeekly.com

19

Traffic Offenses

▦ KEY WORDS AND PHRASES ▦

Absolute speed limit
Blood alcohol content
Cone of vision
Decriminalize
Disabled parking
Drag racing
Driving under
   the influence (DUI)
Driving while intoxicated (DWI)

International accessibility symbol
Low-speed vehicle
Moving violation
Nonmoving violation
Prima facie speed limit
Racing
Reckless driving
Uniform Vehicle Code
Vehicular homicide

19.0 INTRODUCTORY COMMENTS

The tendency is to equate traffic laws with the automobile. Yet prior to the automobile's introduction, laws regulating speed and other aspects of driver conduct were enacted for horse-powered transportation. The invention of the automobile created a need to expand those laws and establish some uniformity.

The first major attempt to develop a set of rules for motorists throughout the nation began in 1926 when the first Uniform Vehicle Code and its "Rules of the Road" were offered. The **Uniform Vehicle Code,** which serves as a basis for state traffic laws, has been adopted in every state with some local modifications. This law has undergone continual revision, taking into account modern technology and modern problems. Not all states have enacted its most recent provisions. Several states have attempted to keep pace.

Because the Uniform Vehicle Code is the most significant traffic law in the United States, we will examine the major traffic offenses through its provisions. The student should consult local statutes and ordinances for differences. Municipalities, subject to state legislative control, have the right to enact their own traffic control ordinances for municipally controlled streets. Generally, cities adopt the Uniform Vehicle Code and add specific regulations to cover local need. However, these local ordinances are subject to careful review because some communities have used the traffic laws not as police regulations but as fundraisers and as substitutes for taxes. State and federal constitutions prohibit taxing under the guise of police power. Thus, a trend has developed whereby cities must get state permission to establish speed limits other than those allowed by state law. In some states, cities are prohibited from using speed-detection devices or are severely regulated as to where, how, when, and who can use them.

19.1 MOVING AND NONMOVING VIOLATIONS: SIGNS AND SIGNALS

Whether one is charged with a parking violation, a **nonmoving violation** a speeding offense, or a **moving violation,** it is essential to the charge that proper notice (signs and signals) were given and, if mechanical, were operating properly. The Uniform Vehicle Code requires obedience to all traffic signs, signals, and markings. However, the sign, signal, or marking has to be in the right place, observable, and properly maintained or working. Cities that allow vegetation to obscure their stop signs and other traffic markers cannot complain when someone runs a sign.

Sizes of signs are sometimes important. For example, in some states motorists have to be warned of radar use by a sign no smaller than 40 inches by 40 inches. If an officer testifies that the sign was posted, it will be presumed to be a valid sign. However, if the defendant proves the sign was considerably smaller, then it violates a prerequisite for a radar-enforced "catch," and the case must be dismissed.

As to the location and size of signs, all states have entered into a compact for uniform sign size and placement. A manual of such requirements is published by the U.S. Department of Transportation and is called *Manual on Uniform Traffic Control Devices for Streets and Highways*. Thus, a sign not placed within a reasonable approximation of the requirements, or the "**cone of vision**," is no sign at all. Stop signs have been observed at the left of an intersection rather than at the right. Why? Road workers have said they could not dig the hole where it was supposed to go. Some signs are placed too far from the intersection; some are placed too high, and some are placed too low. Therefore, it is no idle defense when the defendant proves the improper placement, maintenance, size, color, or even shape of a sign.

19.2 TERRITORIAL APPLICATION OF TRAFFIC LAWS

Of course, traffic laws apply to all streets, roads, and alleys, but what of private roadways? If a private way has become, by custom and usage, a public thoroughfare, then it is subject to regulation. Even shopping center parking lots have been included by the Uniform Vehicle Code and by local ordinances or other state statutes.

There are some offenses under the Uniform Vehicle Code that do not require commission upon a public road. These provisions penalize the conduct no matter where it occurs. Therefore, eluding a police officer, vehicular homicide, reckless driving, hit and run, and driving under the influence do not have to take place on a public street or highway. The philosophy is that the serious nature and consequences of such conduct do not and should not depend upon the place of occurrence.

19.3 WHO MUST OBEY TRAFFIC LAWS

The Uniform Vehicle Code applies to all persons, whether real or corporate and whether driver, pedestrian, or person in control of a vehicle. Thus, nobody is automatically exempted from its provisions. Law enforcement officers, ambulance drivers, and fire personnel, when engaged in a true emergency, can to some extent violate the law. Even then, however, they must use visual and audible warnings, and exercise due care under the emergency circumstances.

The laws regulating the use of public ways must be observed when using any vehicle: motorcycles, cars, trucks, and the like. In many instances, local laws and some Uniform Vehicle Code provisions also include people-powered vehicles. The regulations also apply to pedestrian conduct where appropriate.

19.4 PARKING AND RELATED OFFENSES

In most American cities, the number one traffic problem is parking. There are not enough spaces. Spaces have to be provided for emergency use, access to property, public transportation to safely process passengers, loading zones, disabled persons, and for efficient noncongested flow of traffic.

Despite all these well-defined and, it is hoped, properly marked use zones, people still park illegally or for too long. Parking offenses are usually the simplest of the *mala prohibita* offenses. Yet problems can arise.

What is parking? The Uniform Vehicle Code defines it as the standing of a vehicle, occupied or not, other than temporarily and not actually engaged in loading or unloading property or passengers.

If, therefore, a person is temporarily in a no-parking zone, getting some property to put in a car, the person is stopped but not parked. Unless the zone is properly marked "No Parking or Stopping" rather than just "No Parking," then that person has not parked.

Some of the most bitter parking cases have arisen when a landowner who runs a business has a loading zone established and then parks a vehicle in the spot. The landowner gets ticketed and complains that the zone was for his or her benefit. Unfortunately, the landowner loses. The state or city cannot put the landowner in a more favorable position than it would any other member of the public. The same result confronts the home owner who blocks his or her own driveway when there is an ordinance that prohibits blocking driveways. There can be no favored treatment.

The law presumes that the owner of the ticketed vehicle parked it. It is up to the defendant to show that he or she was not the one who parked the car illegally if he or she wishes to escape personal liability.

The enforcement of parking spaces marked for disabled persons is at the center of current parking concerns. Federal and state laws prescribe the location and number of **disabled parking** places that must be provided at office buildings, restaurants, retail stores, shopping center parking lots, public buildings and so forth. The laws provide for the size of spaces, the color of paint markings, and the shape, size, content, and placement of signs. State statutes generally provide for the issuance and display requirements of placards to be hung from the inside mirror. These laws also govern the issuance of disabled persons' license plates which usually contain the **international accessibility** (wheelchair) **symbol.**

Many state laws require the disabled person to be in the vehicle if a reserved space is to be used. State statutes provide hefty fines for unauthorized parking in a space reserved for the disabled. Nevertheless, there is not one among us who has not observed an unqualified vehicle parked in a reserved space. Local groups representing the disabled call for more enforcement, but it seems that law enforcement can never solve the concerns. Some law enforcement agencies have solicited volunteers who have been given the authority to write citations and specifically look for violations of disabled parking regulations. Time will tell how successful these efforts will prove to be.

19.5 SPEEDING OFFENSES

In the United States, we begin with the proposition that by controlling speed we protect people and property and save lives. This proposition is based on the assumption that most people do not have judgment enough to safely regulate their driving behavior. Thus we have established maximum and minimum speed limits.

There are several types of speed regulations. First is the **absolute speed limit** which, if exceeded, renders the offender guilty. No driving conditions enter into this type of case. Mere proof of the posted legal limit and the defendant's speed is all that is needed. As speeding is a *mala prohibita* offense, mental attitude or intent is not relevant.

Second, the state can adopt a ***prima facie* speed limit,** that indicates that driving at a speed greater than the limit is unsafe and the offender is presumed in violation. However, the defendant can introduce evidence that, under the conditions then present, the actual speed was reasonable.

A third method is to provide an absolute top speed but require drivers to appropriately reduce speed when weather or other conditions are not ideal. Therefore, one might travel at 50 mph in a 55-mph zone and still be in violation because of fog, rain, snow, or other potentially hazardous conditions.

As one can see, a speed less than the maximum limit is not lawful, per se. Similarly, a speed that is reasonable and proper under certain circumstances may be excessive under others. The width of the road, visibility, presence of hills, character of the pavement, number of lanes, intersections, other traffic, and the position of the sun are some conditions used to determine reasonable or appropriate speed. Some courts also look at the condition of the vehicle, its weight, length, and so on, in determining appropriate speed.

In one case, the defendant was traveling at 20 to 25 mph on a snowy winter night. He applied his brakes, skidded, pumped his brakes, and continued skidding until he bumped into the rear of a police car stopped at a traffic light. No one was injured and neither vehicle was damaged. He was charged with operating a vehicle at a speed greater than was reasonable and prudent under the conditions. His conviction was reversed on appeal. The court did not feel that the result indicated the charge.

The slow driver represents a very real danger on the road. For this reason the modern version of the Uniform Vehicle Code, as well as federal law regulating interstate highways, provide punishment for the slow driver. A person violates the law by unnecessarily impeding the normal and reasonable movement of traffic. The federal law provides a 40-mph minimum on interstate highways when conditions are good.

There are exceptions to the slow driver rule. Special exemptions have been made for certain types of equipment moving on highways. Farm vehicles are exempted but are required to display the slow-moving vehicle symbol. Other vehicles involved in commerce are provided some limited exemption. Often slow-moving, oversized equipment has to move on highways. States often require escorts and signs announcing the movement of such equipment.

One of the "old wives' tales" of driving is that it is permissible to exceed the speed limit when passing. In one case, the defendant argued that the excessive speed was necessary and, in fact, enhanced safety in such a passing situation. The court disagreed, saying that, in the absence of a true emergency, safe passing is consistent within the posted limits. The true emergency would arise at the time when, once the passing has begun, out of nowhere a car approaches from the other direction and the passing vehicle is past the point of no return and must go forward rather than drop back.

The state, city, or county, under proper circumstances, can restrict speeds in certain areas or zones due to engineering, social, or natural conditions. State laws governing such variations from the norm are not uniform. The student should consult local law.

The issue of speed raises questions about the operation of golf carts and a new breed of transportation called **low-speed vehicles** which are usually electric and have safety features. In recent years some state laws have permitted municipalities to allow golf carts to operate on municipal streets. This often works in a retirement community with its own streets.

A golf cart was defined as one primarily designed for use on a golf course, with a maximum speed of 20 mph. It does not have lights, a windshield, or other required safety equipment that would allow it to be used lawfully on a state public street or highway.

In mid-1998, the National Highway Traffic Safety Administration (NHTSA) of the U.S. Department of Transportation published a rule in the Federal Register recognizing and authorizing low-speed vehicles. These vehicles must have certain minimum safety equipment as contained in federal regulations and cannot have a maximum speed in excess of 25 mph. For the most part, they are electric. These vehicles are legal for use on the streets and highways of the state and may be required to have more safety features if state law so requires. They are required to be titled and registered and to display a license plate. Of course these low-speed vehicles cannot operate on the interstate highways because they cannot meet the minimum speed requirement.

19.6 RACING

The Uniform Vehicle Code prohibits races, speed competitions or contests, drag races, acceleration contests, physical endurance tests, speed or acceleration exhibitions, or any attempt to make a speed record.

Drag racing can involve one or more automobiles, it is conducted on a course, and either distance or time can be the limit of the course as long as the object is to compare relative speeds or power of acceleration. The course can be a private road, a public street, or a highway.

Racing is more broadly defined; it is the use of one or more vehicles to attempt to (1) outgain, (2) outdistance, (3) prevent another vehicle from passing, (4) arrive first at a given point, or (5) test the stamina of the drivers. What one looks for are determinable, reasonable, and logical inferences from the observable circumstances. Although the speed at which the cars are traveling need not be illegal, it can be one factor. If competitive aspects can be seen in the conduct, it is racing.

Seeing two cars abreast of each other on a four-lane divided highway would not be enough. What is sought is evidence that one driver attempted to prevent the other from passing or overtaking. It is not necessary to show improper lane usage, swerving, dodging, cutting off, or tailgating, even though these are often indications of racing.

The question of speed is relevant. The word race means speed competition, but it does not require that the vehicles be operating at top speed.

Does the fact that those arrested for racing did not know each other before the "race" have any effect on their guilt or innocence? No. There can be a race by total strangers on the spur of the moment. A conviction can be sustained whether or not there was some sort of prearrangement to race.

One does not have to be the driver to be convicted of racing. Anyone in control of the car and driver or who aids or abets such a race can be convicted. However, mere presence at the race or in the car is not enough. The general rules of principals and accessories would be applicable.

19.7   DRIVING UNDER THE INFLUENCE

Unlike most other traffic offenses that measure the manner of driving, **driving under the influence (DUI)** concerns the condition of the operator. This is true even if it is the manner of driving that causes the police to be suspicious of the cause.

Three elements have to be proven. First, there must be a person under the influence of alcohol or drugs. Second, that person must be driving or at least in physical control of a vehicle. Third, the fact of a vehicle has to be proven.

Under The Influence

Some of the earlier attempts to punish the drinking driver were stated in terms of **driving while intoxicated (DWI)**. Intoxication as an absolute was too difficult to prove. Today, the state merely has to demonstrate that drugs or alcohol had an influential effect on reducing the person's ability to control the vehicle and exercise timely judgment to constitute driving under the influence (DUI).

The question of how much was consumed is unimportant. What is important is the effect that the alcohol has had on the individual. To this end the federal government has mandated, and the states have adopted, a presumption that is derived from testing blood alcohol content (BAC). Today, .08% **blood alcohol content** or greater is the level at which the law declares a person to be "under the influence" and unable to control and operate a motor vehicle and exercise timely judgment. This can be tested by analyzing a blood sample or by blowing into a machine that will measure the BAC.

Driving/Physical Control

The second element that needs to be proven for DUI, "drive or operate," has presented difficulty in some states. At first, only the word "drive" was used. This resulted in convictions of those caught while the vehicle was moving and the driver was controlling either the vehicle's speed or direction.

In most states the word "operate" was substituted for "drive." This allows law enforcement to get at the person who obviously drove the car to the location at which it is discovered but who is found slumped over the wheel while the car is no longer moving. However, even the use of "operation" causes some interpretational problems. When is a person operating?

Some states found anyone sitting behind the wheel to be an operator. Others, however, require at least that the keys be in the ignition, whereas still others require the motor to be running. As a result, a few of the states changed "operate" to "in actual physical control." Thus, local statutes and decisions should be consulted.

Vehicle

The vehicle element would not seem to be difficult to prove. Yet if the statute says "motor vehicle," only those that are self-propelled by combustion or electric or diesel power would be included. This definition leaves out bicycles, horse-drawn carts, and others not propelled by a motor. Some states thus use the term "vehicles." But, if the state has the Uniform Vehicle Code, then "vehicles" includes only self-propelled vehicles that do not run on tracks.

The location at which the defendant is caught is usually unimportant. The DUI provision does not require operation upon a public way, because it is the driver's condition that is being judged. However, local statutes should be consulted to see if the legislature has modified the Uniform Vehicle Code to the extent that operation is required on a public way.

The Uniform Vehicle Code punishes only where the driver is rendered incapable of driving safely. It is most important for the officer to test and record the reactions of the stopped driver. The Uniform Vehicle Code clearly states that it is no defense to this charge that the drugs that influenced the driver were prescribed. Chronic alcoholism is also no defense.

19.8 RECKLESS DRIVING

The second of the three most serious traffic offenses is reckless driving. With reckless driving, the manner of driving the vehicle is the most important element. The Uniform Vehicle Code defines **reckless driving** as driving "any vehicle in willful or wanton disregard for the safety of persons or property. . . ."

Reckless driving is a voluntary act done with a conscious disregard of the consequences. It is more than mere inattention or ordinary negligence. Therefore, it takes more than a simple violation of a traffic law. Some states permit criminal or culpable negligence to satisfy the "willful or wanton" aspect of reckless driving.

Speeding alone is not enough. The surrounding circumstances have to show that in addition to the speed there was a disregard for the consequences. Sixty miles per hour in a school zone at night might not be reckless, yet the same speed during school hours would be.

The same would apply to DUI. Being drunk is a heavy factor in some reckless driving cases, yet there are some drunks who drive slowly, at night, on open country roads, closely hugging the berm. Therefore, the totality of the surrounding conditions—population, time, weather, road conditions, and other factors—must be carefully noted by the officer.

A person could be reckless for driving in "pea soup" fog at 35 mph in a 55 mph zone. A person could be reckless if he or she drives knowing that he or she is subject to seizures or comas. People are reckless if they drive when drowsy.

19.9 VEHICULAR HOMICIDE

The third of the more serious traffic offenses is vehicular homicide. Obviously, one could be guilty of murder by using an automobile as a murder weapon. Under modern statutory classes of murder, a person could be guilty of a lesser murder by operating the vehicle in a reckless or heedless manner, thereby causing death. In other states the charge would be a degree of manslaughter.

Because of the stiff penalties generally associated with the general homicide laws, many prosecutors would not prosecute for such deaths. Those that did often discovered that juries were reluctant to find guilt because of the penalties affixed to the general homicide laws.

As an alternative, the drafters of the Uniform Vehicle Code proposed a homicide-by-vehicle statute that reads as follows and clearly defines what constitutes **vehicular homicide:**

> Whoever shall unlawfully and unintentionally cause the death of another person while engaged in the violation of any state law or municipal ordinance applying to the operation or use of a vehicle or to the regulation of traffic shall be guilty of homicide when such violation is the proximate cause of said death.

This provision gives the prosecutor more discretion dependent on the driver's conduct. However, the discretion must be exercised wisely to avoid a double jeopardy claim and to avoid conflict with that part of the greater and lesser included offenses doctrine that holds a conviction of a lesser included offense is the acquittal of the greater included offense (see Chapter 2).

For example, vehicular homicide requires a showing that some vehicular law was violated, and that the violation was the proximate cause of the death. Freda is convicted of drunk driving in Trial 1. Can the state successfully prosecute Freda at a second trial for a death that was proximately caused by her drunk driving? If death had not occurred at the time of the first trial, the second trial would not violate double jeopardy. If the death had occurred before the first trial but Freda had opposed joining the homicide trial with the violation of the vehicular law, then obviously she cannot later bar the trial on the basis of double jeopardy. The more complex issue arises if death had occurred prior to the first trial on the vehicular violation but the state had not charged that offense until after the first trial started. After vacillating on the issue, the U.S. Supreme Court now holds that the "additional element test" must be met to avoid double jeopardy. In the second trial, the state must again prove drunk driving but this time must additionally prove that the death was proximately caused by that drunk driving, thereby avoiding double jeopardy.

Vehicular homicide was not intended to be a strict liability crime. The mere fact that a vehicle driven by the defendant killed a person still requires the resolution of the causation issue. If the defendant feels it was an unavoidable accident, then he or she should be able to introduce evidence to that effect, said the Pennsylvania Supreme Court.

19.10 HIT AND RUN

The Uniform Vehicle Code puts the burden upon drivers involved in accidents to make immediate reports, furnish information, and render aid if death, personal injury, or property damage result.

All the acts required must be done immediately. The duty arises with actual knowledge or in circumstances in which the person should have known he or she was required to stop. The purpose is to keep people at the scene and to prevent them from deliberately leaving.

The requirement applies to all involved in an accident. The statute does not attempt to fix fault. It seeks only to keep those involved from leaving. The accident does not have to be with another vehicle. It can even involve a situation in which someone falls out of the driver's car.

Where no personal injury or death is involved, a duty to report still exists. Many states use a dollar value of damages incurred to determine when such a duty arises. In that event, the amount is what is apparent and not what the actual, later-determined amount is. The amount involved is arrived at by adding all damages together, not by determining the amount of damage to any one vehicle or thing.

19.11 THE DOCTRINE OF LESSER INCLUDED OFFENSES AND TRAFFIC LAWS

Sometimes a defendant charged with DUI will want to plead to reckless driving as a lesser included offense. Unfortunately, to be an included offense an act must be in the chain of crimes sharing always one basic element of the greater crimes.

In murder, for example, the root or base crime is simple assault, which by definition is an attempted battery. A completed assault is at least a battery. When the battery causes death, it is murder, manslaughter, or excusable or justifiable homicide. But at all times the base crime, the assault, is present.

Reckless driving is the act of driving an automobile in reckless disregard for the safety of persons or property. The crime always is a moving violation; it is never a parking violation. It smacks of using the vehicle much as one would use a gun or explosives. The sobriety of the individual is never a question, though it may be a contributing factor. Driving a car on a crowded sidewalk at 15 mph could well be reckless.

A person can be drunk and drive a car in a reasonably safe manner. The way the car is driven is not of any importance as far as a finding of guilt is concerned. The state or condition of the driver is the key and, in essence, the only element. And, except in those states that require the vehicle to be moving, a person whose key is in the ignition or whose motor is running can be found guilty of DUI.

Thus one crime is based on the way in which the vehicle is being operated and the other is based on who is operating and under what conditions the vehicle is being operated. Therefore, an individual could well be guilty of both crimes. Both are listed as separate serious offenses under the Uniform Vehicle Code. So, unless a state has specifically ruled that one is the lesser included of the other, there is no legitimate way in which to accept a plea of reckless driving to a DUI charge unless ample proof of reckless disregard is given.

19.12 CAN THERE BE A DEFENSE FOR *MALA PROHIBITA* OFFENSES?

There are some legitimate defenses to conduct-only crimes such as traffic offenses. Similarly, some excuses offered are merely that and no more. Some excuses that sound good are used by the court in mitigation of the fine or penalty assessed.

Defendants often say, "Everybody does it." Customs or habits, especially bad ones, do not create a defense. As with most crimes, ignorance of the law is no excuse. However, a mistake in law can exist especially when a sign is supposed to be posted and visible and it is not. But, dumb mistakes are no excuse. The fact that somebody else involved in the incident was also wrong is no excuse. As the saying goes, two wrongs do not make a right.

Often a person involved in an accident pays the damages of the other party. Often the defendant wants this restitution to be a defense. It cannot serve as a defense. Entrapment, however, is a defense. There have been communities that purposely rig lights and establish short speed zones that are impossible to obey. What about the brakes that failed? Unforeseen mechanical failures are legitimate excuses. But, the neglectful owner who knew his brakes were going bad cannot avail himself of this defense.

"We had to get to the hospital" is an excuse often seen in cartoons and comedy sketches. This type of actual compelling necessity can be used by the court as an acceptable defense or as mitigation.

Obviously, a person who has a gun pointed at his head may use coercion as a defense. The threat must be real, immediate, and unconditional. The lesser crimes of most states must be prosecuted within a specific time period. Failure to do so within the statute of limitations provides a valid defense.

The statute itself may provide for some acceptable exception or excuse. One exception is emergency vehicles on emergency runs. Individual states and municipalities may have provided others beyond the general provisions of the Uniform Vehicle Code.

19.13 CRIMINAL OR CIVIL: WHICH?

Most states have made violations of their "rules of the road" crimes, providing both a fine and imprisonment as the punishment for violations. This follows the recommendation of the Uniform Vehicle Code. Some have attempted to **decriminalize** minor infractions, leaving them to be pursued as civil matters using civil rules of procedure and evidence. What benefit, if any, is there to decriminalizing most traffic offenses?

A 1973 study at the University of Wisconsin Law School came up with these conclusions: Law as a means of social control would be more effective if the criminal law were made to conform more closely to the popular idea of what that law ought to be, that is, if it labels as crimes only those offenses requiring intent. Second, respect for the law declines as a greater number of crimes are placed on the books. People do not like the stigma of crime on persons not truly criminal. Finally, safeguards of criminal procedure hamper enforcement and prosecution of many of the minor offenses. Thus, the study concluded that using some type of civil procedure would produce better results.

With these conclusions the following recommendations were made to unburden the criminal code:

1. No act which imposes absolute liability should remain criminal (*mala prohibita* offenses).

2. If the only penalty imposed for a present crime is a fine, the act should be removed from the criminal code, for the penalty can be imposed as well by civil action.

3. If a present crime was not a crime at common law, if its commission does not *prima facie* indicate a dangerous personality, and if the act is not popularly regarded as reprehensible, the act should be removed from the criminal code.

4. If the act is of a type that requires the interest and expertness of a specially designated and qualified official for its efficient enforcement, it should not be a crime unless it clearly does not come within Class 3.

Many states have decriminalized many traffic violations with the exception of reckless driving and the DUI violations.

The burden of proof in civil matters is not as stringent as in criminal matters. There is no right to remain silent in civil matters, nor do the rules of Miranda apply. There is no right of arrest in civil matters, nor would there be a right to search the individual incident to that arrest. Depositions of the officer can be used in the civil case, whereas they are often not allowed in criminal cases in which a mere conflict in schedules is the only reason for the officer's not showing up.

Making traffic offenses civil matters does not violate due process, because there would still be notice, a right to be heard, and a hearing on the matter. As the right to counsel attaches only to cases in which a jail sentence is imposed, there would be no such right if the only penalty were a dollar amount.

The only possible problem is that the courts could view these charges and proceedings as quasi-criminal in nature. This would then cause a reattachment of all constitutional criminal rights.

In fact, this is exactly what happened in Nebraska with regard to double jeopardy, which is a defense only in criminal cases. The state lost its case on the civil infraction. As a civil matter, any loser can appeal; in criminal matters, usually only the defendant can appeal. The state, viewing this as a civil matter, appealed, won, and sought retrial. The defense claimed double jeopardy. The court agreed.

On the other hand, changing minor traffic offenses to civil matters would prevent police from making custodial arrests on such charges. The right to make a lawful custodial arrest and search of the individual incident thereto (not allowed in all states) is a valuable enforcement tool. If the police cannot make a custodial arrest, they would not be able to search or impound and inventory a vehicle. This is also a valuable enforcement tool. Law enforcement in states allowing these practices should not act too hastily in championing the reduction of minor traffic offenses to civil matters.

DISCUSSION QUESTIONS

1. While on routine patrol, Officer Jones observes a car with its motor running, parked in a sparsely populated area of the city. Upon investigation, Jones observes a man passed out behind the wheel of the car. The key is in the ignition and the motor is warm. The officer wakes the sleeping man, and, after determining that he is intoxicated, places him under arrest for driving under the influence. Discuss the propriety of the charge.

2. Bill and Don decide that they should have a drag race to determine whose car is fastest. They agree this would be too dangerous to do on an open road. There is an abandoned airbase nearby which is the agreed-upon site. As they start their race the sheriff arrives to see them speed across the runway. After they stop, they are charged with racing under the Uniform Vehicle Code in effect in your state. Are they guilty?

3. Joe has trained his monkey to steer his car. Joe has to press the gas pedal, however. The monkey usually does a good job of steering, but on this day he "rear-ends" a parked car. What offense has been committed and who is responsible?

4. Mary is a diabetic. While driving she lapses into a diabetic coma. Can she be charged with reckless driving?

5. Uncle Dave is 90 years old. He recently was reissued his driver's license. He admits that he does not see well and that he drives from memory and goes places where only right turns are required. Would it be reckless or ordinary negligence for him to drive?

6. While driving recklessly, John hits another car and severely injures the driver. John is cited for reckless driving, to which he pleads guilty. His fine assessed, he serves 90 days. While in jail, the driver of the other car

dies as the direct, proximate result of the accident. Can John be tried for vehicular homicide? What role would the doctrine of lesser and greater included offenses play?

▨▨▨ GLOSSARY ▨▨▨

Absolute speed limit – the posted maximum limit excessive of which renders one guilty.

Blood alcohol content – the percent of alcohol in a person's blood.

Cone of vision – a sign or traffic control device placed with a reasonable approximation of the requirement specified in the *Manual on Uniform Traffic Control Devices for Streets and Highways.*

Decriminalize – making many of the minor traffic infractions civil rather than criminal offenses.

Disabled parking – reserved parking spaces for people certified under state laws to be disabled; federal and state laws prescribe the number and location of parking spaces required.

Drag racing – a race conducted on a course to compare relative speeds or power of acceleration measured either by time or over a measured distance.

Driving under the influence (DUI) – driving or being in physical control of a vehicle while under the influence of alcohol or drugs.

Driving while intoxicated (DWI) – term used to describe an earlier attempt to punish the drinking driver but intoxication as an absolute was too difficult to prove.

International accessibility symbol – the wheelchair symbol used by disabled people, found on license plates and/or placards.

Low-speed vehicle – vehicles that have safety equipment and may be operated on streets, usually electric and have a top speed of 25 miles per hour.

Moving violation – infraction such as speeding.

Nonmoving violation – infraction such as parking violations.

Prima facie **speed limit** – driving at a speed greater than the limit is unsafe, and the offender is presumed in violation of the law.

Racing – the use of one or more vehicles to attempt to outgain, outdistance, prevent another vehicle from passing, arrive first at a given point, or test the stamina of the drivers.

Reckless driving – voluntarily, consciously operating a vehicle with willful and wanton disregard for the consequences or for the rights and safety of persons or property.

Uniform Vehicle Code – serves as the basis for the traffic laws of all states.

Vehicular homicide – in many states carries a potentially lesser penalty than the general homicide laws thus giving the prosecutor the option of charges based on all circumstances.

REFERENCE CASES, STATUTES, AND WEB SITES

CASES

Illinois v. Vitale, 100 U.S. 2260 (1980).

Commonwealth v. Uhrinek, 518 Pa. 532 (1988).

WEB SITE

www.findlaw.com

20

Crimes Affecting Judicial Process

■ KEY WORDS AND PHRASES ■

Affirmation
Criminal contempt
Custody
Embracery
Escape
Failure to appear
False oath
Fugitive
Jumping bail
Jury tampering
Materiality

Oath
Obstructing justice
Perjury
Prison
Prison break
Resisting or obstructing an officer
Rescue
Subornation of perjury
Willful and corrupt
Witness tampering

20.0 INTRODUCTORY COMMENTS

Most, if not all, crimes affect the judicial process in one way or another. Were it not for the commission of crimes, a criminal judicial process would not be needed. There is a certain group of offenses, known to the common law and codified by statute in most jurisdictions, that more directly affect the administration of criminal justice. These offenses directly concern the dignity of the courts and their judicial decisions and the processes by which the criminal justice system is administered. Some of the offenses in this category are covered in other, more appropriate sections of this book. For example, bribery is discussed in Chapter 17. The offenses selected for presentation in this chapter are by no means exhaustive. We have chosen to discuss only those crimes that we feel are of immediate concern to law enforcement officials. Included in this category are perjury, subornation of perjury, embracery, escape, rescue, breaking prison, and obstructing justice.

20.1 PERJURY

The common law misdemeanor of **perjury** consisted of taking a false oath in a judicial proceeding regarding a matter material to that proceeding. A false oath means that the accused willfully and corruptly makes a sworn statement regarding something he or she knew to be false or did not believe to be true. Today, perjury remains a criminal offense in all jurisdictions and in many has been made a felony.

Oath or Affirmation Duly Administered

Before perjury may be properly charged, the testimony must have been given under oath or some legally recognized equivalent. At common law, one test of the competency of a witness was the witness's belief in a supreme being. On the basis of this belief, the witness would swear to tell the truth, or take an **oath.** Thus, taking an oath was one test of determining whether or not a witness was competent to testify. Although taking an oath by swearing to tell the truth is still the prime test today, it is no longer the exclusive test of competency. Today it is generally held that, to qualify as a witness, the person must understand the obligations of an oath and undertake those obligations realizing the penalties of perjury for not telling the truth. If the witness does understand and undertake those obligations without actually swearing to a supreme being, this is called an **affirmation.** Many jurisdictions allow an affirmation instead of an oath. In those states, the affirmation is treated as a legally recognized equivalent of an oath. Thus, a witness who refuses to take an oath may still qualify as a competent witness by affirming to tell the truth.

Aside from religious or other reasons why one would refuse to take an oath, the affirmation is often used in the case of children who are to be witnesses. The fact that the witness is a child will not automatically qualify him or her as a competent witness. In fact, in the case of children, their competency to testify will be more closely scrutinized. A child who may not be mature enough to understand the nature of an oath may still be qualified as a competent witness if, under private examination by the judge, he or she understands the importance of telling the truth and agrees to do so.

There is another ramification of taking an oath or affirmation that must be shown to exist before a perjury charge will be proper. The oath or affirmation must be lawfully administered. This refers to the legality of the oath administered. For an oath or affirmation to have any legal effect, the taking of the oath must be permitted or required by statute. One who has no authority to administer an oath may not legally do so, and false testimony given under such an oath may not be grounds for a perjury charge. This would be true when a person is compelled to give an oath and testify in a "kangaroo" court proceeding run by a group of vigilantes.

False Oath

As stated previously, a **false oath** is a willful and corrupt sworn statement that the accused knows to be false or that the accused does not know to be true. This does not include mistakes, even if the mistake completely contradicts previous testimony on the same subject. Nor does the law make criminal a mistake caused by negligence or carelessness on the part of the witness. As will be seen in the next subsection, perjury requires proof of intent, which is an integral part of this offense.

This element may be satisfied in any number of ways. As the rule indicates, one who swears to the truth of something knowing it to be false has given a false oath. If one testifies a fact is true when he or she is not sure it is true, the witness has given a false oath. Thus, if a witness deliberately testifies that the vehicle seen leaving the scene was a convertible when, in fact, the witness is not sure it was a convertible, he or she may be liable for perjury.

A third manner in which a false oath may be given is by testifying that a fact is true when the witness does not know whether it is true or false. Consequently, if Jones testifies he observed Smith running from the scene of the crime when, in fact, Jones was nowhere near the scene to observe such an event, he has given a false oath. This would constitute perjury even if it later turned out that Jones's statement was correct. Thus, if Smith actually did run from the scene at the time Jones alleged, this would be no defense to a charge of perjury against Jones. In essence, the rule regarding false oath concerns not the testimony itself but, rather, the fact that a lie was given under oath.

Willful and Corrupt

The false oath must be made willfully and corruptly. This requirement supplies the intent element in the crime of perjury. To satisfy this requirement and be considered **willful and corrupt,** the false testimony must be knowingly and deliberately made. This rule eliminates the possibility of holding someone liable for perjury when he or she testifies to something falsely but does so mistakenly or carelessly or under an honest belief that it is true.

Materiality of Testimony

A perjury charge may not lead to conviction if the false testimony was given on a point not really important to the outcome of the case. This does not mean that the testimony must bear directly on the issue of guilt or innocence. As long as the false oath pertains to a point that may have some bearing on the outcome of the trial, it constitutes **materiality.** In each case, whether false testimony, willfully and corruptly given, is material or not is a question of law for the court to decide rather than a question of fact for the jury.

Judicial Proceeding

At common law, perjury was an offense charged when the false oath was given in a judicial proceeding. Thus, the crime was limited to in-court false testimony. There are, however, numerous out-of-court occasions that require that testimony be given under oath or that require that an oath be taken. Because these situations did not conform to the common law requirement for perjury that the testimony be given in a judicial proceeding, a separate offense was recognized by the common law courts. The crime of false swearing was identical to perjury except that it was done in a proceeding other than judicial that also required an oath.

In the majority of jurisdictions today, the crimes of false swearing and perjury have been incorporated into a single statute under the title of perjury, so that taking a false oath in any proceeding, judicial or not, in which an oath is required by law, will constitute the crime.

Historically, disputes were settled in court or privately. With the growth of government, administrative agencies developed to regulate businesses, administer welfare and insurance programs, and collect taxes. These agencies developed regulations that, when apparently broken, necessitated procedures for hearing complaints. They were given the power to subpoena witnesses and records and would sit as a semicourt, take testimony, and render a decision. Thus, the law recognized the need to prevent perjured testimony before these boards. In addition, depositions and interrogatories that may be used in court are taken under oath or affirmation, and false testimony in a deposition will subject a person to prosecution. Affidavits that require an oath, voter registration, and certain license applications requiring an oath can be perjured, subjecting the perjuring party to prosecution.

The word "oath" may convey many meanings to an individual aside from the oath to tell the truth when testifying. For example, when a public official, such as the president of the United States, a senator, or representative, is sworn into office, that official takes an oath swearing to uphold the Constitution of the United States and to perform the responsibilities to the best of his or her ability. May a violation of this oath constitute perjury? Most courts hold that this is not perjury, for that crime pertains to false swearing in a testimonial capacity. However, there are a few cases on record holding to the contrary.

Jurisdiction

A conviction of perjury may be proper if the court or other agency issuing the oath has jurisdiction over the matter tried or the matter sworn to. On the other hand, if the agency or officer administering the oath has no jurisdiction to so do, or if the matter in which the oath is given is beyond the jurisdiction of the agency, no perjury charge will be proper regardless of the amount of false testimony given. For example, if a juvenile court attempts to try an adult for a capital felony, any amount of false oath occurring during the trial will not constitute chargeable perjury, because the lack of jurisdiction voids the legality of the entire trial (see Chapters 3 and 18).

Perjury by a Defendant

An obvious question concerns whether or not a defendant in a criminal case, who takes the stand to testify in his or her own behalf, can subsequently be charged and convicted separately for the crime of perjury, if it can be shown that the person gave a false oath. The answer is yes in most jurisdictions. If the elements of the offense can be proven, the defendant is no less subject to the charge than is any other witness. It must be remembered that the dignity and integrity of the court and of the entire judicial process of our society is downgraded by a perjurer. There is some conflict in this area, however, when a defendant is acquitted of a criminal charge for whatever reason, including his or her own perjured testimony. The federal courts hold that in such a case a perjury charge is not allowed. Most state courts disagree and would allow a charge of perjury to be brought for the reasons just stated. But, when the defendant in a federal case has been convicted of the crime charged, the federal authorities and courts will pursue and allow a charge of perjury.

Requirements of Proof

Because of the many errors and mistakes that witnesses can and do make in criminal trials, whether due to error, mistake, carelessness, or negligence, proof of the crime of perjury must, in fact, be fairly clear-cut. This is not to say that the burden of proof is any greater than is that required for any other criminal case. The law still requires proof only beyond and to the exclusion of every reasonable doubt. What is meant is that a mere conflict in testimony or contradictory statements made by a single witness will ordinarily not be enough to convict. Similarly, the law will not allow a conviction to be based on one person's word against another's. A "your word against mine" situation does not take into account the innocent mistakes that could be made. The law requires that there be independent corroborating evidence to support one person's word against another's or, in the alternative, that there be at least two witnesses to testify against the defendant in the perjury trial.

Consequences of a Perjury Conviction

Perjury today is a felony that in some states carries a maximum penalty of life imprisonment. In addition to these very harsh penalties, perjury, considered an infamous crime from its earliest beginnings, often disqualifies one convicted of the offense from holding public office or from serving on a jury. Some states also provide in their statutes that one who has been convicted of perjury may not be a witness in any judicial proceeding in that state. Thus there are many sanctions imposed on one convicted of this socially degrading crime.

20.2 SUBORNATION OF PERJURY

The crime of **subornation of perjury** is the corrupt procurement of another to commit perjury. If Green procures White to commit perjury, the separate offense of subornation may be charged against Green.

Perjury Committed

The crime of subornation of perjury requires that the crime of perjury actually be committed by the one procured. In the example above, if White does not commit perjury, Green cannot be convicted of subornation of perjury. In light of the previous discussion of the crime of solicitation and the liability of parties to crimes (Chapter 4), the reader may well ask why one who is guilty of procuring another to commit perjury is not an accessory before the fact if the one procured actually commits perjury. The answer to this question lies in the historical background of the crime of subornation of perjury. The common law regarded the procurement of one to commit perjury as a much more socially degrading crime than the actual perjury and, therefore, provided more severe penalties. If the defendant were convicted of subornation, the penalty could be much more severe than if the defendant were convicted as an accessory before the fact to perjury. It is true that perjury was only a misdemeanor at common law. It is also true that subornation of perjury was a common law misdemeanor, but it still carried a more severe penalty. Modern statutes, however, reclassify perjury as a felony with extremely harsh penalties possible. Subornation of perjury usually carries the same penalty, and it too has been elevated to a felony.

If all the other elements of the crime of subornation are satisfied except that the witness procured does not commit perjury, the proper charge would be attempted subornation of perjury. The only qualification affecting this charge is that, had the witness testified as the suborner wanted him or her to, the testimony would have been a false oath, which would have constituted perjury.

By Other Than the Defendant

A brief note is needed to point out that the crimes of perjury and subornation of perjury are inconsistent in the sense that they cannot both be committed in a single incident. The procurer may not be the person who commits the perjury.

Corrupt Inducement by Defendant

To obtain a conviction of this crime, the prosecution must show that the defendant induced the witness to commit perjury. Knowledge on the part of the defendant that the witness intends to, and subsequently does, commit perjury will not support a charge of subornation if unaccompanied by some form of inducement. The inducement must also be corrupt. Thus the suborner must procure another to commit perjury knowingly and willfully.

Defendant Must Have Knowledge of the False Oath

As stated in the preceding paragraphs, mere knowledge that the witness intends to commit perjury will not support a charge of subornation. If, however, this knowledge is accompanied by corrupt inducement, knowledge that the witness intends and does commit perjury becomes an indispensable element of the crime. This element may be satisfied in any of several ways. First, the defendant may know that the testimony of the witness would be given under a false oath. Generally, both the defendant and the witness must know that the testimony is false. If the witness knows the testimony is true, the defendant cannot be convicted of subornation because the witness has not committed perjury. The defendant may, however, still be liable for attempted subornation of perjury. The same rule generally applies if the defendant knows that the testimony given by the witness is true. However, there is an exception to the rule that would not constitute subornation. If the defendant knows the testimony is true and procures a witness to testify to that effect but the witness does not believe the testimony is true, the crime of subornation is complete. This is so because the witness took a false oath, notwithstanding the fact that the testimony given was true.

Matters of Proof

Like perjury, subornation of perjury requires testimony by two witnesses or one witness plus independent corroborative evidence for conviction. However, the element of procuring does not require proof by such a degree of evidence. Testimony of one witness will suffice to establish the act of procurement.

20.3 EMBRACERY

Embracery was a well-known common law offense that today is commonly called **jury tampering** and may be included in other statutory forms instead of as a separate offense. The gist of embracery involves a corrupt attempt to influence the decision of a jury or of an individual juror. It is not necessary that this be done in the form of a bribe, as described in Chapter 17. Of course, if the elements of bribery are well founded, that crime may be prosecuted.

Under our system of law, the various participants in a trial each have assigned functions. It is the responsibility of the judge to apply the law to the facts of the case; the witnesses are responsible for presenting facts; the jury is responsible for drawing conclusions. Each of these functions must be kept distinct to preserve the system. Allowing any of the parties, or a stranger, to influence the decision of the jury destroys the entire concept of the jury system. It is for this reason that the crime of embracery was first recognized.

Unlike the crime of subornation of perjury, it is not essential that a juror actually be influenced by the embracer in reaching a verdict. Embracery is complete when the attempt to influence a juror corruptly is made. Even another juror may be liable for the crime by attempting to influence the verdict by corrupt means. Of course, while a jury is deliberating the fate of an accused person, each member of the jury is attempting to persuade fellow jurors to conform to his or her own opinion as to the guilt or innocence of the accused. This is a perfectly permissible procedure, and it is really the only means by which a jury could reach a unanimous decision. If, however, the juror attempts to influence fellow jurors by promises, threats, money, or other unlawful means not related to the evidence presented in the trial, the crime of embracery has been committed. It has been held that this offense applies not only to trial juries but will also be chargeable when one attempts to influence corruptly the decision of a grand jury.

20.4 ESCAPE, RESCUE, AND BREAKING PRISON

The common law misdemeanor of **escape** was defined as leaving lawful custody without authorization and without the use of force. Escape differs from the common law offense of **prison break,** in which force is an essential element. The crime of **rescue** involves aiding another to flee from custody. An examination of the elements of each of these offenses follows.

Lawful Custody

Our discussion of these crimes relates to custody connected with criminal proceedings, although some states apply these offenses equally to custody commencing from a civil action. The term "custody," in its legal sense, does not imply physical contact, nor does it mean that the prisoner must be behind bars. If the prisoner is deprived of the freedom of mobility, he or she is in **custody**. The fact that the prisoner is in an open area with no guard standing over his or her shoulder does not mean the prisoner is not in custody. If the prisoner is under the general supervision of one who is responsible for and restricts his or her movements, the prisoner is in custody.

The rules regarding custody may have a familiar ring to the reader who has read Chapter 15 on false imprisonment and kidnapping. The distinguishing feature is that in this chapter we are discussing lawful custody. False imprisonment and kidnapping are crimes that involve unlawful custody. Escaping from unlawful custody is not a crime.

In a criminal case, lawful custody may arise in a number of ways: pursuant to an arrest warrant, by authority of a capias or bench warrant issued by a competent court that has jurisdiction and probable cause for believing a felony has been committed, or for an offense committed in the presence of the arresting official or citizen. There are other ways in which a particular agency may gain custody of a person. For instance, the prisoner may be transferred from another institution or may be serving a sentence in an institution. Regardless of the reason as to why the agency or officer has custody of another, the custody must be lawful before an escape will be chargeable if the prisoner departs.

The Act of Departing

To ascertain whether one has departed from lawful custody and thereby committed the crime of escape, it is necessary to determine the limits of the confinement. If we recognize that neither prison walls nor iron bars are required for custody, the explanation is simple. If the prisoner is in lawful custody, regardless of the location or limits, an unauthorized departure will constitute escape. Fred, an inmate at the Super State Prison, is made trustee. He is assigned the task of chauffeuring the warden. One day Fred drives the warden to an appointment at the state capital, 300 miles from the prison walls. While the warden is attending the meeting, Fred decides to leave permanently. Fred has committed escape.

Maybe This Departing Wasn't Such a Good Idea

A prisoner in New Liskeard, Ontario, waiting for his case to come up decided to escape. He ran out of the courthouse with guards right behind him. The suspect ran through the downtown area and out onto the town dock and jumped into the lake. The escaped prisoner started screaming when he remembered he couldn't swim.

The Intent

Escape and breach of prison are general intent crimes, requiring only intent to leave the bounds of lawful custody. There will be times, however, when the circumstances surrounding the departure may be reasonably explained so as to preclude the crime's being charged or a conviction obtained. Consequently, the law will recognize such defenses as mistake of fact or necessity. Referring to the previous example, suppose Fred thought, and could reasonably convince a jury, that he believed the warden told him to see the sights of the capital city while the warden was in conference. In fact, the warden said no such thing. If Fred acted reasonably under the mistake of fact that he honestly believed to be true, this would be a defense to a charge of escape. Would inmate Jones of the same prison be convicted for escape if he was working in a wooded area outside the prison walls with instructions to stay there and he left without intending to leave custody permanently because a forest fire broke out in the area in which he was working? Obviously not.

Force

The existence of the element of force is the only element that distinguishes the crime of prison breach or prison break from the crime of escape. If no force is used to effect the prisoner's departure from custody, the crime is escape. If force is used, no matter what kind or how slight, at common law, the crime was breach of prison. Today, most jurisdictions have combined escape and prison breach into a single statutory offense, but the existence or nonexistence of the element of force is still necessary for proper proof of the crime.

Prison

Any discussion of the crimes of escape and prison break would be incomplete without noting what definition the law attaches to the word "prison." Many of us ordinarily think of the word "prison" to mean the state peni-

---

Too Much Force

Three criminals, being transported to a different jail, made a dash for freedom as they were being unloaded from the back of the sheriff's van. The three were all on one long chain and were connected at the wrists. They only got about ten feet, though. As they ran down the sidewalk, they tried to go in different directions around a telephone pole. The only trouble the police had with the capture was untangling them from each other and the pole.

www. dumbcriminalacts.com

---

tentiary. This is not the interpretation used when referring to the crimes of escape and prison breach. A **prison** is any boundary within which the mobility of a prisoner is restricted. A prisoner is defined as any person who is in lawful custody. Thus, upon being arrested, an individual is in prison. All this leads to the conclusion that, if Sam uses force to effect a departure from the custody of Officer Brown, who has just arrested him on Main Street at high noon, he is liable for prison breach.

Degrees of Liability

Escape without force at common law was a misdemeanor. If force was used, the offense was more serious. Many modern courts have graded the offense in accordance with the crime for which the prisoner has been arrested or convicted. If John escapes from custody after arrest on a misdemeanor charge, his escape is a misdemeanor. If he is arrested on a felony charge, escape will constitute a felony. The same rules apply to an escape after conviction for an offense. In several states, these rules have been altered so that escape after conviction for a crime is a felony regardless of the classification of the crime on which conviction was based. Punishment for escape before conviction is still determined by the classification of the crime for which the arrest was made.

Rescue

The common law crime of rescue is often referred to today as aiding another to escape. Many statutes require that the aid be rendered knowingly. As a general rule, however, the crime of rescue does not require specific intent. Defenses such as mistake of fact will not excuse one for commission of the crime of rescue. Thus, if Joe slips Barney a knife or saw blade to help him escape from lawful confinement, Joe is criminally liable for this offense.

Some Additional Problem Areas

The topics discussed in this section cannot be completed without mentioning three additional areas of concern to law enforcement officers.

Jumping Bail

First, the reader may ask whether jumping bail constitutes escape. In the broadest construction of the term *escape,* **jumping bail,** or failure to appear in court after having been released on bond, is considered a form of escape, but most jurisdictions have a separate statute making this a chargeable offense.

Under the common law it was not always clear whether or not **failure to appear** at a required hearing while under bail was a crime. To remedy the confusion, the Model Penal Code declares such an act a misdemeanor if the failure to appear is without a lawful excuse. The offense can rise to the felony level if the bail jumping involves an appearance on a felony and the defendant flees or hides to avoid the hearing, trial, or punishment.

Fugitive from Justice

Second, Title 18 U.S.C. 1073, a federal statute, declares that crossing state lines to avoid prosecution or confinement for the commission of or attempt to commit any state felony, or to avoid giving testimony in a state felony case, is itself a felony and makes the offender a **fugitive.** If we recognize the separation-of-sovereignties concept discussed under the heading of concurrent and overlapping jurisdiction in Chapter 3, there is nothing to preclude the state from charging escape or prison break even if a federal fugitive warrant has been issued, and if the prisoner is apprehended and returned to the state from which he or she fled. A prosecution in both the federal and state courts would be proper for the different offenses.

Probation or Parole

Third, the question often arises as to whether or not violation of probation or parole by leaving the jurisdiction without permission constitutes an escape. The answer is a definite no. One who is on probation or parole is under the supervision of a probation or parole officer, but the law does not treat these circumstances as custodial and, therefore, there can be no escape.

> ## Isn't Technology Wonderful?
>
> A person can now lawfully be confined to home and be monitored by means of a wrist or ankle band. If a person so confined leaves his or her own home without permission, can escape properly be charged?

20.5 OBSTRUCTING JUSTICE

Many of the crimes previously discussed in this and other chapters were, at early common law, grouped into a general catchall type of offense called **obstructing justice.** Because of the growth in importance of some of them, they were made separate punishable offenses by the later common law and by modern statute. There still exists in many jurisdictions an obstructing justice statute to cover those cases not set apart as distinct offenses. The following is a brief discussion of a few of the facts falling within the purview of this type of statute.

Witness Tampering

In Section 20.3 we discussed the crime of embracery, which consisted of tampering with jurors by attempting to influence their decisions. The crime of obstructing justice includes **witness tampering.** If the defendant in a given criminal case gives the witness $1000 and tells him not to show up at the trial, the crime is complete. Another example would be if the prosecution intentionally hid a witness material to its own or the defendant's case. Most cases of tampering with witnesses have arisen from the use of force or intimidation by one of the parties to the trial. The crime would be chargeable if one of the parties intentionally, by force or threat of force, compelled the witness to stay away from the trial.

Hiding or Destroying Evidence

Another type of conduct giving rise to a charge of obstructing justice involves hiding or destroying evidence. Both parties to a particular case may be charged with this offense. The law applies equally to secretion or destruction of evidence by either the prosecution or defense. Obviously, if the defendant, before trial, destroys material evidence by burning or mutilation, the defendant could be charged with the crime. Likewise, if the defendant rents a safe deposit box in a bank in another town under an assumed name and places in it material that could lead to conviction, which is not discovered by the authorities until after acquittal, the defendant could be charged with the crime.

Resisting or Obstructing an Officer

Many jurisdictions have separate statutes making **resisting or obstructing an officer** in the performance of his or her lawful duties a separate offense. However, some states still include this conduct under the general heading of obstructing justice. Any person who stands in the way of any public officer to prevent the officer from performing the duties of the office is chargeable with this crime. Of major concern to law enforcement officers is obstructing or resisting a peace officer in the performance of legal duties. It is emphasized that the officer must be performing a lawful duty of the office when the obstruction takes place, but, of course, because police officers are sworn officials twenty-four hours a day, this does not mean that the officer has to be on duty at the time. Historically, resisting arrest falls within the purview of this statute if it is not made a separate crime. Obstructing or resisting may be committed either by a person to be arrested or by a third party. For instance, if Officer Smith, armed with a valid arrest warrant for Mr. Brown, attempts to enter the Brown home to execute the warrant but his way is blocked by Mrs. Brown, she is liable for obstructing an officer.

Many courts hold that fleeing an officer to avoid arrest does not constitute resisting or obstructing justice, because no force is employed directly against the officer. Force is an essential element of the offense but only the slightest amount is needed. Even physically barring a passageway will suffice. In some states, fleeing is enough to justify the offense.

Aiding an Officer

One further situation deserves mention in this area. It is the duty of citizens to come to the aid of a peace officer upon the officer's request. Because there is a duty to respond to the officer's call for help, any person refusing is guilty of a misdemeanor. This is true even though the help asked for exposes the individual to greater danger. As a side note, whenever a private citizen is called to the aid of an officer, the law cloaks that individual with the same powers and authority as are possessed by the police officer.

20.6 CRIMINAL CONTEMPT

Another offense closely aligned to obstructing justice, and one that is directly related to the judicial process, is criminal contempt. An act that is disrespectful to the court, calculated to bring the court into disrepute, or of a nature that tends to obstruct the administration of justice is **criminal contempt.** Examples of contemptuous conduct are misbehavior of any person in the courtroom to the discredit of accepted decorum, or disobeyance of a lawful order by a judge for which a punishment may be imposed.

Criminal contempt differs from civil contempt in that the penalty for civil contempt is purely coercive and can be avoided by compliance with the court's order. Thus, in a case in which an attorney was held in contempt for failure to wear a necktie in the courtroom while litigating a case, the penalty could have been avoided if the attorney agreed to wear a necktie. But a waitress who fails to appear to testify in court under orders of a subpoena is liable for criminal contempt. It should be noted that certain conduct can constitute both civil and criminal contempt.

Judicial opinion is split as to whether criminal contempt actually constitutes a crime under the definition given previously in this book. That question is somewhat academic in light of the fact that criminal contempt is punishable as are other criminal offenses.

The most significant problem in this area involves the double jeopardy clause of the Constitution when a person is found in criminal contempt and then the state seeks a subsequent conviction for the commission of the crime(s) that led to the contempt. For example, it is not uncommon for a judge to award bail to a person awaiting trial subject to the condition that the released person commit no crimes. Suppose that Al is released with this condition and then is caught distributing cocaine while out on bail. The judge finds Al guilty of criminal contempt and sentences him to serve jail time. The state later charges Al with the substantive crime of distributing cocaine. Can Al escape punishment for that cocaine distribution charge since he has already been found guilty of criminal contempt by committing the cocaine distribution offense? In a recent case on these facts, the U.S. Supreme Court said it would be double jeopardy to try him for the cocaine distribution offense since at his criminal trial there were no new elements or evidence shown that were not already presented at his contempt trial. This case may have presented a not so common situation. In many instances involving similar facts, the judge, instead of charging criminal contempt, might simply revoke the bail. Al would then be held without bail until disposition of the original charge for which he was arrested. Al may simultaneously be held in jail on the new charge of distributing cocaine.

Changing the above situation a little, if Al had already been convicted of the original offense, been put on probation or some type of community control, and then was charged with distributing cocaine, the normal process would be a filing of a violation of his probation or community control, rather than a contempt of court charge.

Criminal contempt may be classified as either direct or constructive. A direct contempt is one committed in the presence of the court or the judge or so near to the court or judge when the judge is acting judicially as to be interpreted as hindering a judicial proceeding. An indirect or constructive contempt is an act done, not in the presence of the court or of the judge, acting judicially, but under such circumstances and at such a distance that reasonably tends to degrade the court or the judge as a judicial officer, or to obstruct the administration of justice by the court or judge. An example of constructive contempt would be violation of a court instruction to jurors not to discuss a case with outsiders during a trial.

The distinction between direct and constructive contempt is important insofar as the procedure for punishment is concerned. In a direct contempt, the judge has the power to punish summarily (without hearing evidence). In the case of constructive contempt, the judge, not having knowledge of all the facts of the case, is required to hear evidence before making a decision.

It is fairly well accepted that all courts have criminal contempt powers, even courts that do not ordinarily exercise jurisdiction in criminal cases. However, a court may not properly punish for contempt unless the order that was violated was one the court had lawful authority to make in the first place.

■■■■■■ DISCUSSION QUESTIONS ■■■■■■

1. Sam has been charged with a crime, and his trial has been set for a certain day. He contacts Fred and gives him $100 to testify on his behalf. Sam gives Fred five facts that Fred is to testify to under oath. Three of these facts are true, but two are false. The state decides to drop the charges against Sam but learns of the agreement between Sam and Fred. With what crimes, if any, can Sam and Fred be charged?

2. Bill has been charged with the crime of robbery. A jury of twelve and two alternates has been selected. One of the alternates, Al, is Bill's cousin. Neither the prosecutor nor the defense attorney knew this fact. Bill sees Al in the hallway of the courthouse, slips Al $50 and says, "This is for the time you will have to spend on this miserable jury," but says no more. Can Bill be charged with embracery?

3. Jack and Mack were being confined in separate cells in the lower county courthouse jail awaiting trial for their crimes. Neither man knew the other before their day of trial. On the day of trial Mack and Jack are taken from the lower county courthouse cells and put in the back seat of the same car to be transported to the main court house for trial. While en route to the main courthouse Mack attacks the officer driving the car, takes his gun, and forces the officer to stop the car. Mack gets out of the car and hands the gun to Jack, tells Jack to point the gun at the officer's head, and says, "Make this cop drive ten miles; if you don't I'll kill you myself." Out of fear Jack does as he is asked. At the end of the 10 miles he tells the officer to stop, which the officer does. Jack throws the gun away, walks ten steps, and sits down. With what crimes, if any, can Mack and Jack be charged?

▨▨▨ GLOSSARY ▨▨▨

Affirmation – an acknowledgment that a person will tell the truth in an oath without swearing in the name of a supreme being.

Criminal contempt – an act disrespectful to the court, calculated to bring the court into disrepute, or of a nature that tends to obstruct the administration of justice.

Custody – restrained; lawful restraint is a required element before escape or prison break can be properly charged.

Embracery – today called jury "tampering."

Escape – at common law, leaving lawful custody without authorization and without the use of force.

Failure to appear – for a required court appearance; is a separate and distinct offense from the original charge.

False oath – willful and corrupt sworn statement knowing it to be false or not knowing it to be true.

Fugitive – one who hides or leaves a jurisdiction to avoid prosecution; it is a federal crime to cross state lines to avoid confinement or prosecution for a state felony.

Jumping bail – by failure to appear in court when required and while out of jail on a bond.

Jury tampering – the crime of embracery.

Materiality – possibly having some bearing on the outcome of a case.

Oath – swearing or affirming that one will tell the truth.

Obstructing justice – a group of offenses including tampering with a witness, hiding or destroying evidence, obstructing or resisting a law enforcement officer in the performance of his or her duty, and refusing to come to the aid of and when summoned by a law enforcement officer.

Perjury – Taking a false oath in a proceeding regarding a matter material to that proceeding.

Prison – any boundary within which the mobility of a person in lawful custody is restricted.

Prison break – a common law offense of leaving lawful custody without authorization and with the use of force.

Rescue – aiding another to escape from custody.

Resisting or obstructing an officer – standing in the way of any public officer (particularly law enforcement) to prevent the officer from performing the duties of the office.

Subornation of perjury – the corrupt procurement of another to commit perjury.

Willful and corrupt – knowingly, deliberately, with an evil state of mind.

Witness tampering – any activity by the defense to affect the potential testimony of a witness, including the payment of money, intimidation, threats, or force.

■ REFERENCE CASES, STATUTES, AND WEB SITES ■

CASES

Jeffers v. State, 178 N.E.2d 542 (Ind. 1961).

State v. Clawans, 183 A.2d 77 (N.J. 1962).

Edwards v. State, 577 P.2d 1380 (Wyo. 1978).

State v. McCarthy, 38 N.W.2d 679, 686 (Wis. 1949).

Ford v. State, 205 A.2d 809 (Md. 1964).

State v. Walker, 426 S.E.2d 337 (S.C. 1992).

State v. Kiggins, 200 N.W.2d 243 (S.D. 1972).

United States v. Dixon, 509 U.S. 688 (1993).

WEB SITE

www.findlaw.com

CHAPTER 21

Crimes Against Public Order

■ KEY WORDS AND PHRASES ■

Affray
Breach of the peace
Criminal mischief
Disorderly conduct
Disturbing the peace
False public alarm
Gambling
Harassment
Lottery
Malicious destruction of property

Malicious mischief
Nuisance
Riot
Rout
Trespass
Unlawful assembly
Vagrancy
Vandalism
Violation of privacy

21.0 INTRODUCTORY COMMENTS

Much of police officers' working time when not writing reports is spent enforcing the laws against violations of a group of offenses categorized as offenses against public order. Every state has its laws dealing with affrays, disorderly conduct, public profanity, disturbing the peace, vagrancy, nuisance, and so forth. In this chapter we look at the legal composition and enforcement problems that are involved in many of these offenses.

21.1 UNLAWFUL ASSEMBLY, ROUT, AND RIOT

Our English common law heritage recognized the right of people to assemble peaceably for a lawful purpose. This inalienable right was brought to the United States with the colonists and embedded eventually in our basic governing document, the Constitution of the United States. In the First Amendment, this guarantee is maintained. The word "peaceable" is important to the definition, for not every assembly is protected. To be protected, the assembly must be peaceable, lawful, and for a lawful purpose. If it is otherwise, the assembly is not constitutionally protected and the criminal law comes into operation.

Unlawful assembly was a misdemeanor at common law and consisted of the assembly of three or more persons for an unlawful purpose or in an unlawful manner that was violent or riotous in nature or that disturbed the peace of the community. The unlawful assembly itself constitutes commission of this crime, and the unlawful act or unlawful purpose sought to be accomplished by the assembly must not occur. If the unlawful purpose of the assembly is accomplished, the crime of unlawful assembly may not be charged. Such conduct falls within the purview of another offense discussed in the following paragraphs.

The common law misdemeanor of **rout** consisted of any physical conduct that tended toward carrying out the unlawful purpose of the assembly. Thus, if three or more assembled to organize a lynch mob, they were guilty of unlawful assembly. When they left the place of assembly and began to move toward commission of the unlawful act, they were guilty of rout.

Riot, also a common law misdemeanor, was the actual carrying out of the unlawful purpose of an unlawful assembly. If the lynch mob successfully accomplished its goal, the participants were guilty of riot over and above their liability for any homicide they committed. To satisfy the element of riot, the unlawful purpose must have been accomplished in a violent manner or in any manner that would have instilled fear in the minds of the public or that would have constituted a breach of the peace.

Liability for common law rioting required a common purpose by at least three people. This did not mean that all three had to participate physically in commission of the unlawful act. As long as each was participating in such a manner that he or she was able to aid in the accomplishment of a common plan or scheme, there was concerted action. (See Chapter 4 on conspiracy for a complete discussion of concerted action and its ramifications. Also, refer to that chapter for ways in which one may be liable for riot.)

Rioting is a felony in most states today, and it usually requires destruction of or damage to property before it is chargeable. Many statutes also provide for liability of "followers" of a riotous crowd if they fail to obey lawful orders of a police officer or other official to disband.

21.2 AFFRAY

An affray is what we commonly call a fight except that the word "fight," as used in its broadest sense, includes legal as well as illegal conduct. An **affray** is an illegal fight, in which both or all parties (if there are more than two involved) are mutually at fault. For the affray to be chargeable, it must occur in a public place. The mutuality element prevents the offense from covering situations in which a fight is occasioned by one lawfully defending oneself from an unwarranted attack by another person.

Public place is defined as any place in which members of the public could view the affray, regardless of whether the property itself was public or private. An affray was a misdemeanor at common law, and it remains so in most, if not all, jurisdictions today. The characteristic of the offense that puts it in the class of crimes discussed in this chapter is that its commission in a place open to public view constitutes a breach of the peace.

21.3 BREACH OF THE PEACE, DISTURBING THE PEACE, DISORDERLY CONDUCT

Recalling the definition of crime we gave in the beginning of this book as being a public wrong, an act or omission forbidden by law and punishable upon conviction by a prescribed punishment, it is easy to say that all crimes affect the peace and dignity of the people. This concept, however, is theoretical to a certain extent. Certainly all crimes affect all the people, but many crimes are committed in such a way that the immediate impact involves only a few people. Larceny, embezzlement, burglary, and even robbery are not generally committed under circumstances that will arouse terror in the hearts of the people of the community at the time they are committed. In fact, few usually have knowledge of their commission until after the fact. The offenses discussed so far in this chapter differ in the sense that they are committed in public view and under such circumstances as to make the citizenry apprehensive of trouble. The offenses of breach of the peace, disturbing the peace, and disorderly conduct are very general and broad offenses designed to cover these types of situations not otherwise covered by statutes in a particular jurisdiction. Thus if a state has no specific statutes covering affrays, conduct that would constitute an affray would fall within that juris-diction's breach of the peace laws, and so forth.

We use all three offenses interchangeably here, for there is no real dis-tinction between them. Some statutes may call this conduct a breach of the peace, whereas others call it disturbing the peace or disorderly conduct. The same rule applies to municipal ordinance provisions covering such conduct.

To summarize, any conduct that puts fear in the hearts of the citizenry when committed in public view, and that is not dealt with by any other statute in the jurisdiction, may constitute a **breach of the peace, disturbing the peace** or **disorderly conduct.**

It is important to note one further ramification of breach of the peace laws. The common law warrantless arrest powers of private citizens and peace officers alike in misdemeanor cases hinged upon whether the misde-meanor committed in the presence of the arresting person constituted breach of the peace. Private citizens and peace officers had identical arrest powers in misdemeanor cases without a warrant at common law. They could make an arrest in misdemeanor cases only if the misdemeanor was committed in their presence and only if it constituted a breach of the peace. Not all misdemeanors constitute a breach of the peace in accordance with the definitions given earlier. Therefore, arrests could not be made in all cases of misdemeanors committed on view unless a warrant was first obtained. All jurisdictions have enlarged peace officers' authority to arrest without a war-rant for misdemeanors committed in their presence. Generally, a peace offi-cer may now arrest for any offense committed in the officer's presence regardless of whether it constitutes a breach of the peace. The same is not necessarily true concerning the arrest powers of a private citizen. There are still many jurisdictions today that prohibit a private citizen from making an arrest without a warrant for a misdemeanor, unless the misdemeanor is com-mitted in his or her presence and constitutes breach of the peace.

21.4 NUISANCE

There are many times when a law enforcement officer is required to "investi-gate" cases that the officer considers to be a nuisance. Usually, the officer is re-ferring here to situations that the officer thinks are a nuisance to investigate. Nevertheless, there are numerous tasks assigned to law enforcement agencies not directly considered by many to be law enforcement tasks. Every officer has answered, and every future officer will answer, calls in response to a com-plaint of a lost dog or a cat stuck in a tree, or complaints that the neighbor's air conditioner or radio is too loud. These are minor police problems, but they are very real to the people involved and deserve the officer's time and efforts.

All municipal ordinances contain provisions prohibiting the loud play-ing of radios and televisions between certain nighttime hours. It is likewise prohibited for a tavern or bar to emit excessive noise so as to disturb the tranquility of the nearby residential neighborhood, and so forth. These are **nuisance** ordinances and are well within the realm of the enforcement powers of a peace officer. As indicated, many nuisances are specifically covered by statute or ordinance. In addition, many jurisdictions have created separate laws to encompass those conditions not otherwise covered and have termed them nuisance laws. A nuisance may take one of two forms under civil law that are equally applicable to the "criminal" provisions. A nuisance may be either private or public. This depends on whom and how many people the nuisance affects. The distinction will also determine the extent of action that may be taken to abate the nuisance. In the case of a tavern, a warning will usually suffice as "a word to the wise." Sometimes, however, it might be necessary to secure a court order to close the establishment or take other legal action to abate the nuisance.

21.5 MALICIOUS MISCHIEF, MALICIOUS DESTRUCTION OF PROPERTY, VANDALISM, CRIMINAL MISCHIEF

Again we have a situation in which four offenses are used interchangeably. There is no real distinction between **criminal mischief, malicious mischief** and **malicious destruction of property. Vandalism** is sometimes more narrowly construed to include only malicious destruction of property that is considered a treasure or object of art.

The foundation of the offense is that there be some actual damage or destruction of property belonging to someone other than the accused and that the damage or destruction be caused voluntarily, not negligently or accidentally. The offense requires no specific intent but must be committed consciously with a realization of its effect. Malicious mischief may be committed against real or personal property. If John throws paint on Bob's house, he is liable for the offense. If Sam, a high school dropout, throws bricks through the local high school windows, he has committed the offense, and when Archie smashes the windows in the school principal's car, he, too, is criminally liable for malicious mischief.

To constitute the crime, there must be some damage or destruction of the property. The damage need not be great but it must be material, usually requiring some amount of repair.

Malicious mischief was a misdemeanor at common law and remains so in some jurisdictions today. There are many states, however, that now recognize, by statute, that the malicious mischief may cause a great deal of damage or destruction and have, therefore, graded the offense as either a felony or misdemeanor, depending on the type of property damaged or destroyed, the method by which the damage or destruction is done, and the extent of the damage or destruction.

> The FBI's Uniform Crime Reporting (UCR) program defines vandalism as willful or malicious destruction, injury, disfigurement, or defacement of any public or private property, real or personal, without the consent of the owner or persons having custody or control.

This problem has been specifically dealt with by the Model Penal Code. The code punishes anyone who causes or risks a catastrophe. According to the code (§220.2):

1. Causing Catastrophe. A person who causes a catastrophe by explosion, fire, flood, avalanche, collapse of building, release of poison gas, radioactive material or other harmful or destructive force or substance, or by any other means of causing potentially widespread injury or damage, commits a felony of the second degree if he does so purposely or knowingly, or a felony of the third degree if he does so recklessly.

2. Risking Catastrophe. A person is guilty of a misdemeanor if he recklessly creates a risk of catastrophe in the employment of fire, explosives or other dangerous means listed in Subsection 1.

3. Failure to Prevent Catastrophe. A person who knowingly or recklessly fails to take reasonable measures to prevent or mitigate a catastrophe commits a misdemeanor if:

   a.   he knows that he is under an official, contractual or other legal duty to take such measures; or

   b.   he did or assented to the act causing or threatening the catastrophe.

21.6   TRESPASS OF REAL PROPERTY

The word "trespass" is used throughout the law to represent a variety of legal conditions. In its general usage, **trespass** implies unauthorized violation of the person or property of another. Consequently, assault and battery is a trespass against the person of the victim. The taking of property in a larceny case without permission is a trespass, and so forth. In this section, we are concerned only with trespass as it applies to setting foot on the real property of another without permission.

Trespass is universally recognized as a civil wrong for which a tort action may be brought. The civil action in no way involves the state as a party. The only participation of the state in civil cases is to provide the forum in which the parties may litigate their problems to a successful and, it is hoped, just conclusion.

Most jurisdictions also provide that trespass onto the real property of another may be a criminal offense. The seriousness of the offense will depend on the circumstances surrounding the trespass, including the manner in which the trespass occurs and what the trespasser does after entering the land of another without authorization.

As a general rule, one who enters the land of another either intentionally or otherwise, and does nothing more, is not subject to criminal liability for trespass. Most jurisdictions require that to complete the offense, the trespasser must first be advised that he or she is unwelcome or advised that it is forbidden to enter the property before trespass may be charged. Statutes taking this position are generally referred to as trespassing after-warning laws. In specific instances, when a trespasser enters the land of another to engage in hunting or fishing without the permission of the owner, trespass may not be charged unless the owner has posted signs in a number of conspicuous places around the enclosure advising against trespassing. It is usually further required that such signs be posted for a certain time before any change in a previous custom of allowing people to enter is made. A typical statute of this type may read as follows: "Six days' notice by poster to be posted in at least four different and conspicuous places around the enclosure shall be given by parties wishing to avail themselves of the benefit of this section" (referring to the provisions making trespass to hunt or fish a violation).

The grade of the offense and the severity of punishment for trespass will vary from state to state and often from case to case. Depending on the types of activities the trespasser engages in after entering the land, the crime may be a misdemeanor of the most minor nature or may be punishable as a felony deserving a long prison sentence.

In addition to the cases mentioned in the preceding paragraph, trespass may involve acts such as tearing down fences, cutting or carrying away timber, severing and taking other property that is part of the real estate, or picking vegetables, flowers, or fruit. Each state legislates its own punishment for acts such as these. A word of caution is in order at this point. If the act the trespasser commits after entry on the land would constitute any other crime such as larceny, that crime would be chargeable instead of trespass. But, as in the case of hunting or fishing after trespassing on another's property, no other separate and distinct offense would be chargeable. If another offense such as larceny is committed along with the trespass, the former is the proper charge if it can be proved. Trespass becomes a lesser included offense of the larceny.

21.7 VAGRANCY

In Chapter 4, vagrancy was discussed in relation to its character as a status or an act. There are authorities in the criminal law field who would disagree with our conclusion on that point, but to go into detail at this juncture would be to dwell on the theoretical. Whichever view the reader chooses to accept, it is agreed that vagrancy is an offense punishable by the statutes of many jurisdictions.

Vagrancy, where it exists, is generally a misdemeanor and encompasses many and varying forms of conduct such as wandering and loitering.

The prime basis on which the existence of vagrancy statutes hinges is founded in theories of crime prevention rather than crime repression. Vagrancy is, and may be used as, a tool by law enforcement officers to prevent the commission of serious offenses. It is believed that removing the vagrant from the streets reduces the opportunity to commit crime. Of course, this supposition rests on the premise that it is the vagrant who will commit the more serious offenses. If this justification is true, as has been indicated time after time by the courts, it is well for the reader to recognize that the people, the legislature, and the courts are placing an enormous burden and an exceptional amount of power in the hands of law enforcement officers. In no other areas of either substantive or procedural criminal law is such unrestricted discretion given to law enforcement. Nevertheless, to retain the authority to arrest for vagrancy, law enforcement officers must not use or abuse the statutes indiscriminately. Even though there are a large number of acts or conditions that are incorporated within any particular vagrancy statute, the courts have said that these provisions are specific enough to allow people to understand the provisions of the statutes and reasonably to conform their conduct to the requirements of the law. Therefore, by implication, if not by express mandate, the offense must be capable of being proved by the arresting officer and prosecutor. If the vagrancy laws are used as a catchall or for harassment, law enforcement officers may find themselves without this authority.

Because of the breadth of vagrancy-type statutes, which include conduct such as loitering, they have often been attacked as being unconstitutional, basically on the grounds of the "void for vagueness" doctrine of constitutional law. Throughout the years, many courts have continually upheld the validity of vagrancy laws. At this writing, it is again under attack in the highest court in the land. The U.S. Supreme Court considered the validity of a Chicago gang loitering ordinance which was held invalid by the Illinois Supreme Court. The case of *Chicago v. Morales* again raised the issue of "void for vagueness" along with "substantive due process" because it restricted freedom and personal liberty to engage in harmless activities. The ordinance defined loitering as remaining in one place with no apparent purpose. Apparent to whom? As anticipated, the U.S. Supreme Court affirmed the Illinois Supreme Court decision and held the ordinance invalid. This does not mean that all vagrancy or loitering laws are invalid. If the statute objectively specifies the prohibited conduct that is understandable so that people can conform their behavior to the requirements of the law, such a statute can be valid.

21.8 FALSE PUBLIC ALARMS, HARASSMENT, AND VIOLATION OF PRIVACY

Two of the three newly identified crimes in this area as recognized by the drafters of the Model Penal Code are false public alarms and harassment.

False Alarms

We have seen the false fire alarm grow into the false bomb scare. Both activities are increasing. Such conduct is not limited to juveniles seeking a thrill. In one state, the state office buildings have been emptied on such scares an average of once a month. In one instance the "scare" was caused by a disgruntled claimant against the workers' compensation commission. Such activity is deemed a criminal **false public alarm** if the person makes such a report knowing that it is false or baseless and that it will cause public inconvenience, alarm, or the evacuation of a building.

Harassment

The second crime, harassment, presents a unique response to an age-old problem. Some people do not want communication or dialogue; they merely want to harass others. The Model Penal Code makes such conduct a petty misdemeanor. If a person with the purpose to harass calls someone without a legitimate purpose to communicate, as when the number is rung and the caller hangs up upon the receiver's answer, then the crime is committed. The act requires only one such call.

This offense is also complete if a person purposely harasses by insults or challenges that are likely to provoke a violent or disorderly response. Repeated anonymous communications at inconvenient hours or in offensive, coarse language are also prohibited. The Code prohibits **harassment,** or any course of harmful conduct serving no legitimate purpose of the actor. There is often an interrelationship between harassment and what has become a more serious offense—stalking (see Section 7.7, Stalking Crimes, for a more complete discussion of that offense).

Privacy

The third offense is truly a twentieth century creation. The use of "bugs," tapes, and other modern surveillance devices has increased and is part of the private sector. Husbands spy on wives, bosses on employees, neighbors on neighbors. Add to this the use made by the public sector, and one can see why concern with the individual's privacy rights is rising. Therefore, concern about **violation of privacy,** or illegal and unauthorized surveillance of private life, is also growing.

21.9 GAMBLING AND RELATED OFFENSES

The common law did not recognize **gambling,** or betting, as an offense against society. The only way one could get in trouble for gambling was by conducting oneself in such a manner that the conduct constituted a public nuisance. This possibility exists in many states today.

Thus, unless a state has, by its constitution or statutes, declared gambling illegal, it is legal. Most states do prohibit gambling or some aspect of it. These states have declared that gambling is against the public welfare, and they use the police power to prohibit it.

State-Authorized Gambling

A number of states permit gambling under certain circumstances and under certain conditions. This is often referred to as legalized gambling. Some states, for example, permit highly regulated betting at race tracks but only at the tracks through the parimutuel system. Betting on the same races outside the system through a bookmaker is illegal in these same states.

Some states permit limited gambling at certain religious, charitable, or public events, such as at fairs. Even at these places, the types of games and the amount at stake or the duration of the game is regulated.

Every few years additional states are allowing state-regulated casino gambling.

A few states that prohibit gambling of all types exempt the "weekly" private poker or other game at a private house as long as it is unconnected with organized crime and is not conducted so as to constitute a public nuisance.

Those states that permit gambling, either in a limited form such as a parimutuel betting or in all forms as is done in Nevada, require that any person conducting such a gambling operation be licensed by the state and adhere to state regulations. Conduct of a gambling operation outside the statute will render the operator chargeable for illegal gambling.

Illegal Lotteries

One form of gambling often prohibited by statutes is the lottery. There are three ingredients to a **lottery:** a chance, a prize, and a consideration paid for a chance at the prize. If any one of these is missing, there is no illegal lottery.

"No purchase required" is a term we have all seen with regard to national advertising of certain companies' big giveaways. If this were not the case, if one had to purchase oil or gas products to get a chance at the prize offered by the oil company, it would be a lottery and would be prohibited in most states because it would require the payment of something of some value for the chance at the prize. Game programs that we see on television are not lotteries, because the consideration element is lacking.

The element of chance looks to a fortuitous result, a result not of skill or fixed rule-following but merely "luck." If the skill of the participant is the real determining factor in the outcome of the contest or game, then it is not a lottery.

The federal government prohibits the use of the mail for the purpose of conducting a lottery or for transporting records, lists, number slips, and the rest of the paraphernalia through the mail.

Some states, both now and in the past, have a state lottery for the purpose of raising state revenues. The number continues to grow. Lotteries operated by the states are not illegal.

Illegal Gambling Houses

A number of states, by statute, make it a crime to maintain or keep a gambling (gaming) house. For the crime to be chargeable, there has to be more than one incidence of gambling, but none of the statutes set forth a particular length of time that the gambling house must be in operation.

It is not essential that the gambling house be open to the general public. Usually, it is enough if the house is maintained for a certain restricted clientele, which includes members of an organization and their invited friends. The gambling house can be hidden. It does not have to be open to the view of the public.

The type of building is unimportant. As a matter of fact, the entire building does not have to be devoted to gambling. It could be a back room in a restaurant or drugstore, for instance. It could be a spare room in a private house, hotel, or motel. Legitimate activities could be, and usually are, conducted simultaneously with the illegitimate gambling operation.

Owners, employees, and others who benefit from the operation of a gambling house may be prosecuted as keepers. This is true whether there is active participation or tacit approval of the use of one's facilities.

One of the main questions is whether the gambling that goes on in a private home is subject to the gambling laws of the state. There are at least three positions taken by state statutes. The first is that all gambling is prohibited, no matter where it is carried on. The second exempts gambling in private homes. The third type permits it in private homes as long as the home is not commonly used for gambling purposes. It is essential to be thoroughly familiar with the local statutes and ordinances. Even in a state in which the statute allows gambling in private residences, if a residence becomes a public place, then the gambling that goes on there may be stopped and the participants arrested. Thus, there is a distinction between commercial and organized gambling and occasional gambling. One state prohibits gambling in all places, but provides a lesser fine for gambling in the private home.

Related Enforcement Problems

Perhaps here as in no other area does the question of selective law enforcement come up. There is no possibility of full enforcement of all criminal law. Human resources and funds are not available. Additionally, many people would prefer law enforcement officers to stay with major crimes, such as homicide, robbery, and burglary, and leave the vices alone. Yet, too often the vices are the breeding grounds for the so-called major crimes. This puts law enforcement on the oft-quoted horns of a dilemma. Some compromise is usually reached. Law enforcement often seeks only to prevent organized crime efforts in the area of gambling, while looking the other way when "locals" are involved in an occasional way. Similarly, law enforcement officers tend to enforce against groups whose gambling activities spawn crimes of violence and crimes against property while not touching the penny-ante bridge clubs. There are those who suggest that law enforcement should concentrate only on flagrant public gambling and leave the private gambling alone.

These questions are moot in a few states in which the statute provides a penalty for neglect of duty with regard to gambling laws. In these states no discretion is left to law enforcement officials. They must enforce the law. However, the prosecutor's discretion is left intact. The prosecutor can take a *nolle prosequi* (refuse further prosecution), or fail to seek an indictment or issue an information. This can only frustrate law enforcement officials. The answer to the dilemma is found in the need to balance priorities.

▓▓▓ DISCUSSION QUESTIONS ▓▓▓

1. John Builder, president of the Builder Land Developing Corporation, has opened a subdivision in a suburban area of Big City. John and his public relations firm decide to kick off the sales of property in the subdivision with a very attractive offer. Everyone who buys a piece of property will have the recording number of the deed written on a piece of paper that is placed in a fishbowl. After all the property in the subdivision is sold, a drawing will be held and the person whose deed number is drawn will have his or her mortgage paid off, or, if the buyer paid cash for the property, the purchase price will be refunded. Builder puts the plan into effect, sells the first lot, and places the first number in the fishbowl. Has Builder committed any crimes?

2. Jay Hanson is shopping in a local department store. While walking down an aisle in the store, Jay observes a man shoplifting merchandise. The shoplifter was taking a $2 imitation ruby ring. May Jay, as a private citizen, arrest the shoplifter?

GLOSSARY

Affray – an unlawful fight in which both or all parties are at fault.

Breach of the peace – conduct that puts fear in the hearts of the citizenry when committed in public view.

Criminal mischief – voluntarily damaging or destroying property belonging to another person.

Disorderly conduct – conduct that puts fear in the hearts of the citizenry when committed in public view.

Disturbing the peace – conduct that puts fear in the hearts of the citizenry when committed in public view.

False public alarm – false fire alarms, bomb scares, etc. knowing the alarm to be false.

Gambling – betting.

Harassment – nagging or annoying phone calls or unwanted repeated personal contact.

Lottery – combination of a chance, a prize, and a consideration.

Malicious destruction of property – voluntarily damaging or destroying property belonging to another person.

Malicious mischief – voluntarily damaging or destroying property belonging to another person.

Nuisance – a group of minor offenses like loud music and noises that are often violations of ordinances.

Riot – carrying out the unlawful purpose of an unlawful assembly.

Rout – physical conduct tending toward carrying out the purpose of an unlawful assembly.

Trespass – setting foot on the real property of another without permission.

Unlawful assembly – assembly of three or more people for an unlawful purpose or in an unlawful manner.

Vagrancy – encompasses varied forms of conduct such as wandering without visible means of support and loitering.

Vandalism – voluntarily damaging or destroying property belonging to another person; sometimes only applied to art objects or treasures.

Violation of privacy – unauthorized, illegal surveillance of one's private life.

REFERENCE CASES, STATUTES, AND WEB SITES

CASES

City of Birmingham v. Richard, 203 So.2d 692 (Ala. App. 1967).

City of Chicago v. Morales, 527 U.S. 41, 119 S.Ct. 1849, 144 L.Ed.2d 67 (1999).

STATUTES

Alabama: Ala. Code §. 13A-11-7 (2001).

California: Ca. Pen. Code §. 415 (2001).

District of Columbia: D.C. Code ann. §. 22-3502 (2001).

Idaho: Idaho Code §. 18-3802 (2000).

Maryland: Md. Code Ann. art. 33 §. 16-204 (2001).

Michigan: Mich. Comp. Laws §. 432.202 (2001).

Minnesota: Minn. Stat. §. 609.725 (2000).

Mississippi: Miss. Code Ann. §. 97-35-37 (2001).

Montana: Mont. Code Ann. §. 45-8-101 (2000).

New Mexico: N.M. Stat. Ann. §. 30-14-6 (2001).

WEB SITE

www.findlaw.com

22

Organized, White Collar, and Commercial Crimes

■ KEY WORDS AND PHRASES ■

Antitrust laws
Blue-sky laws
Computer crimes
Contraband
Environmental crimes
Extortionate credit transaction
False advertising
Food and drug acts
Forfeiture

Loan sharking
Mail fraud
Money laundering
Monopoly
Racketeer-Influenced Corrupt Organization (RICO)
Tax evasion
Wire fraud

22.0 INTRODUCTION

Every law enforcement officer should read the *Task Force Report: Organized Crime*, written by the President's Commission on Law Enforcement and Administration of Justice in 1967. As the report states, the "core of organized crime activity is the supplying of illegal goods and services . . . to countless numbers of citizen customers." However, organized crime even gets into the legitimate areas, such as business and labor unions. Achieving its ends requires pressure, force, and the corruption of public officials. The influence of organized crime can cause prices to rise. But it is not the direct effect on prices that causes so much worry. The accumulation of wealth by organized crime means that it can be a "strong motive for murder" and may "corrupt public officials." By avoiding taxes, organized crime causes each taxpayer to pay more than he or she should.

Too often the public is indifferent. Most people do not know the extent to which organized crime affects their daily lives. The principal activities of organized crime are gambling, loan sharking, narcotics, labor racketeering, and infiltration of business. Prostitution and bootlegging are no longer the mainstays of organized crime's illicit incomes.

Although only an estimate, legal betting on horse races grosses only $5 billion a year, whereas illicit betting taps $20 billion with a profit of more than $7 billion a year to organized crime. **Loan sharking** is the lending of money at interest rates greater than the law allows. The Task Force found that the average interest rate is 20 percent a week, although it can run as high as 150 percent a week. The profit here has been estimated in the multi-billion-dollar range.

Today, the largest source of income is the distribution of narcotics. Most of the activity here is restricted to importing and first-level wholesale distribution. The lesser wholesale transactions and street transactions are left to independents. The profit to organized crime is estimated to be billions of dollars annually.

A "legitimate business," the report says, allows the acquisition of "respectability." The Illinois Crime Commission said, "There is a disturbing lack of interest on the part of some legitimate business concerns regarding the identity of the persons with whom they deal. This lackadaisical attitude is conducive to the perpetration of frauds and the infiltration and subversion of legitimate business by the organized criminal element." A 1957 meeting in New York State by 75 racket leaders found that 9 were in the building industry, 10 owned grocery stores, 17 owned bars and restaurants, 16 were in the garment industry, and so on. One criminal syndicate owned over $300 million in real estate investments.

There are four ways in which organized crime enters existing businesses. It invests hidden illicit profits. It accepts business interests in payment of the owner's gambling debts. Usurious loans are foreclosed. Or, it uses various forms of extortion. Organized crime in labor can prevent unionization in some industries. It can steal from union funds. It can extort money from business through the threat of possible labor strife. Finally, it can use union funds for syndicate business ventures.

It is no exaggeration that organized crime is everywhere. However, it is most firmly entrenched in the major population areas. Perhaps the most vile of all the activities carried on by organized crime is the corruption of public officials. Organiz?d crime is currently directing its efforts to corrupt law enforcement at the chief or middle supervisory level. Who will investigate the investigators? Even more effective is the effort at corrupting the political leaders who "tied the hands of the police."

What was said in the late 1960s and the 1970s, for the most part, remains true today. Although some inroads have been made—several members of organized crime going public, the conviction and imprisonment of some high-level "family" members, and some fairly effective legislation—organized crime has not gone away. It remains a threat to our society.

As the laws in the fight against organized crime were developing, so were white-collar crimes and commercial crimes. The spectrum of problems required the law to develop new prohibitions against certain business practices because it became clear that some businesspeople who were not necessarily connected to organized crime would avoid the law when possible. Standard common law and statutory business crimes of bribery, embezzlement, forgery, bad-check violations, and false pretenses were often difficult to prove. Extortion, perjury, obstructing justice, and the like were equally difficult to demonstrate. Therefore, specific crimes were created to enforce the stated public policy to ensure the goals of the new laws. Mail fraud, of course, is one of those laws that addresses both organized crime and white collar crime. Others include tax evasion, false advertising, food and drug law violations, monopoly, antitrust, and security law crimes. Computer crimes, environmental offenses, and the procedural contraband forfeiture have been added. This area of white collar crime will continue to evolve through both legislative actions and court decisions. There seems to be no slowdown in legislative activity and the regulation of business affairs. Safety enactments in both the consumer products field and in manufacturing are further evidence of this type of growth. Environmental concerns will continue to spark new legislation despite the economic burdens thereby imposed. Computer crimes will continue to grow and expand. What can be seen, from an enforcement perspective, is the growth of highly specialized law enforcement needs that can be satisfied only by specially trained law enforcement officers with significant enforcement and business, economic, scientific, computer, or social history backgrounds.

I. ADDRESSING ORGANIZED CRIME

22.1   RICO AND CONTRABAND FORFEITURE

One of the more effective legislative creations was the **Racketeer-Influenced Corrupt Organization (RICO)** section of the Organized Crime Control

Act of 1970 which was indeed aimed primarily at the Mafia, La Cosa Nostra, and similar organizations. Because Congress is required to write general criminal laws and not single out one or two known criminals or criminal groups, the language ultimately has affected other individuals and other organizations engaged in criminal activity in violation of federal law. The success of the federal law in its application to labor unions and consortiums of business groups, such as power companies, led most states to adopt their own civil and criminal RICO statutes. A number of high-level organized crime figures have been convicted under the federal criminal RICO statute. The civil portion of the law permits the seizure of property used in or used for a criminal racketeering activity. RICO statutes have been and continue to be highly successful investigative aids to law enforcement at both the federal and state levels.

Initially, it was assumed that there had to be an economic motive underlying the criminal activity in order to apply the federal RICO statute. The question was asked of the United States Supreme Court whether a corrupt organization could be punished if their motive was not economic. The case that raised the question had specific application to antiabortion groups. The Court held that economic motive was not the determining factor. The Court defined racketeering as a pattern of illegal activity to achieve any goal. If the illegal acts are committed and the pattern is provable, racketeering has occurred and RICO is applicable.

Following the Supreme Court ruling, federal state legislation has been expanded and strengthened to allow for the application of RICO statutes to a greater variety of offenses, often using a judicial process called "contraband forfeiture." Many states have enacted laws that allow the government to seize personal property, including vehicles and monetary assets, related to the sale of illegal drugs. Such **forfeiture** laws allow law enforcement officers to possess any property that they have probable cause to believe is used in the unlawful manufacture, growth, processing, transport, distribution, sale, concealment, or storage of a controlled substance or the manufacture or sale of drug paraphernalia. Thus, if a defendant is apprehended transporting a substantial quantity of cocaine in his truck, the government may seize the defendant's truck in addition to prosecuting him for the drug crime. These laws also allow the government to seize any property offered in exchange for the performance of such acts or any revenues generated from such acts. In actuality, many of the statutes are not limited to narcotic transactions, rather, they might include a number of specified serious illegal activities. Drug cases happen to be the most prevalent.

In addition to proving they have probable cause to believe that the property is related to the illegal act, the government must prove, when suing to seize **contraband** property, that either the defendant's possession of the property is unlawful or that the defendant was involved in or knew of the illegal activity. Many jurisdictions hold defendants liable if they should reasonably have known of the illegal activity, even though actual knowledge could not be established. Some states will allow a defendant to assert a defense that he or she took all reasonable steps to abate the illegal activity and took all reasonable steps to prevent the property from becoming involved in the illegal activity or, that the defendant is actually an innocent owner or co-owner of the property who had no knowledge or role in the illegal act.

As significant as RICO statutes are, they, and contraband forfeiture, are not the only criminal statutes addressing organized and white collar crime.

22.2 EXTORTIONATE CREDIT TRANSACTIONS

Congress found that a substantial part of the income of organized crime is generated by extortionate credit transactions. **Extortionate credit transactions** are characterized by the use, or the express or implicit threat of the use, of violence or other criminal means to cause harm to a person, reputation, or property as a means of enforcing payment for a high-interest loan (loan sharking). Congress even said that when extortionate transactions are purely intrastate in character, they nevertheless directly affect interstate and foreign commerce.

Thus, a person who uses force or the threat of force to another person or to a property, or the reputation of another in order to force payment on a loan with an illegal interest rate, can be charged under this act and fined $10,000 or imprisoned for 20 years, or both. However, proof of this offense requires the government to show that the rate of interest on the loan is in excess of 45 percent. The act also punishes the "banker," or money source.

22.3 MONEY LAUNDERING

The primary piece of federal legislation addressing money laundering is the Money Laundering Control Act of 1986. This legislation was enacted because Congress finally recognized that the proceeds from money laundering had become a most lucrative and sophisticated business activity of organized crime. Without the ability to move and hide their wealth, large-scale criminal activities would not be able to operate at their current levels. **Money laundering** is defined as a process by which one conceals the illegal source of income or conceals the illegal application of income to other activities and disguises the income to make it appear to be legally obtained.

The act prohibits a person from being involved in a financial transaction when the individual knows that property involved in the transaction represents proceeds from certain specified illegal activities. Those illegal activities are the primary activities which constitute the source of big money in organized crime including drugs, loan sharking, etc. In order for an individual to be charged, the person must engage in the transaction either with the intent to promote the conduct of the specified unlawful activity or with the knowledge that the transaction is done for the purpose of avoiding the requirements of the law. A second part of the law prohibits an individual from engaging in, or attempting to engage in, a monetary transaction which involves property valued at more than $10,000 when the individual knows that the property is derived from the specified illegal activities.

There are five elements to the crimes. They are: (1) knowledge, (2) existence of a specified unlawful activity, (3) a financial transaction, (4) proceeds, and (5) intent. An example of the application of this statute occurred in a case in which a defendant had taken the proceeds of drug transactions and purchased an automobile using cash and cashier's checks as the methods of payment. In this case it was shown that the defendant acted willfully and with knowledge of the statute. The drug transaction from which the money was derived was among the specified unlawful activities. The purchase of the vehicle constituted a financial transaction in which the defendant obtained a vehicle and his intent was to use the money to conceal the existence of the source of the funds.

Closely related to the Money Laundering Control Act is the antistructuring law, which requires banks and other financial institutions to file reports whenever they are involved in a cash transaction that exceeds $10,000. The law makes it illegal to "structure" transactions, that is, to break up a single transaction that is above the $10,000 threshold into two or more separate transactions for the purpose of evading the financial institution's reporting requirements. Anyone who willfully violates this provision is subject to criminal penalties. Proof of the willfulness of the violation requires the government to establish that the defendant acted with knowledge that the conduct was unlawful. In a federal case arising in Nevada, the defendant had lost a great deal of money to a casino and was given a short period of time in which to come up with the $160,000 debt. Shortly thereafter the defendant came into the casino with $100,000 cash. When the defendant was advised that the casino could not accept a cash transaction in excess of $10,000 without reporting it to state and federal authorities, the casino had a limousine drive the defendant to a number of different banks in which the defendant procured cashier's checks, each for less than $10,000. The defendant was charged with violation of the antistructuring law in order to evade the bank's obligation to report cash transactions exceeding $10,000. However, the case was overturned on appeal because the government could not prove that the defendant knew that what he was doing was unlawful.

22.4 MAIL AND WIRE FRAUD

Federal law makes it a felony, termed **mail fraud** or **wire fraud,** to use the mail or any interstate electronic communication network to further any scheme or artifice to defraud or to obtain money or property by means of a false or fraudulent pretense. The misrepresentation need not be of a present fact but may also include suggestions and promises for the future. The conduct does not have to be a common law fraud. The offense is using the mail or interstate communication networks and not the scheme itself. The material sent through the mail or over the wire must have a specific relation to a scheme to defraud.

Originally the mail fraud statute only applied to the illegal use of the United States Postal Service, but in 1994 the statute was amended to include private interstate delivery services such as United Parcel Service and Federal Express. Although mail fraud and wire fraud are slightly different forms of communication, the elements of the crimes are essentially the same and include: (1) a scheme to defraud, (2) committed with the intent to defraud, and (3) use of mail, private interstate commercial carrier, or electronic network to further the fraudulent scheme. Decisions of the courts have indicated that the federal government is not required to prove that the scheme to defraud was successful but is required only to show that the scheme existed in which the use of the mail or the wire was reasonably foreseeable and that actual mailing or use of the wire did occur in furtherance of the scheme. The only essential difference between the mail fraud and wire fraud statutes is that in mail fraud there is no requirement that the mail actually cross state lines; however, in the wire fraud statute, it is necessary to show that the communication did cross state lines before there can be conviction. Radio and television transmissions are included in the federal wire fraud statute.

II. ADDRESSING WHITE COLLAR AND COMMERCIAL CRIME

22.5 TAX EVASION

The gravamen of **tax evasion** is that the taxpayer has intentionally cheated the government of tax moneys by preparing a false report, using phony deductions, or intentionally refusing to pay taxes. Tax evasion was difficult to label in a common law context because it was a type of embezzlement, yet the government did not have the appropriated or kept money; it was a false pretense but only in the methods employed. It was not a "pure" larceny. Sometimes it was a fraud. To correct the shortcomings, the government wisely created the crime of tax evasion. This crime was used successfully against some of the more infamous criminals of the 1930s, such as Al Capone, to achieve convictions where other criminal statutes failed.

22.6 FALSE ADVERTISING

Of course, the smart salesperson will know how to advertise his or her wares without violating the false pretenses law. Yet the public can still be hurt. Some activity sought no specific exchange of goods for money but sought only to get customers to come in. Therefore, "come-on" prices for goods not in stock were used. It became obvious that some method was needed to protect a gullible public. To be convicted of **false advertising,** it is not necessary to show that anyone was cheated or defrauded; the offense is complete if the advertisement is false and if it was purposefully circulated to get the public to buy property or pay for the service offered.

22.7 SECURITIES ACTS

Like the federal government, the states have enacted laws to protect people from stocks and bonds frauds. These are usually called **blue-sky laws.** These statutes punish for any scheme or attempted scheme with an intent to defraud. The scheme, therefore, does not have to succeed or injure anyone. Any act that represents an effort to get stock subscriptions or that promotes the sale of the sham security is sufficient.

22.8 FOOD AND DRUG ACTS

When one thinks of drugs and law enforcement, the attention is usually focused on the illicit drug trade of controlled substances. This is but one facet of the consumable problem. People have suffered illness or death through the use or ingestion of adulterated foods and noncontrolled substances. Both state and federal laws have taken aim at such practices as selling impure food and harmful drugs and cosmetics by enacting **food and drug acts.** This type of regulation helps provide for the health and safety of people. Such acts are necessary because people must purchase prepackaged foods manufactured some distance from them. While people really have no way of protecting themselves, inspection, labeling, and packaging requirements for foods and drugs aid in the effort.

In all these requirements, there is usually an accompanying penal provision. According to many of these statutes, some type of intent is required for penal sanctions. There are some exceptions, especially where a person sells adulterated food. Here the seller should know better, but more than that, the act itself and not the intent is important. Thus some of these offenses are *mala prohibita.*

22.9 MONOPOLY AND ANTITRUST LAWS

Whenever a person seeks to "hog" the market and does so by a series of unfair trade practices and not by head-to-head competition, the individual is said to be attempting to create a **monopoly.** The public loses when competition dies. Similarly, when businesses combine together and avoid competition by price fixing and other acts, they are said to be in violation of **antitrust laws.** Intent is not important. The acts that have the desired effect are enough. Thus cartels such as OPEC would in fact be illegal in the United States.

Any act in which the actors limit production or supply is a violation. Some manufacturers, for instance, have refused to supply dealers who would not keep prices at the level at which the manufacturer wanted them. One clothing manufacturer would not allow its dealers to put the articles on "sale." This was found to be a violation.

Antitrust laws are aimed at preventing any number of "dirty tricks." So intricate and devious are some of the schemes that those who are not law enforcement officers in the traditional sense handle these enforcement problems. Special enforcement units had to be created requiring the services of economists, accountants, marketing specialists, and attorneys.

22.10 COMPUTER CRIMES

The commission of crimes using the Internet can be subtler and more damaging than the same acts committed in a traditional fashion. On the Internet, hate mail needs no paper, no envelopes, and no stamps; pornography needs no film screen, projector, or printer. Manufacturers can now ship most mass market computers Internet ready with sophisticated communication software that can send and receive documents, videos, photographs, audio and computer programs.

Although there are many hackers operating in the cyber world, the FBI Law Enforcement Bulletin identified employees as the most significant threat to any organization's computer systems because they have access to internal networks and data that a hacker can only dream of. An employee can be dangerous if the employee has personal interests, motives, and grievances that conflict with his or her employer.

With so many different methods of communication available on the computer, including e-mail transmissions, newsgroups, chat rooms, etc., the possibility of initiating or conducting criminal conduct is virtually boundless. The simultaneous presence of the Internet in all states and countries and its ability to let users make millions of copies from a single posted document, photograph, video, or song poses unique problems of criminal, civil, and ethical liability. Key laws governing Internet crime focus on the offending user rather than the owner of the communication system. Nevertheless, even an innocent owner can end up being arrested because of a mistaken belief about the way computers store data.

In many ways laws treat a computer used in Internet-based crime as a weapon, much like a gun or a knife. Its use can therefore lead to additional charges or a heightened degree of severity. The national and worldwide accessibility of cyberspace creates risks of legitimate multiple prosecutions in several jurisdictions simultaneously (see the sidebars in Chapter 3).

A typical computer crime statute is the Alabama Computer Crime Act that punishes as an offense against intellectual property the willful, knowing, and unauthorized accessing, modification, destruction, disclosure, use, or taking of data and computer programs or supporting documentation from a computer system. The act punishes such behavior as a misdemeanor, but it punishes such behavior as a felony if the act is committed for the purpose of fraud or obtaining other property or if the behavior results in damage to the intellectual property, interrruption of public utilities, or physical injury to a person not involved in the act. Essentially, **computer crimes** encompass offenses in which a computer is used for identity theft, employee theft, or making pornography available to children.

Some states treat a computer crime as having occurred in their jurisdiction if the process of transmission used any computer in their jurisdiction. Consequently, the transmission of a single e-mail message, for example, could conceivably violate the laws of many jurisdictions. Prosecutors might interpret some states' laws to treat, as a separate offense of "distribution," each download of an obscene photograph.

Besides the federal attempt to protect children from pornographic material on the Internet (see sidebar in Chapter 18), the Computer Fraud and Abuse Act is the federal government's core deterrent against crimes made possible or easier because of computers. The act makes it a crime for an authorized user to access a computer that is federally owned or is a "protected computer" (where the conduct would affect a financial institution or the government or the computer is used in interstate or foreign commerce or communication) for the purpose of obtaining records from a bank, credit card issuer, or consumer reporting agency, or to commit fraud or extortion, or transmit a destructive virus or command, or to traffic in stolen passwords, or to threaten damage to a computer system to extort money or other things of value.

The Electronic Communications Privacy Act provides criminal penalties for intercepting e-mail. The act makes it a federal felony to intercept electronic communications or to disclose or use the contents of an electronic communication notably e-mail.

New laws and court decisions will continue evolving these areas of substantive criminal law at both the state and federal levels.

22.11 ENVIRONMENTAL CRIMES

Criminal prosecutions for acts that affect the environment have increased dramatically in recent years. Although laws have been on the books for some time, the enforcement of criminal penalties and the use of environmental crime violations to address other criminal activities is only of recent origin. Both at state and federal levels, environmental crime statutes have been used, as was tax evasion in the 1930s, to address those who could not necessarily be convicted of other violations.

At the federal level there are a number of laws that contain criminal penalties, including the Clean Air Act, the Safe Water Drinking Act, the Clean Water Act, the Toxic Substance Control Act, and the Resource Conservation and Recovery Act. The criminal provisions of these and many state laws are primarily directed at industries and business to ensure that their activities are not conducted in a manner that violates the welfare of the environment. Disposal of toxic or hazardous waste into the ground or failure to dispose of it at all (thus constituting an **environmental crime**) are the types of behavior that statutes seek to prohibit in order to protect the environment. Businesses, sometimes by neglect, sometimes accidentally, and sometimes intentionally, violate these laws. The violations can be used in many cases to seek both civil and criminal sanctions and have proved useful in many cases where other suspected unlawful activity was not able to be established to the point of bringing criminal charges. As an example, owners and operators of motor vehicle salvage yards have been criminally cited and charged for environmental crimes while under investigation for auto theft–related activities but without sufficient evidence upon which to make arrests and seek prosecutions. In some cases it has resulted in the business being closed down, which has the effect of preventing the continuation of other types of criminal activity.

Many states have joined the federal government in enacting criminal penalties for a variety of environmental concerns. For example, Montana punishes defendants who knowingly violate its air pollution measures, misrepresent facts on required forms, or disable air pollution monitoring devices, with fines up to $10,000 per violation and prison terms of up to two years. California has enacted measures that punish a person who knowingly disposes of hazardous wastes at a facility that the defendant should have known lacked required permits, or who knowingly transports wastes he or she should have known to be hazardous, with a prison term of up to one year. The statute authorizes courts to impose a fine upon a person convicted of between $5,000 and $100,000 for each day that the hazardous waste remains in violation of the state hazardous waste laws, and it authorizes the courts to impose additional and consecutive prison terms and fines of up to $250,000 if the illegal transport or disposal causes great bodily injury or a "substantial probability that death could result."

DISCUSSION QUESTIONS

1. Research applicable criminal and civil RICO statutes in your jurisdiction. To what conduct do these statutes apply? What are the criminal penalties? Do the statutes require an economic motive underlying the criminal activity?

2. What is money laundering and why is it critical to large-scale criminal activities?

3. A title loan law of the state of Excelsior allows companies to charge up to 264% annual interest on loans secured by the title to the borrower's car. If any payment is over 60 days late, the vehicle can be sold, and payment on the loan is still due and payable. In view of the federal law on extortionate credit transactions, how can Excelsior's law be justified?

4. Smithfield goes to Joe Barnaby Motors and wants to purchase a brand-new, top-of-the-line Yamuki sport utility vehicle. The negotiated price is $36,000. Smithfield wants to pay cash. What are the legal obligations and responsibilities of both parties?

5. Billy Joe Jefferson starts a chain letter on the e-mail of his home computer that requests that recipients send a dollar to the person whose name appears at the top of the list, then send the e-mail to six friends after putting their name at the bottom. Jefferson promises that, if the chain is unbroken, participants could receive thousands of dollars within weeks. He sends the e-mail to friends all over the country. In turn, many of the recipients send the e-mail to their friends throughout the country and the world. At the same time, Jefferson mails hard copies of the e-mail to other in-state friends. Of what crimes is Jefferson guilty? Why?

GLOSSARY

Antitrust laws – designed to prevent business monopolies.

Blue-sky laws – designed to protect people from stocks and bonds frauds.

Computer crimes – offenses involving sexually explicit materials directed at or available to children; employee theft and identity theft.

Contraband – any item, the possession or use of which is unlawful.

Environmental crimes – offenses involving the disposal of toxic or hazardous waste or other laws designed to protect the environment.

Extortionate credit transaction – the use of force or threats to enforce the payment of a high interest loan.

False advertising – false advertisement purposefully circulated to get the public to buy property or service.

Food and drug acts – inspection, labeling, and packaging requirements to protect against adulterated foods and noncontrolled drugs or other substances.

Forfeiture – a judicial procedure by which the government seizes ownership of property used in a criminal enterprise or fruits of the unlawful activity.

Loan sharking – the use of force or threats to enforce the payment of a high interest loan.

Mail fraud – use of the mail to further a scheme to defraud by false pretenses.

Money laundering – a process by which one conceals the illegal source of income or conceals the illegal application of income by making it appear to be legally obtained.

Monopoly – the absence of competition in a type of business.

RICO – Racketeer-Influenced Corrupt Organization.

Tax evasion – preparing a false report, using phony deductions, or intentionally refusing to pay taxes.

Wire fraud – the use of an interstate electronic communication network to further a scheme to defraud by false pretenses.

▓▓ REFERENCE CASES, STATUTES, AND WEB SITES ▓▓

CASES

National Organization for Women v. Scheidler, 510 U.S. 249 (1994).

STATUTES

United States: 18 U.S.C. 1341 and 1343.

Alabama: Code of Ala. 13A-8-102(d) (2001).

California: Cal. Health & Saf. Code §. 25189 (2001).

Colorado: Colo. Rev. Stat. 16–13–503 and 504 (2000).

Montana: Mont. Code Anno., §. 75-2-412 (2000).

WEB SITE

www.findlaw.com

Index